Image Processing in Agriculture and Forestry

Image Processing in Agriculture and Forestry

Special Issue Editors

Gonzalo Pajares Martinsanz
Francisco Rovira-Más

MDPI • Basel • Beijing • Wuhan • Barcelona • Belgrade

MDPI

Special Issue Editors

Gonzalo Pajares Martinsanz
University Complutense of Madrid
Spain

Francisco Rovira-Más
Polytechnic University of Valencia
Spain

Editorial Office
MDPI
St. Alban-Anlage 66
Basel, Switzerland

This is a reprint of articles from the Special Issue published online in the open access journal *Journal of Imaging* (ISSN 2313-433X) from 2015 to 2017 (available at: http://www.mdpi.com/journal/jimaging/special_issues/ImageProcessing_Agriculture_Forestry)

For citation purposes, cite each article independently as indicated on the article page online and as indicated below:

LastName, A.A.; LastName, B.B.; LastName, C.C. Article Title. *Journal Name* **Year**, *Article Number, Page Range.*

ISBN 978-3-03897-097-2 (Pbk)
ISBN 978-3-03897-098-9 (PDF)

Contents

About the Special Issue Editors

Gonzalo Pajares Martinsanz received a Ph.D. in Physics from Distance University in Spain in 1995, discussing a thesis on stereovision. Since 1988, he has worked at Indra in critical real-time software development. He has also worked at Indra Space and INTA in advanced image processing for remote sensing. He joined the University Complutense of Madrid in 1995 on the Faculty of Informatics (Computer Science) in the Department of Software Engineering and Artificial Intelligence. His current research interests include computer and machine visual perception, artificial intelligence, decision-making, robotics, and simulation. He has authored many publications, including several books, on these topics. He is the director of the ISCAR Research Group. He is an Associated Editor in the indexed journal of Remote Sensing and serves as a member of the Editorial Board for the following indexed journals: Sensors, EURASIP Journal of Image and Video Processing, and Pattern Analysis and Applications. He is the Editor-in-Chief of the Journal of Imaging.

Francisco Rovira-Más received a degree in Agricultural Engineering in 1996 from the Polytechnic University of Valencia, Spain, where he was an Assistant Professor from 1997 to 2000. He obtained his Ph.D. in 2003 from the University of Illinois at Urbana-Champaign (USA). He has been a member of the Intelligent Vehicles System group at the John Deere Technology Center in Moline (USA), and a Research Associate with the Department of Agricultural and Biological Engineering in the University of Illinois, conducting research at the John Deere Intelligent Vehicle Systems unit in Urbandale (USA). Currently, Francisco is a Professor at the Polytechnic University of Valencia in Spain, founder and director of the Agricultural Robotics Laboratory, and Head of the International Programs Office at the School of Agricultural and Biological Engineering (ETSIAMN). His research interests include autonomous vehicles, machine vision, controls, stereoscopic vision, off-road equipment automation, robotics, and artificial intelligence.

Preface to "Image Processing in Agriculture and Forestry"

This book contains high quality works from the Special Issue entitled "Image Processing in Agriculture and Forestry" that demonstrate significant achievements and advances in these areas. They are self-contained works addressing different imaging-based procedures and applications.

Images are the main source of information in perception systems in agriculture. These images can be captured from machine vision systems on-board autonomous or non-autonomous ground vehicles, unmanned aerial vehicles and satellites, as well on manually operated and static systems. Several devices with complementary metal-oxide-semiconductor (CMOS) or charge-coupled device (CCD) technologies have been used, operating in the visible spectrum (red-green-blue (RGB)-based) or in multiple spectra (RGB and infrared). Satellite imagery has been also used, including Synthetic Aperture Radar (SAR) operating in the L-band, as well as hyperspectral devices from the Landsat 5 Thematic Mapper (TM), Landsat 8 Operational Land Imager, and WorldView-2.

Regarding image processing methods and algorithms, this book contains artificial neural networks (including convolutional), machine learning and classifiers, specific segmentation approaches (including vegetation indices), and three-dimensional (3D) reconstruction.

Regarding agricultural and forestry applications, several different methods are addressed and the order of papers in the book is determined according to the application of these methods. A brief description is provided in this preface for each paper. The first paper provides guidelines for selecting machine vision systems for different tasks (navigation and plant detection). The second and third papers are focused on navigation and obstacle avoidance. The fourth to sixth papers deal with canopy-based approaches on trees. Aboveground biomass (AGB), phenotyping, chlorophyll, water content, leaf area index (LAI), and nutrient contents are addressed in the seventh to tenth papers. Land cover changes are addressed in the eleventh paper and soil analysis is covered in the twelfth and final paper. A brief description of the papers in the order they appear in the book follows:

1. Guidelines are provided for selecting machine vision systems for optimum performance in navigation, crop detection, specific treatments, and different tasks on board autonomous and non-autonomous agricultural vehicles. This implies the analysis of spectral bands (visible and infrared), imaging sensors and optical systems (including intrinsic parameters), and geometric visual system arrangement (considering extrinsic parameters and stereovision systems).
2. Simple cross-correlation functions are used to establish high-resolution navigation control, in the centimeter range, for secure and robust navigation in small agricultural robots. Any color CMOS or CCD cameras with near infrared (NIR) sensitivity (~800 nm to 900 nm) can be used.
3. A deep convolutional neural network is fine-tuned for the detection of a specific obstacle based on RGB-images captured with an iPhone 6.
4. Three-dimensional (3D) reconstruction, based on feature descriptors and matching, was applied for characterizing tree canopies in croton and jalapeno pepper plants and lemon trees based on a binocular stereo (two high-definition cameras assembled on a wooden box with a parallel optical axis).
5. Contrast stretching for image enhancement and thresholding was applied for detecting peach blossoms on trees using an off-the-shelf unmanned aerial system (UAS), equipped with a multispectral camera (near-infrared, green, blue).
6. Artificial neural networks were used for early yield prediction in apple orchards, where canopy images captured with a RGB camera were segmented into fruit, foliage, and background categories.
7. Non-parametric MaxEnt and Random Forests machine learning algorithms were employed to model AGB retrieval from Synthetic Aperture Radar (SAR) L-band backscatter intensity and coherence images.
8. K-means clustering and minimum-distance classification were used on red-green-blue (RGB) images of sorghum plants as a means to measure stalk thickness (plant phenotyping) as an essential

step for crop breeders to improve sorghum for energy production.

9. Twenty spectral vegetation indices (chlorophyll, water content, and LAI) from off-nadir viewing satellite imagery (WorldView-2) were obtained by combining different bands to estimate the stress-related biophysical variables of capital intensive orchard crops.

10. Partial least squares regression was applied to achieve field measurements and vegetation indices derived from WorldView-2 images to produce canopy nutrient (nitrogen) levels, AGB, and LAI maps in order to determine the biophysical variations of a mangrove forest.

11. A Random Forest classifier (pixel-based and object-based) was used to determine land cover change on Assateague Island as a result of Hurricane Sandy from Landsat 5 Thematic Mapper (TM) and Landsat 8 OLI sensors for pre- and post-processing [11].

12. A neural network model was used for estimating soil structure, soil texture, bulk density, pH, and drainage category using color and texture descriptors from images captured with a smartphone with GPS.

Industrial evolution and future connected systems are demanding cutting-edge technologies in all fields, including agriculture and forestry. In this regard, readers can find an excellent source of resources in this book regarding the development of the research and industrial activity involving imaging and processing procedures focused on agricultural and forestry applications.

The international scientific, farming, and industry communities worldwide are also indirect beneficiaries of the abovementioned developments. Indeed, this book provides insights and solutions for the many different problems addressed. Also, it lays the foundation for future advances in the face of new challenges. In this regard, new imaging sensors, technologies, and procedures contribute to the solutions of existing agricultural and forestry problems. Likewise, the need to resolve certain problems demands the development of new imaging technologies and associated procedures.

We are grateful to everyone involved in the publication of this book. Without the invaluable contributions of the authors together with the excellent help of the reviewers, this book would not have come into fruition. More than 50 authors have contributed to this book.

Thanks are also extended to the Journal of Imaging and the team of people involved in the production of this book for their support and encouragement.

Gonzalo Pajares Martinsanz and Francisco Rovira-Más
Special Issue Editors

Journal of
Imaging

MDPI

Technical Note

Machine-Vision Systems Selection for Agricultural Vehicles: A Guide

Gonzalo Pajares [1,*], Iván García-Santillán [1], Yerania Campos [1], Martín Montalvo [1], José Miguel Guerrero [1], Luis Emmi [2], Juan Romeo [1], María Guijarro [3] and Pablo Gonzalez-de-Santos [2]

[1] Department Software Engineering, School of Computer Science, University Complutense of Madrid, José García Santesmases, 16, 28040 Madrid, Spain; ivandgar@ucm.es (I.G.-S.); yeraniac@ucm.es (Y.C.); mmontalvo@ucm.es (M.M.); jmguerre@ucm.es (J.M.G.); jromeo01@ucm.es (J.R.)
[2] Center for Automation and Robotics, UPM-CSIC, 28500 Madrid, Spain; luis.emmi@car.upm-csic.es (L.E.); pablo.gonzalez@csic.es (P.G.-d.-S.)
[3] Department of Computer Architecture and Automatic, School of Computer Science, University Complutense of Madrid, 28040 Madrid, Spain; mguijarro@ucm.es
* Correspondence: pajares@ucm.es; Tel.: +34-1-394-7546

Academic Editor: Francisco Rovira-Más
Received: 12 September 2016; Accepted: 15 November 2016; Published: 22 November 2016

Abstract: Machine vision systems are becoming increasingly common onboard agricultural vehicles (autonomous and non-autonomous) for different tasks. This paper provides guidelines for selecting machine-vision systems for optimum performance, considering the adverse conditions on these outdoor environments with high variability on the illumination, irregular terrain conditions or different plant growth states, among others. In this regard, three main topics have been conveniently addressed for the best selection: (a) spectral bands (visible and infrared); (b) imaging sensors and optical systems (including intrinsic parameters) and (c) geometric visual system arrangement (considering extrinsic parameters and stereovision systems). A general overview, with detailed description and technical support, is provided for each topic with illustrative examples focused on specific applications in agriculture, although they could be applied in different contexts other than agricultural. A case study is provided as a result of research in the RHEA (Robot Fleets for Highly Effective Agriculture and Forestry Management) project for effective weed control in maize fields (wide-rows crops), funded by the European Union, where the machine vision system onboard the autonomous vehicles was the most important part of the full perception system, where machine vision was the most relevant. Details and results about crop row detection, weed patches identification, autonomous vehicle guidance and obstacle detection are provided together with a review of methods and approaches on these topics.

Keywords: machine-vision; spectral bands; imaging sensors; optical systems; geometric arrangement; 3D/2D mapping; crop rows detection; weed control; guidance; obstacle detection

1. Introduction

The incorporation of machine vision systems in agricultural environments is becoming more and more common, and is undergoing a period of continuous boom and growth, particularly onboard agricultural vehicles (autonomous and non-autonomous), but not limited to this case. These systems can be used for different agricultural tasks, including crop (patches, rows) detection, weed identification for site-specific treatments, monitoring or canopy identification, among others, where precise guidance is required and the security and surveillance in the area of influence become crucial issues.

With progress, machine vision systems become imperative in autonomous vehicles and very useful for driver assistance in non-autonomous vehicles, considering that they work under adverse outdoor conditions from the point of view of image processing, required for such purposes. A review of control systems in autonomous vehicles was carried out in [1–3], where four subsystems were identified: guidance, weed detection, precision actuation and mapping. Imaging sensors were used for different tasks, including guidance, weed detection or phenotyping analysis. Crop rows detection and weeds identification are common tasks in precision agriculture where image processing techniques are used for site-specific treatments [4–16], guidance based on crop lines following [17–20], obstacle detection for security purposes [21–25] or mapping the environment in olive trees [26], among others. Recent technological advances have allowed the incorporation of vision systems onboard Unmanned Aerial Vehicles (UAVs), which can also be considered as agricultural vehicles, demanding priority attention [27].

Once a vision system is to be installed onboard an agricultural vehicle, either for assistance, autonomy or specific applications, with the purposes expressed above, several questions are to be considered, namely: what is the system? What specifications should it have it? Where should we place it onboard the vehicle? How should it be oriented it towards the 3D scene? Answers to these and other problematic questions are crucial, and they are to be considered for any engineering design, affecting the machine vision system and its integration onboard the vehicle. The following main issues are to be addressed, without forgetting the economic costs:

1. Tradeoff between vision system specifications and performances. Operating spectral ranges are to be identified, i.e., multispectral, hyperspectral, including visible, infrared, thermal or ultra-violet. Spectral and spatial sensor's resolutions are also to be considered including the intrinsic parameters.

2. Definition of the region of interest and panoramic view. Apart from the spatial resolutions mentioned above, the optical system plays an important role in acquiring images with sufficient quality, based on lens aperture. At the same time, lens distortions and aberrations are to be determined. The field of view, in conjunction with the sensor resolution, must also be determined.

3. Vision system arrangement with specific poses onboard the vehicles (ground or aerial). All issues concerning this point are related to the vision system location: height above the ground, distance to the working area or region of interest, rotation angles (roll, yaw and pitch). Extrinsic parameters are involved.

Thus, regarding the above considerations this paper addresses three main issues concerning the machine vision systems onboard agricultural vehicles, namely: (a) spectral-band selection; (b) imagers sensors and optical systems and (c) geometric system pose and arrangement. The main contribution of this paper involves such issues, which are to be considered before a machine vision system is selected to be installed onboard an agricultural vehicle for specific tasks in agriculture.

This paper is organized in two parts. The first one comprises three Sections 2–4. Section 2 describes the spectral band selection. Section 3 is devoted to imaging sensors and optical systems. Section 4 deals with the geometric system pose. Illustrative examples in agricultural contexts are also provided to clarify the related issues. The second part comprises Section 5, which describes a case study, based on the RHEA (Robot Fleets for Highly Effective Agriculture and Forestry Management) project [28]. In the corresponding subsections of Section 5 we explicitly indicate the link with Sections 2–4. Finally, an additional appendix provides the basic concepts for camera system geometry.

2. Spectral-Band Selection

2.1. Visible Spectrum

Most agricultural tasks using machine vision systems require image processing techniques with the aim of identifying specific spectral signatures. Vegetation indices allow the extraction of spectral features by combining two or more spectral bands, based on reflectance properties produced by the vegetation [29,30]. Some of them use only the three visible spectral bands, i.e., Red (R), Green (G) and Blue (B), where the goal is to enhance some specific band, accentuating the spectral signatures (color) of interest. In this regard, if the greenness is the interest, the G band values are to be enhanced, when soil segmentation is the interest, the R band values should be enhanced, excess green and excess red are two well-known indices for such purposes [11]. The first one is applied for detecting green plants, including crop patches and crop rows, weed patches, leaves and other vegetative parts. The second one is used for other purposes such as soil analysis (organic composition, moisture, etc.). CCD (Charge Coupled Device) or CMOS (Complementary Metal Oxide Semiconductor) are two common technologies used in imaging sensors devices. They are both based on the photoelectric effect to produce digital intensity values from incident light over specific picture elements (pixels), which are the smallest units, conveniently arranged in matrices with specified horizontal (H) and vertical (V) sizes or linearly as an array of pixels. Section three is devoted to sensors.

The greater the intensity of light, the more electrons are produced [31]. Light consists of photons (discrete particles), but a light source produces photons randomly throughout time. This causes noise in the perceived intensity of the light and this magnitude is equivalent to the square root of the number of photons generated by the source of light (Shot Noise) measured in electrons (e^-). Ideally, every photon would be converted in one electron, so that this conversion is governed by physical laws. Nevertheless, there are factors altering the ideal conversion, which produce what is known as noise, such as the read-out noise due to electronic operation, camera processing noise or dark current shot noise, among others, leading to discrepancies between the ideal and real performance. The electrons generated are stored within each pixel in the Well and the number of electrons that can be stored is known as Saturation Capacity or Well Depth (measured in e^-), so that if the Well receives more electrons than the saturation capacity no additional electrons are stored. The charge measured in the Well is called the Signal and the error due to this measurement is known as Temporal Dark Noise (TDN) or Read Noise (measured in e^-). After this, the grey imaging value (Grey Scale) is obtained by converting the signal value expressed in electrons into pixel values in bits (8, 16 or others) through Analog to Digital Units (ADUs). The ratio between the analog signal value and the digital grey scale value is known as Gain (measured in electrons per ADU) which differs from the analog to digital conversion. Manufacturers provide information about the ratio between the ideal and real situations measured in terms of Signal to Noise Ratio (SNR) in decibels (dB) or equivalently in bits of data, applying the conversion expression bit = log2(10SNR/20). Typical values of SNR are around 50–60 dB which can be determined through specific calibration processes. This is a quality measurement of camera performance between the ratio of noise versus the signal together with Dynamic Range, this last one also measured in dB or bits. The difference is that Dynamic Range considers only the TDN, while SNR also includes the root mean square summation of the Shot Noise. There is another metric known as Absolute Sensitivity Threshold, which is the minimum number of photons required to obtain a signal equivalent to the noise produced by the sensor. Below this threshold value no significant signal is produced. Sometimes, light density (photons/μm^2) against signal (e^-) or SNR are available and the best sensor is the one with the highest signal/SNR values for the same light densities. The above is valid for both CCD and CMOS devices.

CCD and CMOS are blind to color, so that when color is to be generated a band-pass filter is placed in front of each sensor to allow the incidence of light according to the input radiation. Depending on the type of system, i.e., with a unique CCD or several, different technologies are used. A typical arrangement in imaging sensors with a unique CCD is the known as Bayer's filter. Alternating red-green and blue-green pixels are conveniently placed to obtain RGB (Red, Green, Blue) images, complementary color's filters (cyan, magenta, yellow) can also be used to produce CMY images. Software-based image processing techniques allow the direct/reverse transformation between the two colors models. In CCD devices, the charge produced on the pixels by the incident light is transferred, using vertical shift registers, to a node or nodes where the charges are converted to voltage, buffered and sent out as an analog signal, which is amplified and digitalized by an analog to digital (A/D) converter through the ADU. In CMOS devices, each pixel contains its own converter from charge to voltage, sometimes including amplifiers, noise reducers and electronic digitization. Because of this, the output uniformity is greater in CCD than in CMOS, giving high image qualities but with higher noise. In contrast, CMOS technology produces lower levels of noise with faster read-out, and lower power consumption.

Manufacturers of camera-based sensors (CCD, CMOS) provide for each device a data-sheet containing information (sometimes graphical) about the sensor sensitivity measured in terms of absolute Quantum Efficiency (QE) or Relative Response (RR) [31]. QE is the percentage of photons converted to electrons at a specific wavelength, expressed in percentage. The Signal (as a measure of the charge, as mentioned above) is computed as the product of LightDensity (LD), expressed as the number of photons/μm^2, the pixel area (pixel size, PS) and QE as follows,

$$Signal = LD \times PS^2 \times QE \tag{1}$$

Figure 1a displays an illustrative generic graph representing a RR against wavelengths for a RGB sensor. Figure 1b also displays the QE against wavelengths for a three spectral RGB sensor with response in the near infrared and beyond. If the sensor is monochrome, a typical profile could be the one represented in Figure 1c, also against wavelengths.

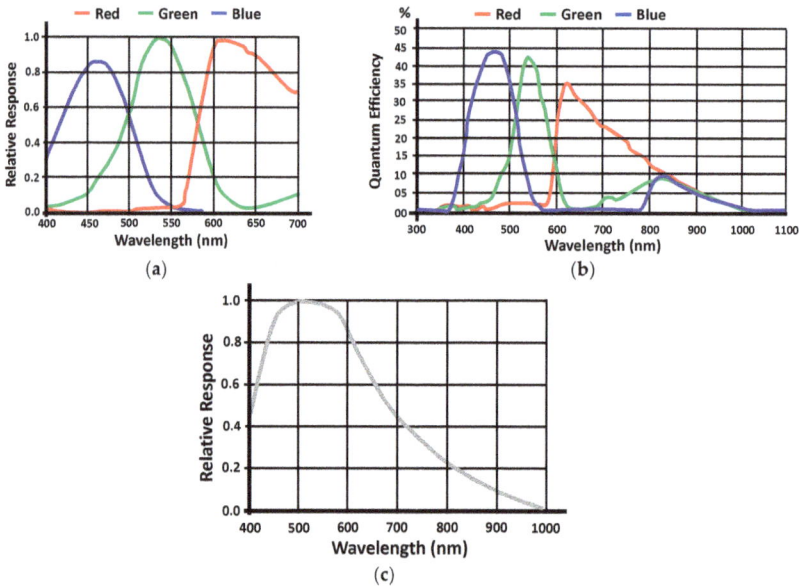

Figure 1. Generic spectral responses: (**a**) Relative Response (RR) for a RGB sensor; (**b**) Quantum Efficiency (QE) for a RGB sensor; (**c**) RR for a monochrome sensor.

2.2. Spectral Corrections: Vignetting Effect and White Balance

In agricultural outdoor environments the machine vision system works in adverse conditions where the natural illumination contains high NIR and UV spectral components (radiation). Generally, imaging sensors are highly sensitive to NIR radiation starting at 760 nm and to a lesser extent to UV, below 400 nm. Indeed, based on the spectral responses displayed in Figure 1b, the NIR heavily contaminates the three spectral channels (R, G and B), mainly the red channel in the range 760–800 nm, producing images with hot colors. This makes identification of green vegetation unfeasible. To avoid this undesired effect, cut-off filters are required, such as a Schneider UV/IR 486 [32]. Its operating curve specifies that wavelengths below 370 nm and above 760 nm are blocked, i.e., both UV and NIR radiations. Figure 2a displays just a corrupted image acquired without the UV/IR 486 cutting filter and Figure 2b equipped with such filter. As mentioned above, without such a filter the contamination is obvious and the undesired effect is clearly minimized with the filter. These images were acquired with a CCD-based sensor with the corresponding optical system onboard the tractor dedicated to maize crops belonging to the fleet of robots in the RHEA project [28]. Details about this system are provided in section five.

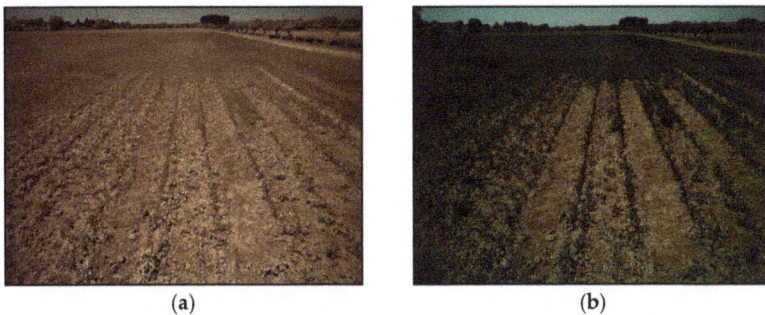

(a) (b)

Figure 2. Effect of the UV/IR cutting filtering: (**a**) without filter; (**b**) with filter.

Despite the blocking filtering, a vignetting effect still remains, requiring correction. As specified by the manufacturer, the Schneider UV/IR 486 cut-off filter is based on what is known as thin-film technology containing more than thirty coats on one of its sides and a multi-resistant coating on the opposite one. The incidence angle of rays in the periphery of the filter is greater than in the center and they must travel longer distances along the different layers of interference. This effect is more pronounced the lower is the focal length of the lens, i.e., lenses with wide-angles. These cutting filters, particularly IR filters, are generally incorporated by the manufacturer on off-the-shelf digital cameras, because its selection for a specific agricultural application is unnecessary. The vignetting effect causes important anomalies on the spectral features. Indeed, because of the larger distances travelled by these rays, the IR wavelengths are filtered with higher intensity in areas far from the image center than in the central part of the image. By proximity of IR and Red (R) wavelengths in the spectrum, this last one is also affected with an excess of filtering at the expense of Green (G) and Blue (B) bands introducing an excess of G with respect to R, expressed with higher greenness at the external parts of the image and particularly at the corners. Figure 3a displays an image with greenness segmentation by applying the ExG index [8,11]. It is clear that an excess of green plants are segmented. Two approaches can be considered to correct this undesired effect. The first consists on the installation of UV/IR cutting filters just in front of the sensor (CCD, CMOS), with the aim of minimizing the distances traveled by the rays. As mentioned before, in off-the-shelf digital cameras this filter is built-in at the factory and most of the time no additional actions are required. In the second approach, when the first fails or it is not possible, specific spectral bands (R,G and B) corrections are required via software. For each pixel (x,y) a normalized distance ranging in $[0,1]$ is computed as follows,

$$d(x,y) = \frac{(x - x_c)^2 + (y - y_c)^2}{(x_d - x_c)^2 + (y_d - y_c)^2} \qquad (2)$$

where (x_c, y_c) and (x_d, y_d) are the coordinates of the image center and a corner point respectively, Figure 3b. Thus, the following intensity corrections can be applied,

$$R'(x,y) = R(x,y) + \mu_R d(x,y); G'(x,y) = G(x,y) + \mu_G d(x,y); B'(x,y) = B(x,y) + \mu_B d(x,y) \quad (3)$$

The corrected spectral values R', G' and B' for each pixel location at (x,y) are obtained by adding to the original spectral values R, G and B (normalized in the range [0,1]) a term which is a function of the normalized distance $d(x,y)$ and multiplied by the corresponding correction factor μR, μG and μB ranging in [0,1]. In this example, only R is to be increased but not the green and blue, because the greenness segmentation is intended. Figure 3c displays the corrected image by applying the following correction factors $\mu R = 0.3$, $\mu G = \mu B = 0.0$; as can be seen, the excess of greenness has been considerably reduced with the unique emphasis on R.

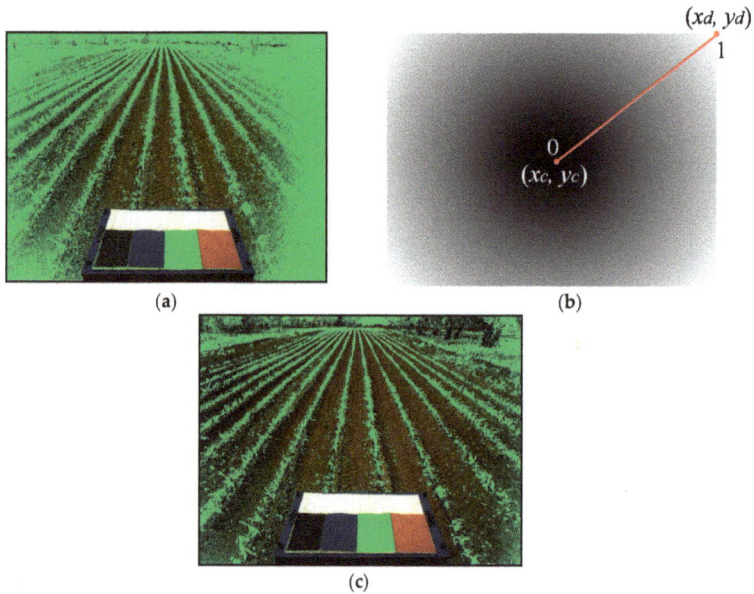

Figure 3. Vignetting: (a) effect with emphasis on the external parts; (b) correction mask; (c) corrected image.

The B spectral band is also affected by proximity to the UV band when a cutting UV/IR filter is used. In this regard, a blue correction could be suitable in order to increase intensity values in the blue band. Nevertheless, because in agricultural applications the greenness is usually the interest, as in the above example, the blue correction is unnecessary.

White balance is another option for improving image quality, based on the correction with reference to known spectral values. Assume we have a reference white panel with nominal spectral white values R, G, B as (255, 255, 255) or equivalently (1, 1, 1) for normalized values. Considering a region on the known white reference panel with sizes of 50 × 50 pixels as an example, the average values R_W, G_W, B_W are computed for such a region and the white balance correction is applied as given by Equation (4). Figure 4 displays in (a) an original image with balance correction in (b).

$$\begin{bmatrix} R' \\ G' \\ B' \end{bmatrix} = \begin{bmatrix} 255/R_W & 0 & 0 \\ 0 & 255/G_W & 0 \\ 0 & 0 & 255/B_W \end{bmatrix} \begin{bmatrix} R \\ G \\ B \end{bmatrix} \quad (4)$$

The problem with the application of white balance is that the black area must be correctly located and free of additional effects, such as projection of shades affecting exclusively to such region but not to other parts in the image. For example, a shadow from the cabin on the reference panel causes anomalies on the spectral correction in the rest of the image.

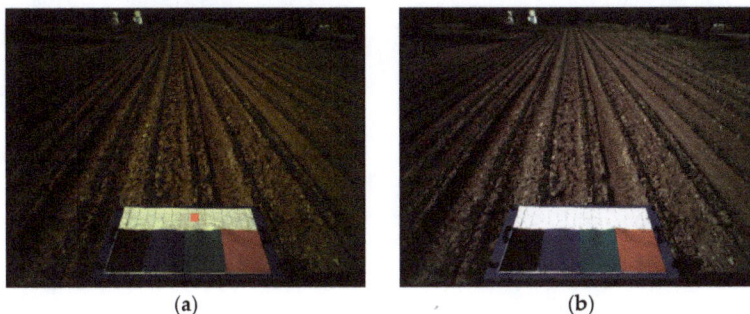

(a)   (b)

Figure 4. White balance: (**a**) original image; (**b**) corrected image.

2.3. Infrared Spectrum

It is well-known in remote sensing applications [33], where green vegetation is to be identified from sensors onboard airborne or satellite platforms equipped with multi(hyper)-spectral imagery sensors, that near infrared is a useful band for plant identification and phenotyping because green vegetation produces high reflectance in the NIR band due to chlorophyll activity and absorption [34,35]. In this regard, according to the agricultural application to be developed, the best approach consists of determining the matching between the agricultural objects to be detected and the sensor spectral response. Figure 5a displays typical reflectance spectra profiles at different wavelengths for crop and soil, which are roughly drawn from the information provided in [34], where the maximum reflectance is achieved between 700 nm and 1300nm. Thus, considering that NIR corresponds to wavelengths falling within the 760 to 1400 nm range, the best sensor for capturing crop reflectance should be the one with the higher response inside this range. There exist sensors based on Indium Gallium Arsenide (InGaAs) technologies covering different infrared ranges, roughly Short-Wave infrared (SWIR, ~1400–3000 nm), Mid-Wave infrared (MWIR, ~3000–8000 nm), and Long-Wave infrared (LWIR, ~8000–15,000 nm). Figure 5b displays two responses covering two spectral ranges corresponding to two respective versions of the Bobcat-640-GigE sensor [36]. This sensor contains a detector based on InGaAs (Indium/Gallium/Arsenic) as the substrate to build the focal plane array with two readout integrated circuit (ROIC) modes (Integrate Then Read, ITR and Integrate While Read, IWR) and noise level of 90 e^- and 640 × 512 pixels. Other substrates are also possible for NIR-based devices, covering different spectral ranges, such as Indium/Antimonide (InSb), Mercury/Cadmium/Tellurium (HgCdTe) among others with different sensibilities.

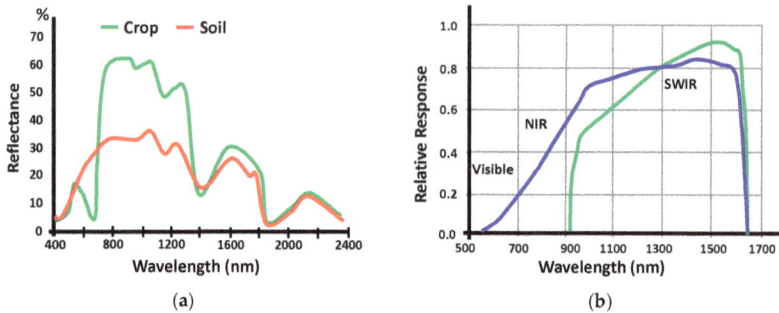

Figure 5. (a) Typical spectral reflectance profiles for crops and soil roughly drawn from the information provided in [34]; (b) Relative response from two generic sensors covering Near-Infrared (NIR) and Short-Wave infrared (SWIR) spectral ranges.

So, if we want to detect crop reflectance below 900 nm the most appropriate sensor is the one covering the range between 550 to 1700 nm, otherwise, if the crop reflectance is above 900 nm, the sensor covering the range from 900 to 1700 nm should be acceptable.

Table 1 summarizes different ranges of wavelengths (λ), expressed in nm, and related to the spectral bands (S) commonly used in agricultural applications, particularly for greenness identification. They cover Ultra-Violet (UV), Visible with Blue (B), Green (G), Red (R) and Infra-Red (IR) split on Near-Infrared (NIR), Short, Mid and Long waves.

Table 1. Spectrum (S) and wavelengths λ (nm).

S	λ (nm)	S	λ (nm)	S	λ (nm)	S	λ (nm)
				Blue	450–500		Near	760–1400
UV	1–380	Visible	380–780	Green	500–600	IR Short-Wave (SWIR)	1400–3000
						Mid-Wave (MWIR)	3000–8000
				Red	600–760	Long-Wave (LWIR)	8000–15,000

2.4. Illustrative Examples and Summary

Assume we have a sensor with the spectral specifications displayed in Figure 1a, where the agricultural application consists in the crop row detection of green plants for guiding purposes in maize fields, where typical reflectance values are around 560 nm. Wavelengths for green reflectance is around 500–570 nm, thus the sensor response according to Figure 1a provides a relative red reflectance r = 0.20 and a relative green reflectance g = 0.80 and the Green Red Vegetation Index (GRVI) [33], GRVI = (g − r)/(g + r), results in 0.60. Nevertheless, if the reflectance sensor profiles are the ones provided in Figure 1b, r = 0.02 and a relative green reflectance g = 0.35 and GRVI is 0.89, then the sensor represented in Figure 1b is more efficient in this kind of situation. The best sensor for greenness identification, where wavelengths range from 500–570 nm, is the one with a green spectral response covering this range with tails being the minima out of such a range. In contrast, if the red spectral response in the range of 500–570 nm is null, the GRVI achieves maximum values. In short, the best sensor for greenness identification will be the one with high green spectral responses in 500–570 nm and null for the red ones, i.e., with minimum overlapping between the spectral R and G bands. Regarding a monochrome sensor with its relative response displayed in Figure 1c, we can see that for 560 nm its response is close to 1.0, i.e., with a good performance for the intended greenness identification. Sometimes, during tilling operations, perhaps for automatic guidance [37], the goal is the identification of spectral responses from the soil. Consider that we are interested in the segmentation of dry clay soils with reflectance values around 650 nm. According to Figure 1a,b GRVI values are respectively −1.0

and −0.9; again the sensor represented by Figure 1a provides the best performance. Table 2 displays values for different vegetation indices [8,11] based on r, g and b values for 560 nm according to the RR and QE spectral responses in Figure 1a,b respectively. The best performances are achieved with the maximum values marked in bold.

Table 2. Vegetation indices values for RR and QE for a wavelength of 560 nm.

Vegetation Indices	RR for 560 nm, Figure 1a r = 0.20; g = 0.80; b = 0.01	QE for 560 nm, Figure 1b r = 0.02; g = 0.35; b = 0.03
GRVI = (g − r)/(g + r)	0.60	0.89
ExG = 2 g − r − b	1.39	0.65
ExR = 1.4 r − g	−0.52	−0.32
ExGR = ExG − ExR	1.91	0.97
CIVE = 0.441r − 0.811 g + 0.385b + 18.78745	18.23	18.52
VEG = gr^{-a}b$^{(a-1)}$ with a = 0.667 which was defined in [38]	10.85	15.29

There exist commercial 2CCD (bi-channel) [39] or 3CCD (three-channels) [40] devices capturing simultaneously visible RGB in raw Bayer or separated together with NIR, respectively. Visible and NIR spectra are separated by the dichroic coatings of the prism with a separation wavelength of about 760 nm in the 2CCD device and about 600 nm and also 760 nm for the separation of the green, red and NIR in the 3CCD device.

Sometimes, a band pass NIR filter can provide a solution by placing it in front of the optical system in the visible imager. In this regard, based on the visible spectral responses displayed in Figure 1, we must consider that the sensor is still active with sufficient responses for wavelengths inside the infrared range so that the CCD or CMOS cells are activated with wavelengths crossing the NIR filter. This was the solution proposed in [41,42] in the context of stereovision systems intended for autonomous navigation.

Another solution is the one proposed in [35], where the IR cutting filter in the visible camera, if any, is removed, allowing the input of NIR so that the RGB spectral channels contain an amount of NIR, i.e., R + NIR, G + NIR and B + NIR. With a filter blocking the blue wavelengths, placed in front of the lens or immediately in front of the sensor, the blue channel should be exclusively impacted with NIR exclusively providing the NIR component. Subtracting the blue channel (containing only NIR) from the other two, R, G and NIR spectral responses are obtained. Nevertheless, because the responses from all devices are real, and differ from the nominal or ideal, this procedure requires an extra effort in order to define the best cutting blue filter and also the combination of bands to obtain the required R and NIR real responses to derive vegetation indices by using R and NIR channels. A calibration and estimation is carried out in the laboratory with a tunable monochromatic light source spectrometer.

Active sensors are used for phenotyping studies based on Normalized Difference Vegetation Index (NDVI) and canopy densities [43]. A monochrome CCD camera (5 MPix) is mounted in a position two meters above the canopy surface inside a box with a LED light panel also inside the box illuminating the surface to produce nine spectral wavelengths (465, 500, 525, 590, 615, 625, 660, 740 and 850 nm) as the active light source for multispectral images.

Plant phenotyping represent an important challenge in agriculture applications where wavelength band selection plays an important role for determining some specific parameters such as morphology, biomass, leaf forms, fruit characteristics, yield estimations, water content, photosynthetic activity or stress. Different machine vision systems are to be considered because of the advances in imaging techniques, involving spectroscopy (multi-hyper), thermal infrared, fluorescence imaging, 3D imaging, and recently tomographic imaging (Nuclear Magnetic Resonance Imaging, Positron Emission Tomography, X-ray Computed Tomography) for seeds, roots or transport analysis [44].

Under the above considerations, the specifications and features for a machine vision system in outdoor applications and particularly for agricultural tasks can be summarized as follows:

1. Broad spectral dynamic range with adjustable parameters to control the amount of charge received by the sensor, considering the adverse environmental conditions that cause high variability on the illumination in such outdoor environments. In this regard, specific considerations are to be assumed depending on the vehicle (ground, aerial) where the machine vision system is to be installed onboard. Of particular relevance is the effect known as bidirectional reflectance, which appears in sunny days due to angular variations, which may become critical in aerial vehicles [45].

2. Ability to produce images with the maximum spectral quality as possible, avoiding or removing undesired effects such as the vignetting effect.

3. A system robust enough to cope with adverse situations and with responses as deterministic as possible.

3. Imaging sensors and Optical Systems Selection

3.1. Imaging Sensors

An important step during the machine-vision selection process is to consider the imaging sensors and the optical system together. As mentioned before, CCD or CMOS devices consist of pixels conveniently arranged in matrices with specified horizontal (H) and vertical (V) sizes or linearly as an array of pixels. The variety of sensors with different $H \times V$ sizes or linear sizes is certainly high, commercially there is a high variety of ranges, from 160×120 to 6576×4384 or above, UV and NIR-based systems contain the lower resolutions. The product $H \times V$ determines what is called MegaPixel (MP) in terms of millions of pixels, for example, a device with resolutions of 3376×2704 is measured as 9.1 MP. Physically each pixel has its own horizontal (h) and vertical (v) sizes, typical values range from 3.45 to 7.50 µm. The nominal sensor sizes, horizontal (S_h) and vertical (S_v), can be computed as $S_h = h \times H$ and $S_h = v \times V$; nevertheless, real dimensions are a bit larger in size because the required alignments and arrangements. In summary, CCD and CMOS chip sizes vary considerably [46]. Historically, these dimensions come from Vidicon TV cameras with their imaging tubes projecting the image in a circle with a given diameter. CCD/CMOS chips are designed with rectangular dimensions with their corresponding diagonal lengths. The following association (type, diagonal) is established between the type of sensor and its diagonal in length units expressed in mm: (1/8", 2.0), (1/6", 3.0), (1/4", 4.0), (1/3.6", 5.0), (1/3.2", 5.68), (1/3", 6.0), (1/2.7", 6.72), (1/2.5", 7.18), (1/2.3", 7.7), (1/2", 8.0), (1/1.8", 8.93), (1/1.7", 9.50), (1/1.6", 10.07), (2/3", 11.0), (1/1.2", 13.3), (Super 16 mm, 14.54), (1", 16.0), (4/3", 21.6), (Canon APS-C, 26.7), (Pentax Sony Nikon DX, 28.4), (Canon APS-H, 34.5), (35 mm, 43.3), (Leica S2, 54.0), (Kodak KAF 3900, 61.3), (Leaf AFi 10, 66.57) or (Phase One P 65+, 67.4). A specific sensor is assigned to the type whose diagonal is at least as large as the one of the sensor. For example, assume a chip with sizes 11.6×9.5 mm^2, with diagonal of 14.99 mm, it will be assigned to the type 1". This will lead to some geometric constraints of particular interest in machine vision systems in agriculture and particularly for stereovision systems and also when choosing the optical lenses, as we will see later.

In terms of agricultural applications, the choice of a sensor will be determined by its potential features. So, if poor illumination conditions are expected, such as the ones carried out at dawn or dusk the most suitable should be a CMOS technology. CMOS is also appropriate when timing is critical, for example when the time between image acquisition and actuation is extremely low. This could be the case during weeds removal by applying herbicide, based on nozzle sprayers, where the camera is attached to each single nozzle sprayer and weeds are identified for immediate spraying. Cameras in zenithal positions with respect to the region of interest, CMOS should be suitable as it provides rapid responses. Nevertheless, in most agricultural applications, involving image processing, time is critical but not extreme. Moreover, illumination conditions cause problems because of high variability in days with alternating periods of sun and clouds with rapid and frequent changes. Also, problems can appear in days with high/low lighting intensity due to sunny/cloudy days in the outdoor agricultural environments, but these problems are never critical enough to require the use of a CMOS. Here, CCD-based sensors could be appropriate, conveniently connected to real time processors under efficient HW/SW architectures [47]. In this regard, Giga Ethernet (GigE, sometimes including dual ports), Camera Link, USB 2.0, USB3 Vision (USB 3.0) or IEEE 1394a,b (FireWire) are appropriate interfaces to guarantee sufficient data (images) transmission rates. A description about specific features and reasons for choosing the right camera bus are given in [48] based on throughput, cable length, standardized interface, power over cable, CPU usage, I/O synchronization and also effective cost, where relative rankings are provided for each bus.

Additionally, to deal with the adverse illumination conditions in outdoor agricultural environments we have still available two resources: exposure time and aperture. They can be controlled either by the optical system, by applying external control via HW/SW, or based on image processing or both, to achieve sufficient qualities avoiding images with over/under-exposure [49].

Exposure time is the time that the sensor is continuously receiving the light until the signal is produced. The higher the exposure time the greater is the light received by the sensor and vice versa. The exposure time values depend on several factors, including the type of sensor, such values are specified by manufacturers, generally varying from 3 μs to 60 s as maximum values in internal control mode and to ∞ for external control.

A trade-off must be achieved between the exposure time and aperture. The Exposure Value (EV), Equation (5), has been defined to combine both magnitudes so that different combinations give the same exposure value.

$$EV = \log_2 \left(F^2/t \right) \tag{5}$$

where F is the *f-number* (defined in Section 3.2) and t is the exposure time.

EV is used in professional photography where there exists a broad knowledge about the more appropriate values for specific scenes, so that by fixing one of them the other can be obtained from Equation (5), once the EV is determined, based on existing look-up tables. In agricultural outdoor environments, as far as we know, there are no evidences about such values. In this regard, in optical systems with manual aperture, i.e., when the *f-number* must be set before the agriculture task, the best option is to apply a control via image processing. This is the case in the RHEA project [28] for weeds and crop row detection, where a Region Of Interest (ROI) was selected (Section 5), which is the area where specific treatment is to be applied and also the area containing the crop rows used as reference for guiding the autonomous vehicle. The image brightness on the ROI is processed, based on histogram image analysis, and the exposure time is conveniently increased or decreased depending on first order statistical histogram values, such as the mean and standard deviation. An image processing procedure was designed in [50] to automatically set the exposure time.

Another issue concerning the selection of imaging devices is the capability to capture frames, measured as frame rates per second (*fps*). Depending on the sensory technologies and spatial resolutions, currently *fps* can vary from 7 to 1300 or above; so that, in general, CMOS-based technologies allow the sensors to achieve higher *fps* than CCD-based. In this regard, from the point of view of agricultural applications, it is required to determine the best *fps* choice for performance. Common operation speeds in autonomous ground agricultural vehicles can range from 3 km/h (0.83 m/s) to 8 km/h (2.22 m/s) or higher. This means that the ROI to be processed, once it is mapped on the image plane, must be defined with sufficient length to guarantee that it can be processed inside the specified time limits, when the autonomous vehicle moves forward. In this regard, the *fps* and the tasks allocated to the imaging processor must be considered, because the processor is probably in charge of other different processes coming from other sensors [47].

3.2. Optical Systems

The amount of radiance received by the sensor is controlled by the optical system consisting of the following main features and elements [51]:

1. *Set of lenses*, which is the main part of the optical system. Manufacturers provide information about the focal length (f) and related parameters. Sometimes includes a manual focus setting or autofocus to achieve images of objects with the appropriate sharpness. Systems with variable focal length exist, based on motorized equipment with external control. The focal length is a critical parameter in agricultural applications which is to be considered later for geometric machine vision system arrangement.

2. *Format*. Specifying the area of the sensor to be illuminated. This area should be compatible with the type of imaging sensor, specified above. An optical system that does not illuminate the full area creates severe image distortions. Figure 6 displays a sensor of type 2/3″ and a lens of 1/2″, i.e., the full sensor area is greater than the area illuminated by the lens.

3. *Iris diaphragm* automatic or manual. This consists of a structure with movable blades producing an aperture which controls the area where the light, traveling towards the sensor, passes. Manufacturers specify it in terms of a value called the f-stop or f-number, which determines the ratio of f, to the area of the opening or more specifically the diameter (A) of the aperture area, i.e., $N = f/A$. The aperture setting is defined as steps or f-numbers, where each step defines a reduction by a half of the intensity from the previous stop and consequently a reduction in the aperture diameter of $2^{-1/2}$. Figure 7 displays a lens aperture according to the f-number which is minimum in (a) with 16 and maximum in (b) with 1.9. Depending on the system, the scale varies, represented in fractional stops. So, to compute the scaled numbers in steps of $N = 0, 1, 2, \dots$, with the scale s, the following sequence is normally used: $2^{0.5(Ns)}$. The scales are defined as full stop ($s = 1$), half stop ($s = 1/2$), third stop ($s = 1/3$) and so on. The following is an illustrative example, if $s = 1/3$ the scaled numbers are: $1, 1.1, 1.3, \dots, 2.5, \dots 16, \dots$

4. *Holders and interfaces*. With the aim of adapting the required accessories, filter holders are specified. The type of mount (C/F) is also provided by manufactures.

5. *Relative illumination and lens distortion*. Relative illumination and distortion (barrel and pincushion) are provided as a function of focal distances.

6. *Transmittance (T)*: Fraction of incident light power transmitted through the optical system. Typical lens transmittances vary from 60% to 90%. A T-stop is defined as the f-number divided by the square root of the transmittance for the lens. If T-stop is N the image contains the same intensity as the ideal lens with transmittance of 100% and with f-number N. Relative spectral transmittance with respect wavelengths is also usually provided. Special care should be taken to ensure the proper transmission of the desired wavelengths toward the sensor.

7. *Optical filters*. Used to attenuate or enhance the intensity of specific spectral bands, they transmit or reflect specific wavelengths. To achieve the maximum efficiency, their different parameters should be considered, including central wavelength, bandwidth, blocking range, optical density, cut on/off wavelength [52]. A common manufacturing technique consists of a deposition of layers alternating materials with high and low index of refraction. An example of a filter is the Schneider UV/IR 486 cut-off filter [32].

Figure 6. Imaging distortion caused by a sensor of type 2/3″ and a lens of 1/2″.

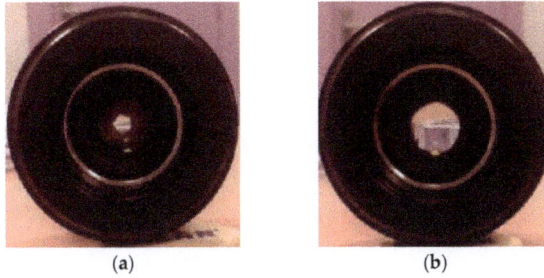

(a)   (b)

Figure 7. Lens aperture according to the f-number: (a) minimum with 16; (b) maximum with 1.9.

The choice of the optical system for agricultural applications is of special relevance in order to guarantee a correct performance oriented toward the acquisition of images with sufficient quality. In this regard, the image must be correctly focused (manually or with autofocus) because feature extraction depends highly on focus. Plants and structures that are out of focus do not provide appropriate features for discrimination. A compatible format between sensor and lens is mandatory in order to avoid distortions. An iris diaphragm could be automatic for self-adjusting, although a manual diaphragm could sometimes be suitable such that it can be controlled for a sufficient amount of illumination, which together with the exposure time control and image analysis allows the correct control for acquisition of images with the required quality. Transmittance and optical filters should be chosen properly to minimize undesired effects, such as vignetting. In agricultural applications the focal length selection is crucial for defining the most appropriate ROI. The next section is devoted to this issue.

3.3. Focal Length Selection

An important subject concerning the optical system is the selection of the focal length [53]. Depending on the field of view, the working distance where objects of interest are placed and the sensor sizes, the focal length requires a convenient selection. As mentioned before, it is well-known that the main element in the optical systems is the lens with its corresponding focal length, f, where in a converging lens all incoming rays parallel to the optical axis intersect. Figure 8 displays the basic elements of a generic converging optical system. H represents the field of view in the scene, h is the sensor size, D is the working distance and d is the distance from the lens to the image plane, i.e., the focus distance where the object appears focused on the image plane.

The Gaussian lens expression and magnification factor (m) are given as follows,

$$\frac{1}{f} = \frac{1}{D} + \frac{1}{d}; \quad m = \frac{h/2}{H/2} = \frac{d}{D} \tag{6}$$

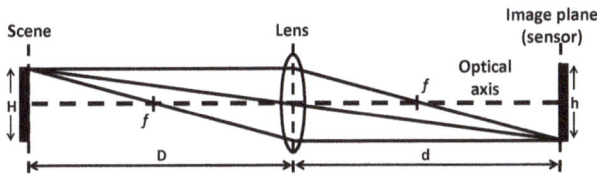

Figure 8. Optical system setup.

By combining both expressions, the following relation can be derived,

$$f = \frac{hD}{h + H} \tag{7}$$

For example, consider an agricultural machine vision application based on the Kodak KAI 04050 M/C sensor, specified in Section 5.1, with horizontal size 2336 pixels × 5.5 μm/pixel = 1285 mm. The ROI is 3 m wide or a tree is 3 m height, i.e., H = 3 m and the working distance is D = 5m. Under these considerations, applying the Equation (7) the required f results in 10.68 mm, which is a reference for selecting the focal length.

4. Geometric Visual System Attitude

4.1. Initial Considerations

Once the above issues have been considered, the next action, oriented toward the visual system selection in agricultural applications, is the geometric system arrangement. The main goal in this regard consists of determining the vision system pose, particularly onboard autonomous ground vehicles, where a set of specific 3D extrinsic parameters, involving translation and rotation matrices, are critical. These parameters combined with the also critical intrinsic parameters (focal length, sensor dimensions) allow us to determine how the 3D scene in the field is to be projected on the image plane. This represents an important challenge; particularly during the vision system selection process. Indeed, there are several tasks with specific requirements. The following is a list of examples:

1. *Crop row detection*: sometimes a fixed number of crop rows are to be detected for crops and weeds discrimination for site-specific treatments or precise guiding [5,7,10–16]. Depending on the number of crop rows to be detected or to follow during guidance, the vision system must be conveniently designed such that the required number of rows, considering the inter crop row spaces, can be imaged with sufficient image resolutions.
2. *Plants leaves, weed patches, fruits, diseases*: different applications have been developed based on sizes of structures. In [54] morphology of leaves is used for weed and crop discrimination based on features by applying neural networks. Apples are identified and counted on their context on the trees in [55]. Fungal or powdery mildew diseases are identified in [56,57]. The machine vision must provide sufficient information and the structures (leaves, patches fruits) must be imaged with sufficient sizes and dimensions to obtain discriminant features for the required classification or identification. In this regard, small mapped areas could be insufficient for such a purpose.
3. *Tracking stubble lines*: machine vision systems for tracking accumulations of straw for automatic baling in cereal has been addressed in [58], where a specific width is required to guide the tractor dragging the baling machine.
4. *Spatial variations*: plant height, fruit yield, and topographic features (slope and elevation) have been studied in [59], where specific machine vision system arrangements are studied.
5. *3D structure and guidance*: stereovision systems are intended for 3D structure determination and guidance [20,21]. Multispectral analysis is carried improving the informative interpretation of crop/field status with respect to the 2D image plane. The panoramic 3D structure obtained must contain sufficient resolution for such interpretation and also provide a map where the autonomous ground vehicle applies path planning and obstacle avoidance for safe navigation. A variable field of view setup has been experimented for guidance in [22]. An adapted NDVI was used in [60] for distinguishing soil and plants trough a camera-based system for precise guidance in small vehicles.

4.2. System Geometry

The above are illustrative examples where the correct definition of intrinsic and extrinsic parameters will determine the machine vision effectiveness. The process to select a machine vision system, assuming image perspective projection, consists of the following steps:

1. Fix the position of the machine vision Cartesian system onboard the vehicle.
2. Take as reference the central point of the sensor *o*, i.e., the point where the two diagonals in the image plane intersect. This point will be the origin of the secondary coordinate system *oxyz*, with axes (*x,y,z*).
3. Fix the origin *O* and associated Cartesian axes (*X,Y,Z*) of the primary world coordinate system *OXYZ*. This is an imaginary system where the 3D points in the scene are to be referenced. Its positioning must be conveniently set as to facilitate the agricultural tasks.

Given a point *W*(*X,Y,Z*) with its corresponding spatial coordinates, the goal is to define the mapping of this point onto the image plane to obtain its coordinates (*x,y*) with respect to the system *oxyz*, either expressed as length or pixels units. Under the image perspective projection, the problem becomes a transformation between two 3D Cartesian coordinate systems, namely *OXYZ* and *oxyz*. To do that the following steps are required, where at each step an elemental homogenous transformation matrix is applied as follows [61]:

1. Initially the systems *OXYZ* and *oxyz* are both coincident, including their origins.
2. Move the origin of *oxyz* to a new spatial position located at $W_0(X_0,Y_0,Z_0)$, which is the point chosen to place the central point of the image plane, i.e., the origin of the *oxyz* system. This operation is carried out by applying a translation operation through the matrix *G*.
3. Rotate the axes *x*, *y* and *z* with angles α, β and θ respectively. These rotations produce the corresponding elementary movements to place the image plane oriented toward the 3D scene (ROI) to be analyzed. These operations are carried out by applying the following respective operations R_α, R_β and R_θ.
4. Once the image plane is oriented toward the scene, the point *W*(*X,Y,Z*) is to be mapped onto the image plane to form its corresponding image. This is based on the image perspective projection by applying the perspective transformation matrix *P*.

The point *W*(*X,Y,Z*) is mapped onto the image coordinates *x* and *y* through the following composition of elementary matrices in homogenous coordinates as defined in Appendix A.

$$\begin{pmatrix} x \\ y \\ z \\ k \end{pmatrix} = PR_\theta R_\beta R_\alpha G \begin{pmatrix} X \\ Y \\ Z \\ 1 \end{pmatrix} \tag{8}$$

The sizes of the sensor are measured in length units as expressed above as S_h and S_v, thus considering the origin of the *oxyz* reference system placed at the central point of the sensor device, the endpoints of the sensor are located at $(-S_h/2, +S_h/2)$ and $(-S_v/2, +S_v/2)$ for axes *x* and *y* respectively. The coordinates *x* and *y* are also expressed in length units with values in the following ranges: i.e., $-S_h/2 \le x \le +S_h/2$ and $-S_v/2 \le y \le +S_v/2$. Thus, to express *x* and *y* in pixel coordinates, x_p and y_p respectively the following transformation is applied,

$$x_p = \frac{(S_h/2 + x)\,H}{S_h} \quad y_p = \frac{(S_v/2 + y)\,V}{S_v} \tag{9}$$

Given a vision system setup, we can determine the imaging mapping of pixels in the 3D agricultural scenario allowing efficient analysis focused on secure specific operations. The following is a list of issues that can be established under the vision system setup for its correct selection:

1. *Mapping of specific areas*: to determine the number of pixels in the image, which allows us to determine if the imaged area is sufficient for posterior image processing analysis, such as morphological operations where the areas are sometimes eroded. For example, it is very important to determine if such areas can provide discriminatory information based on shape descriptors for dicotyledons against monocotyledons or other different species. Maximum and minimum weed patches dimensions should be also of interest [6,7,10–14,16,62].
2. *Crop lines in wide row crops*: determination of the maximum number of crop lines that can be fully seen widthwise. Maximum resolution that can be seen along with discriminant capabilities. Separation between crop lines to decide if weed patches can be distinguished or they could appear overlapped with the crop lines. Crop lines width and coverage [6,7,10].
3. *Fruits*: sizes of fruits for robust identification [63], where the imaged dimensions determine specific shapes based on sufficient fruit's areas.
4. *Canopy*: where plant's heights or other dimensions can be used as the basis for different applications, such as for plant counting to determine the number of plants of small young peach trees in a seedling nursery [64].

Illustrative examples are provided in section five in the context of the RHEA project [28], where the goal is to determine the best camera system arrangement for crop rows detection.

The machine visual system geometry represents an important issue to be considered in machine vision systems for agriculture:

1. The loss of the third dimension when the 3D scene is mapped onto 2D requires additional considerations in order to guarantee imaged working areas (ROIs) with sufficient resolutions and qualities.
2. Camera system arrangements onboard agricultural vehicles, together with the definition of the sensor's resolutions and optical systems, are to be considered.
3. It is appropriate simulation studies to determine the best resolutions, based on geometric transformations from 3D to 2D.

4.3. Stereovision Systems

Stereovision systems, based on conventional lenses, are specifically dedicated to build 3D maps for different purposes in agriculture [65], including vehicles navigation, operator-assisted and autonomous systems [41], precision agriculture [42], recognition of fruits [66] or for obstacle avoidance for safety purposes [67]. Following the Barnard and Fishler [68] terminology, the problem of stereovision consists of the following steps: image acquisition, camera modeling, image matching and depth determination. The key step is the image matching, that is, the process of identifying the corresponding points in 3D scene. A set of constraints are generally applied for solving the matching problem, as explained in [68–70]: epipolar, similarity, uniqueness or smoothness.

Epipolar: derived from the system geometry, given a pixel in one image its correspondence in the other image will be on the unique line where the 3D spatial points belonging to a special line (epipolar) are imaged. *Similarity*: matched pixels have similar attributes or properties. *Uniqueness*: a pixel in the left image must be matched to a unique pixel in the right one, except for occlusions. *Smoothness*: disparity values in a given neighborhood change smoothly, except at a few discontinuities belonging to the edges, such as borders on trunks or obstacles.

Consider two image planes, I_L and I_R associated to two stereo-cameras with parallel optical axes and projection centers O_L and O_R respectively and separated a baseline B, Figure 10a. The world coordinates system is defined by $OXYZ$, with the *effective focal length, f*, which is assumed to be identical in both optical systems. Let $P(X,Y,Z)$ a 3D point expressed in $OXYZ$, which is projected onto the images planes on $P_L(X_L, Y_L, Z_L)$ and $P_R(X_R, Y_R, Z_R)$ with respect the image coordinates systems $O_L X_L Y_L Z_L$ and $O_R X_R Y_R Z_R$. The projected rays PO_L and PO_R define the *epipolar plane*, whose intersections with image planes define the *epipolar line*. Given the projected point P_L in the left image, its corresponding point P_R in the right image lies on the epipolar line, which defines the epipolar constraint for stereo matching. The difference $d = X_L - X_D$ is known as disparity. By applying triangulation and the similar triangles principle, once d is known by applying stereo correspondence, the depth, Z, for the point P can be established and hence the 3D determination. Figure 10b and Equation (10) display the similar triangles and the depth derivation.

$$\left. \begin{array}{l} O_L : \frac{\frac{B}{2}+X}{Z} = \frac{X_L}{f} \\ O_R : -\frac{\frac{B}{2}-X}{Z} = \frac{X_R}{f} \end{array} \right\} \Rightarrow \left. \begin{array}{l} X_L = \frac{f}{Z}\left(X + \frac{B}{2}\right) \\ X_R = \frac{f}{Z}\left(X - \frac{B}{2}\right) \end{array} \right\} \Rightarrow d = X_L - X_R = \frac{fB}{Z} \Rightarrow Z = \frac{fB}{d} \qquad (10)$$

Once both f and B parameters have been fixed, the main issue is the computation of the disparity for each pixel or for specific features (edges, regions, interest points), this is known as the correspondence problem, which has been addressed broadly, although in different robotics contexts [71], but equally valid in agricultural settings.

In this regard, consider the following example, where we want to design a stereovision system with the following specifications and requirements: baseline 10 cm, the spatial coverage in the X direction should be at least 30 m for a distance Z of 60 m, and f of 10 mm. From Equation (10) we can obtain: $X_L = \frac{10mm}{60 \times 10^3 mm}\left(30 \times 10^3 mm + \frac{100mm}{2}\right) \approx 5.01mm$ and $X_R = \frac{10mm}{60 \times 10^3 mm}\left(30 \times 10^3 mm - \frac{100mm}{2}\right) \approx 4.99mm$, i.e., as an example, the CCD Kodak KAI 04050 M/C sensor, described in section five, with image resolutions of 2336 $\times$ 1752 pixels and 5.5 $\times$ 5.5 µm pixel-sizes suffices for this purpose. Indeed, $X_L/5.5$ and $X_R/5.5$ result in 910.61 pixels, falling inside the image resolutions.

Precision in stereovision systems in agricultural applications becomes an important issue, because sometimes the ratio between 3D parameters and measurement errors becomes very significant. Indeed, assume the goal is to determine plant heights with few centimeters, if the systematic error introduced by the stereovision system is also of centimeters, the results could be dramatic and the system performance will be limited. This issue has been conveniently addressed in [72] under different system settings. Part of these limitations arises from the arrangement of the cells in the CCD/CMOS sensor device [73]. Assume the device contains n pixels (elements) along the horizontal X direction defined by its width p, Figure 9a, we can thus deduce the following relationship expressed in Equation (11),

$$tg\beta = \frac{np}{2f} \approx \beta (radians); \text{ for very small angles} : \frac{p}{f} = \frac{2\beta}{n} \qquad (11)$$

where β determines the Field of View (FOV) angle.

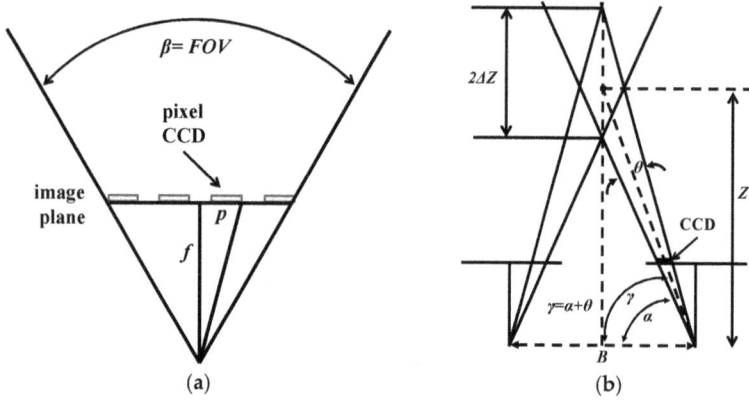

Figure 9. Precision in stereovision systems (images from [73]): (**a**) geometric setting and parameters defined by the CCD; (**b**) geometric relations on triangles from the 3D mapping.

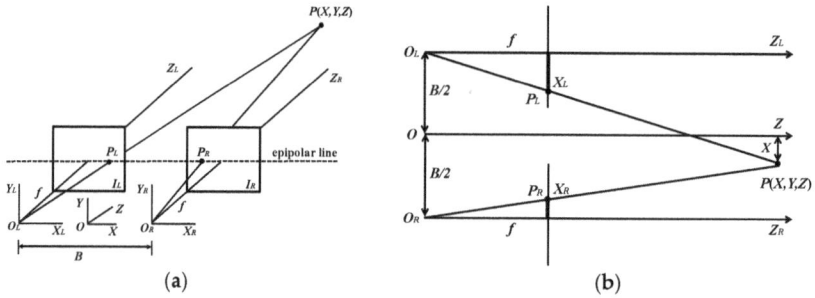

Figure 10. Stereovision system geometry with parallel optical axes (images from [73]): (**a**) mapping of 3D point *P(X,Y,Z)* onto image planes; (**b**) geometric parameters on similar triangles.

From the geometric relations in Figure 9b the following equations can be derived,

$$tg\gamma = \frac{Z + \Delta Z}{B/2}; \quad tg\alpha = \frac{Z}{B/2} \tag{12}$$

$$Z + \Delta Z = \tfrac{B}{2} tg\gamma$$
$$\Delta Z = \tfrac{B}{2} tg\gamma - Z = \tfrac{1}{2} Btg(\alpha + \theta) - Z = \tfrac{B}{2} tg\left(tg^{-1}\left(\tfrac{2Z}{B}\right) + \theta\right) - Z \tag{13}$$

where ΔZ determines the accuracy in terms of the distance Z and the baseline. The Equation (13) can be expressed as a function of Z per baseline units as follows,

$$\frac{\Delta Z}{B} = \frac{1}{2} tg\left(tg^{-1}\left(\frac{2Z}{B}\right) + \theta\right) - \frac{Z}{B} \tag{14}$$

As an illustrative example, let a stereovision system with baseline B = 30 cm and f = 10 mm where each pixel is 5 μm as defined by the manufacturer. According to Equation (14) we need to know θ, which can be inferred from Figure 9, under the following assumption $\theta \approx tg^{-1}(p/f) \approx tg^{-1}(5 \times 10^{-3} mm/10mm) \approx 5 \times 10^{-4} rad$. Once obtained, the inaccuracy for a distance of 4m can be derived from Equation (13), as: $\Delta Z = \frac{30cm}{2} tg\left(tg^{-1}\left(\frac{800cm}{30cm}\right) + 5 \times 10^{-4} rad\right) - 400cm \approx 5.4cm$, this means that the system must be validated with this inaccuracy to be considered as feasible or unfeasible.

5. A Case Study: Machine Vision Onboard an Autonomous Vehicle in the RHEA Project

The RHEA project [28] was envisaged for precision agricultural tasks in maize (*Zea mays* L.), wheat (*Triticum aestivum* L.) and olive trees (*Olea europaea* L.), and the experiments were performed over four years with a final demo (May, 2014) in two fields located in Arganda del Rey, Madrid, Spain, (40° 18′ 50.241″, −3° 29′ 4.653″ for wheat and 40° 18′ 57.924″, −3° 29′ 3.7134″ for maize and olive trees). A fleet of autonomous vehicles (ground and aerial) equipped with different sensors, all including a machine vision system, were the innovative elements used for such purpose. This case of study is focused on the machine vision system, installed onboard an autonomous ground vehicle based on a commercial tractor chassis, Figure 11a, used for weed detection and its removal in maize fields (wide row-crops). Weed detection is based on crop rows detection with respect the ground vehicle that allows the location of weed patches, at the same time it acts as an aid for guiding the vehicle. This study describes the full machine vision system onboard a tractor, considered as a whole, oriented toward a specific agricultural application. The full system contains a specific description related to the main issues addressed in the first part on this paper, i.e., spectral-band selection (Section 2), imaging sensors and optical systems (Section 3) and geometry (Section 4). This is explicitly stated.

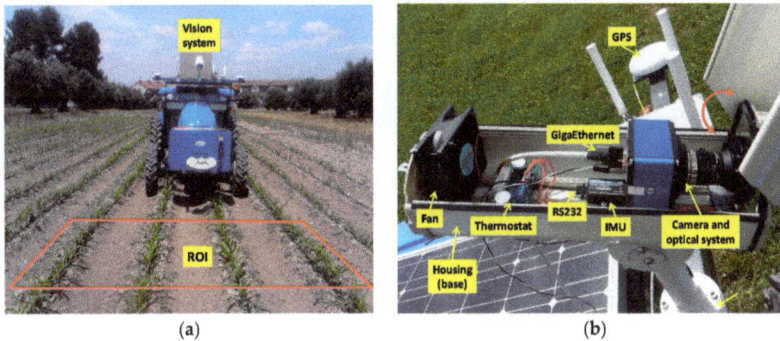

(a)   (b)

Figure 11. Machine vision system: (a) onboard the autonomous vehicle; (b) camera and optical systems and other elements in a housing system. Images adapted and taken from [47] respectively.

5.1. Machine Vision System Specifications

The main components in the machine vision system were a camera-based with its optical system and an IMU (Inertial Measurement Unit), both embedded into a housing system with a fan controlled by a thermostat for cooling purposes, assuming that some agricultural tasks are conducted under high working temperatures, above 50 °C, Figure 11b. The housing system is IP65 protected to work in harsh environments (exposure to dust, drops of liquid from sprayers, etc.). The goal was to apply specific treatments in the ROI in front of the vehicle, which was a rectangular area 3 m wide and 2 m long, Figure 11a. It covers four crop rows in the field, as specified in RHEA. This area starts at 3 m (Section 5.2) with respect to a virtual vertical axis traversing the center of the image plane in the camera, i.e., where the scene is imaged, Figure 12.

The IMU, of LORD MicroStrain® Sensing Systems (Williston, VT, USA) is a 3DM-GX3®-35 high-performance model miniature Attitude Heading Reference System (AHRS) with GPS [74]. It is connected via RS232 to the processor and provides information about pitch and roll angles. These angles were used as aid for estimating the crop rows in the image, based on the geometric imaging projections from 3D to 2D, as described in Section 5.2.

Specific considerations about spectral-band selection (Section 2), imaging sensors and optical systems specifications (Section 3) are provided below. The camera-based sensor, Figure 13a, is the SVS4050CFLGEA model from SVS-VISTEK [75] and is built with the CCD Kodak KAI 04050M/C sensor with a GR Bayer color filter; its resolution is 2336 × 1752 ($H \times V$) pixels with a 5.5 by 5.5 μm pixel size. The manufacturer provides a data sheet for this device, with additional specifications, namely: frame rate (16.8 fps), sensor size ($h \times v$ =12.85 × 964 mm), type sensor format (1"), optical diagonal (1606 mm), minimum/maximum exposure times (6 μs/60 s or ∞ external), Red/Green/Blue gains modes (manual and auto), SNR (58 db/9 bit), internal memory (64 MB), manual/automatic white balance, lens mount (C-Mount), information about the operating temperature. The RR covers typical ranges in the visible spectrum, see Figure 1a as reference, starting at 300 nm with tails above 760 nm, i.e., receiving the impact of UV/IR radiations. The camera is Gigabit Ethernet compliant connected to the main processor. This processor consists of a CompactRIO-9082 [76], with a 1.33 GHz dual-core Intel Core i7 processor, including an LX150 FPGA with a Real-Time Operating System. LabVIEW Real-Time, release 2011, from National Instruments [77], was used as the development environment. On average, each image was processed on 400 ms.

The optical system, Figure 13a,b, consists of a lens with focal length of 10 mm, *f-number* varying from 1.9 to 16 covering maximum and minimum aperture respectively, format of 1" (as required by the sensor format) and transmittance of 86%; it is equipped with an external UV/IR 486 filter with cutting wavelengths below 370 nm and above 760 nm, as described in Section 2.2.

In RHEA the *f-number* was fixed to 8 (intermediate value) and the exposure time was controlled by applying the procedure described in [50], which was based on the histogram analysis of the ROI. Vignetting correction was applied, as described in Section 2.2. No white balance was used because of the problems with shadows mapped onto the reference panel, described in Section 2.2. The frame rate was fixed to 3 fps, which was sufficient. Indeed, the maximum speed of the vehicle during the working operation was fixed to 6 Km/h, so that the vehicle requires 1.8 s to travel the 3 m length of the ROI, i.e., we had available about 5 frames, allowing us to discard possible failed images.

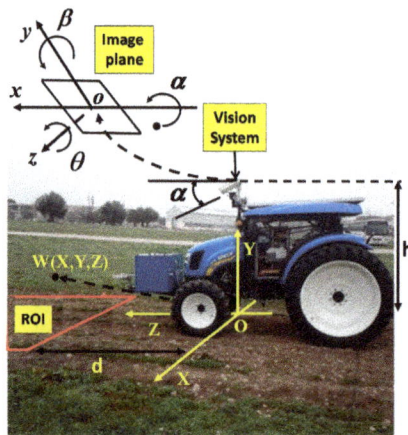

Figure 12. Camera system geometry. Image from [47].

Figure 13. Charge Coupled Device (CCD) sensor, lens and UV/IR cut filter: (**a**) assembled; (**b**) separated.

5.2. 3D Mapping onto 2D Imaging

Figure 13 displays the camera system geometry, based on the considerations addressed in Section 4. *OXYZ* is the reference frame located in the ground with its axes oriented as displayed; *h* is the height from *O* to the origin *o* of the reference frame *oxyz* attached to the camera; roll (θ), pitch (α), and yaw (β) define the three degrees of freedom of the image plane with respect to the referential system; *d* is the distance from the beginning of the ROI to the X axis.

As an illustrative example for defining the vision system geometry, consider the camera-based sensor and optical system specified in Section 5.1. based on the geometric scheme described in the Appendix A. The ROI is imaged onto the image plane as displayed in Figure 14a. Six crop rows are specified (which is a number different from the four crop rows in RHEA) separated from each other 0.75 m; eight horizontal strips are considered with a separation of 50 cm. The ROI is placed on the ground with 4.5 × 4 m^2 (wide and long), placed 3 m ahead of the tractor with reference to the origin of the world coordinate system *OXYZ*, i.e., with XYZ coordinates (0,0,3) m, respectively. The extrinsic camera parameters are: $(X_0,Y_0,Z_0) \equiv (0,2,0)$ m and $(\alpha,\beta,\theta) \equiv (20°,0°,0°)$. Figure 14b displays the same ROI imaged with the same arrangement but with a different θ, i.e., $(\alpha,\beta,\theta) \equiv (20°,0°,+5°)$. As we can see the image becomes distorted in the second case. The asterisk displayed in both images is the mapping of a reference point with coordinates $(X,Y,Z) \equiv (0,1,1)$ m.

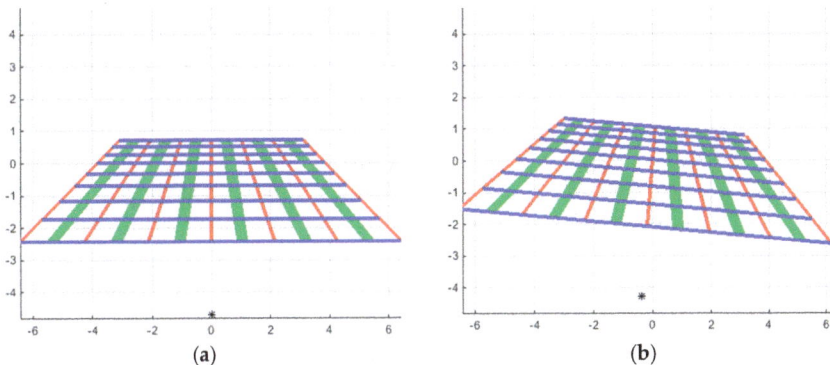

Figure 14. Geometric imaging projections of a Region Of Interest (ROI) for the SVS4050CFLGEA sensor with two different settings at $(X_0,Y_0,Z_0) \equiv (0,2,0)$ m: (**a**) with $(\alpha,\beta,\theta) \equiv (20°,0°,0°)$; (**b**) with $(\alpha,\beta,\theta) \equiv (20°,0°,+5°)$.

Assume the same sensor SVS4050CFLGEA placed at $(X_0,Y_0,Z_0) \equiv (0,2,0)$ m, given a simulated patch with size 20 × 20 cm^2 placed onto the ROI described above at different distances from the center

O in the world coordinate system $OXYZ$. The imaged areas of this patch are measured in pixels and displayed in Table 3 as a function of the distances from the center, i.e., with Z values of 3, 4, 5 and 6 m, $Y = 0$ and $X = \pm10$ cm; with two α values (15° and 20°), β and θ both fixed to 0° and also for the following four focal lengths (3.5, 8.0, 10.0, 12.0) mm.

Table 3. Imaged areas in pixels for a patch of 20×20 cm^2 at different distances form the origin in the coordinates of the world system, α angles and focal lengths.

Distances from O (m)	$\alpha°$	f (mm)	Area (pixels)	Distances from O (m)	$\alpha° f$ (mm)	Area (Pixels)
3	15	3.5	756	5	3.5	212
		8.0	3901		8.0	1070
		10.0	6136		10.0	1675
		12.0	8840		12.0	2415
	20	3.5	710		3.5	188
		8.0	3645		8.0	1007
		10.0	5684		10.0	1596
		12.0	8296		12.0	2320
4	15	3.5	371	6	3.5	120
		8.0	1900		8.0	637
		10.0	2934		10.0	1026
		12.0	4312		12.0	1428
	20	3.5	342		3.5	110
		8.0	1806		8.0	620
		10.0	2818		10.0	998
		12.0	4128		12.0	1388

We can see that the maximum/minimum areas are 8840/110 pixels, which corresponds respectively to imaged patches of 94×94 and 11×10 pixels^2 for the same patch on the 3D ROI. This allows the evaluation of the vision system configuration in order to discriminate shapes or for posterior processing such as morphological operations. For example, if binary morphological erosion is applied over the above areas with a 3×3 structuring element these areas are reduced to 8464/72 pixels representing reduction rates of 4.2%/34.5%. This means that the best arrangement is the first one as the largest area allows for better subsequent discrimination based on area analysis. Designers can use the vision system geometry for different simulations. However, in addition, the following robot simulators could be used for previous analysis on agricultural environments [78,79].

5.3. Crop Rows Detection and Weed Coverage

Three methods were tested in RHEA for crop rows detection [5,10,80], the vignetting effect, produced by the use of the Schneider UV/IR 486 cut-off filter, was compensated based on the approach proposed on subsection (2.2). No white balance was required because this action was replaced by histogram analysis on the ROI, as explained in the above references. Alignments of row pixels were identified in [5] along specific directions defining the crop rows. Maximum accumulations corresponding to the number of expected crop rows define the crop rows. This approach, inspired by the human visual perception, is a simplification of the Hough transform [16]. Linear regression was applied in [10,80], where greenness was identified based on the computation of vegetation indices [11] followed by automatic thresholding. The IMU provides pitch and roll angles, which together with the remainder intrinsic/extrinsic parameters and Equation (8) the expected crop rows are drawn on the image, and then linear regression (least squares and Theil-Sen respectively) was applied for adjusting the expected crop rows to the real ones on each image.

More than 3000 images were analyzed belonging basically to three groups according to the growth stage of the crop: Low (5 cm), Medium (15 cm), High (30 cm). The images were acquired over different days under different illumination conditions, i.e., cloudy, sunny days, and days with high light variability. With the set of images analyzed and considering the maize crops at the above-mentioned three growth stages (low, medium, high), the averaged percentage of successes displayed in Table 4 were obtained.

Table 4. Percentage of success for crop lines detection for three (low, medium, high) maize growth stages.

	Crop Lines Detection		
Maize growth stage	Low	Medium	High
% of success	95	93	90

For each image, a density matrix of weeds associated with each ROI was computed. This matrix contains low, medium, and high density values. Figure 15 illustrates two consecutive images along a sub-path. They contain three types of lines defining the cells required for computing the density matrix as follows:

1. Once the crop lines are identified, they are confined to the ROI in the image (yellow lines).
2. To the left and right of each crop line, parallel lines are drawn (red). They divide the inter-crop space into two parts.
3. Horizontal lines (in blue) are spaced conveniently in pixels so that each line corresponds to a distance of 0.25 m from the base line of the spatial ROI in the scene.
4. The above lines define 8×8 trapezoidal cells, each trapezoid with its corresponding area A_{ij} expressed in pixels. For each cell, the number of pixels identified as green pixels was computed, G_{ij}, (drawn as cyan pixels in the image). Pixels close to the crop rows were excluded, with a margin of tolerance which represents 10% of the width of the cell along horizontal displacements. This is because this margin contains mainly crop plants but not weeds. The weed coverage for each cell is finally computed as $d_{ij} = A_{ij}/G_{ij}$, expressed in percentage. The different d_{ij} values compose the elements of the density matrix.

Figure 15. Consecutive images along a sub-path with the detected crop lines (**yellow**); parallel lines to the left and right crop lines (**red**); horizontal lines covering 0.25 m in the field. Images taken from [47].

From a set of 500 images, obtained during the test campaigns mentioned above and also with the different growth stages, the weed coverage was classified according to three levels (Low, less than 33%, Medium, between 33% and 66%, and High, greater than 66%), associated to the Liquefied Petroleum Gas pressure levels of the physical weed controller used for weed removal in RHEA. These percentages are checked against the criterion of an expert, who determined the correct classification. The results are summarized in Table 5.

Table 5. Percentage of success for weeds detection for three (low, medium, high) maize growth stages.

	Maize Growth Stage		
Weed densities	Low	Medium	High
% of success	92	90	88

As before, the worst value corresponds to maize fields with a high growth stage, which is consistent with the real situation because of the reasons expressed above. Part of the inaccuracy comes from the incorrect crop lines detection.

5.4. Guidance

Nowadays, GPS systems are commonly used, being a well-known approach for autonomous guidance. However, in RHEA, tractor's guidance was achieved by combining GPS and machine vision systems.

In RHEA a Mission Manager software-based was developed for handling the multi-robot system. It is responsible for generating global trajectories determining the path planning [81], which are previously established for each vehicle before the mission starts [82]. Regarding the global path following planned for the tractor, parallel routes that move alternately from one extreme of the field to the other are planned, following the crop row direction and turning in the headlands (the outer areas of the field). At the first stage, the GPS was used to provide information to the tractor to place it at the beginning of the crop rows, points belonging to the plan, as aligned as possible.

Once the tractor is placed and aligned, the tractor starts moving along the crop rows following the planned path by using GPS information. Specifically, a RTK-GPS (Real Time Kinematics-Global Positioning System) sensory system was used consisting of two GNSS (Global Navigation Satellite Systems) rover antennas, one for XYZ positioning and the other for heading calculations [47,81], where the correction signal is produced locally generated with a reference (local) base station, providing localization errors of below ± 2 cm. Precise guidance was applied for controlling deviations from the planned path, based on the machine-vision system.

The system determines the *diagonal* (D) line equidistant to the two central crop rows detected in the image. Considering the bottom line in the image defining the ROI, two principal points are identified. The first is exactly the central point (P_c) in the horizontal row crossing the full image and overlapped with the bottom line in the ROI. The second (P_i) is the intersection point between D and the same bottom line in the ROI.

The difference between the x-horizontal coordinates of P_i and P_c determines the deviation with respect the correct trajectory. This difference (positive, negative or null) transformed from image pixels to length measurements was used for trajectory correction.

When P_i and P_c match, no correction was required; otherwise, the appropriate correction with respect the planned path (line-of-sight) was applied. In order to assume incorrect information provided by the machine vision system because of failures during the crop row detection, lower and upper limits were established considering that deviations greater than ± 3 cm are ignored and that the path following continues with the GPS following the line-of-sight. The limit of ± 3 cm represents the 8% with respect the half of the distance of 75 cm existing between adjacent crop rows.

Figure 16a,b display two consecutive images acquired during the execution of a straight trajectory from the line-of-sight with their processed images and crop rows detected in the ROI (weeds are also identified around the crop lines). The tractor in 16a undergoes a slight deviation from the correct trajectory. Indeed, the upper right corner in the box, belonging to the tractor, is very close to the rightmost crop row and that this box is misaligned with respect to the four crop lines detected in the image displayed in 16c. This misalignment is corrected and can be observed in Figure 16b where the box is better centered relative to the central crop rows, Figure 15d. This situation was very common on rough maize fields because they contain abundant irregularities.

Figure 16. Alignment of the vehicle along the crop rows. Images adapted and taken from [47]: (a) original image with deviation; (b) original image after correction; (c) misalignment of the tractor with respect the crop rows; (d) misalignment corrected.

For testing purposes, a set of 400 images were randomly selected. Corrections ordered by the machine vision system were checked. After each correction, the position of the vehicle with respect the crop rows in the next image was verified. A correction has been demanded for 30% of the images (120 images). From these, the tractor was correctly positioned on 89% of the subsequent images. For the remaining images, the correction was erroneously demanded. In these cases, the following path was exclusively based on GPS for guidance. Figure 17 illustrates the comparison between the use of the information provided by the machine vision system and the use of the information provided exclusively by the GPS for crossing the maize field, where it is noteworthy that the row detection system slightly improves the row following, taking into account that the theoretical path to be followed using only the GPS system corresponds to the center of the row by which the two results are compared. It is worth noting that the crop rows at the end of the experimental field were slightly damaged (the last 10 m), due to the large number of tests performed, and in this area, the vision system for row detection produced a large number of errors.

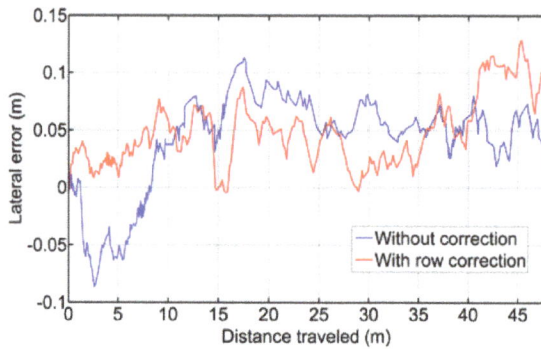

Figure 17. Comparison of the vehicle guidance in a maize field, represented as the lateral error of the rear axle with respect to the theoretical center of the rows. Image from [47].

5.5. Security: Obstacle Detection

Spatial and temporal analyses were applied in video sequences in obstacle detection for safety purposes in [25]. The spatial analysis is based on the b^* channel in the CIELAB color space where most objects can be distinguished from the main structures (plants and soil). When objects contain high red and/or white components L^* and a^* channels were used. Texture information for each pixel is also computed, based on differences between maximum and minimum gray level values in a neighborhood environment around the pixel. Binary images were obtained at each step and combined with the logical *and* binary operation to obtain a final binary image containing potential objects in the environment. The temporal analysis is based on the difference between two consecutive frames where significant differences are obtained where objects appear and a new binary image is computed. The matching of the binary image obtained based on spatial analysis was compared to the one obtained for temporal differentiation. A comparison is established between the two binary images to verify/discard binary matches, which determine the presence of objects. Figure 18 displays illustrative examples with three persons and a vehicle coming from the front containing dangerous situations on the working agricultural scenario. New trends and methods are currently being tested based on deep learning approaches [67] following the ISO/DIS 18497 which is a standard for safety of highly automated agricultural machines, including tractors.

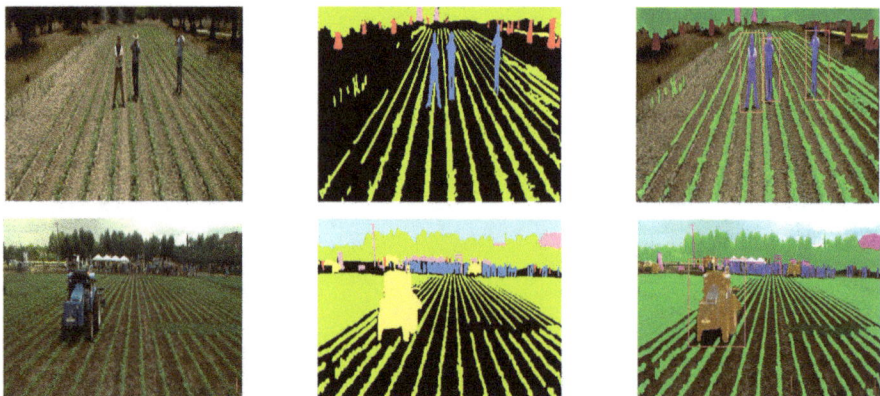

Figure 18. Peoples and a vehicle identified as obstacles in the working environment.

6. Conclusions

Machine vision is a relevant system in agricultural vehicles (autonomous and non-autonomous) for different tasks, including UAVs. An appropriate choice of such systems is an additional guarantee the successful performance of tasks in outdoor environments. In this regard, this paper has addressed the following three main topics for a correct selection in agricultural environments: (a) spectral band for identifying significant elements (plants, soil, objects); (b) imaging sensors and optical systems for mapping the scene onto images with sufficient quality and (c) geometric system pose and arrangement for mapping specific areas. A general overview, with detailed description and technical support, has been provided for each topic with illustrative examples focused on specific applications in agriculture. This represents a set of guidelines with sufficient details and descriptions, so that future engineers have sufficient basis for designing machine vision systems in agricultural applications, which represents a compilation and condensation of scattered ideas in the teeming area of applications in agriculture based on machine vision systems. The way is open for the incorporation of new incoming technologies, particularly 3D systems such as the ones based on Time of Flight (ToF) technologies.

A case study is provided as a result of research in the RHEA project (funded by the European Union) for effective weed control in maize fields (wide-rows crops) where many of the technical issues described in the paper have been applied with successful results.

Acknowledgments: The research leading to these results received funding from the European Union's Seventh Framework Programme [FP7/2007–2013] under Grant Agreement No. 245986. Part of this work has been carried out by the first author funded by Universidad Politécnica Estatal del Carchi (Ecuador) and the second author funded by the National Council of Science and Technology of Mexico (CONACyT) for the doctoral grant number 210282 to undertake doctoral studies. Authors are grateful to the referees for their suggestions and constructive criticism of the original version of this paper.

Author Contributions: All of the authors contributed extensively to the work presented in this paper. Gonzalo Pajares has coordinated the work and participated on all sections, on his role as main researcher in RHEA on the University Complutense of Madrid. Iván García-Santillán provided specific support for defining contents related to spectral bands, imaging sensors, optical systems specifications and crop rows detection. Yerania Campos was in charge of obstacle detection, including image segmentation. Martín Montalvo, José Miguel Guerrero and Juan Romeo contributed equally in the design and testing of image processing methods for quality improvement, crop rows detection and weeds identification. Luis Emmi was in charge on autonomous guidance and its performance. María Guijarro revised the manuscript and supplied ground-truth images for assessment. Pablo Gonzalez-de-Santos, on his role of European project coordinator in RHEA, revised the results described in the case study and designed the mechanical system for the installation of the machine vision system onboard the vehicle.

Conflicts of Interest: The authors declare no conflict of interest.

Appendix A. Camera System Geometry

The point $W(X,Y,Z)$ is expressed in the 3D space with respect to the $OXYZ$ world reference system. The origin o of the image plane is displaced with respect O according to the vector w with coordinates (X_0,Y_0,Z_0). The elementary translations and rotations as described in Section 4 are expressed as follows, including the focal length (f),

$$
\begin{pmatrix} x \\ y \\ z \\ k \end{pmatrix} = PR_\theta R_\beta R_\alpha G \begin{pmatrix} X \\ Y \\ Z \\ 1 \end{pmatrix}
\tag{A1}
$$

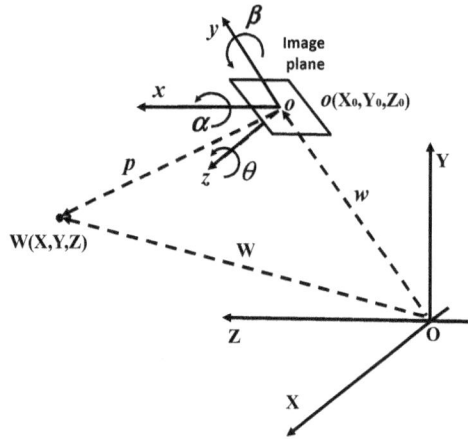

Figure A1. Reference systems and relations.

The elementary matrices involved are defined as follows, where $CX \equiv cosX$ and $SX \equiv sinX$.

$$G = \begin{pmatrix} 1 & 0 & 0 & -X_0 \\ 0 & 1 & 0 & -Y_0 \\ 0 & 0 & 1 & -Z_0 \\ 0 & 0 & 0 & 1 \end{pmatrix} \quad P = \begin{pmatrix} 1 & 0 & 0 & 0 \\ 0 & 1 & 0 & 0 \\ 0 & 0 & 1 & 0 \\ 0 & 0 & -\frac{1}{f} & 1 \end{pmatrix}$$

$$R_\alpha = \begin{pmatrix} 1 & 0 & 0 & 0 \\ 0 & C\alpha & S\alpha & 0 \\ 0 & -S\alpha & C\alpha & 0 \\ 0 & 0 & 0 & 1 \end{pmatrix} \quad R_\beta = \begin{pmatrix} C\beta & 0 & -S\beta & 0 \\ 0 & 1 & 0 & 0 \\ S\beta & 0 & C\beta & 0 \\ 0 & 0 & 0 & 1 \end{pmatrix} \quad R_\theta = \begin{pmatrix} C\theta & S\theta & 0 & 0 \\ -S\theta & C\theta & 0 & 0 \\ 0 & 0 & 1 & 0 \\ 0 & 0 & 0 & 1 \end{pmatrix}$$

(A2)

$$R = R_\theta R_\beta R_\alpha = \begin{pmatrix} C\theta & S\theta & 0 & 0 \\ -S\theta & C\theta & 0 & 0 \\ 0 & 0 & 1 & 0 \\ 0 & 0 & 0 & 1 \end{pmatrix} \begin{pmatrix} C\beta & 0 & -S\beta & 0 \\ 0 & 1 & 0 & 0 \\ S\beta & 0 & C\beta & 0 \\ 0 & 0 & 0 & 1 \end{pmatrix} \begin{pmatrix} 1 & 0 & 0 & 0 \\ 0 & C\alpha & S\alpha & 0 \\ 0 & -S\alpha & C\alpha & 0 \\ 0 & 0 & 0 & 1 \end{pmatrix} =$$

$$\begin{pmatrix} C\theta C\beta & C\theta S\alpha S\beta + C\alpha S\theta & -C\alpha S\beta C\theta + S\theta S\alpha & 0 \\ -S\theta C\beta & -S\theta S\alpha S\beta + C\alpha C\theta & S\theta C\alpha S\beta + C\theta S\alpha & 0 \\ S\beta & -S\alpha C\beta & C\alpha C\beta & 0 \\ 0 & 0 & 0 & 1 \end{pmatrix}$$

(A3)

The composition of the elementary rotation matrices derive in a composed rotation matrix as follows,

$$RG = \begin{pmatrix} C\theta C\beta & C\theta S\alpha S\beta + C\alpha S\theta & -C\alpha S\beta C\theta + S\theta S\alpha & -X_0 C\theta C\beta - Y_0\left(C\theta S\alpha S\beta - C\alpha S\theta\right) - Z_0\left(-C\alpha S\beta C\theta + S\theta S\alpha\right) \\ -S\theta C\beta & -S\theta S\alpha S\beta + C\alpha C\theta & S\theta C\alpha S\beta + C\theta S\alpha & -X_0 S\theta C\beta - Y_0\left(S\theta S\alpha S\beta - C\alpha C\theta\right) - Z_0\left(-C\alpha S\beta S\theta + C\theta S\alpha\right) \\ S\beta & -S\alpha C\beta & C\alpha C\beta & -X_0 S\beta - Y_0\left(-S\alpha C\beta\right) - Z_0\left(C\alpha C\beta\right) \\ 0 & 0 & 0 & 1 \end{pmatrix}$$

(A4)

$$PRG = \begin{pmatrix} C\theta C\beta & C\theta S\alpha S\beta + C\alpha S\theta & -C\alpha S\beta C\theta + S\theta S\alpha & -X_0 C\theta C\beta - Y_0\left(C\theta S\alpha S\beta - C\alpha S\theta\right) - Z_0\left(-C\alpha S\beta C\theta + S\theta S\alpha\right) \\ -S\theta C\beta & -S\theta S\alpha S\beta + C\alpha C\theta & S\theta C\alpha S\beta + C\theta S\alpha & -X_0 S\theta C\beta - Y_0\left(S\theta S\alpha S\beta - C\alpha C\theta\right) - Z_0\left(-C\alpha S\beta S\theta + C\theta S\alpha\right) \\ S\beta & -S\alpha C\beta & C\alpha C\beta & -X_0 S\beta - Y_0\left(-S\alpha C\beta\right) - Z_0\left(C\alpha C\beta\right) \\ -\frac{1}{f}\left(S\beta\right) & -\frac{1}{f}\left(-S\alpha C\beta\right) & -\frac{1}{f}\left(C\alpha C\beta\right) & -\frac{1}{f}\left(-X_0 S\beta - Y_0\left(-S\alpha C\beta\right) - Z_0\left(C\alpha C\beta\right)\right) + 1 \end{pmatrix}$$

(A5)

Finally, the projections on the image plane are expressed as,

$$x = -f\frac{(X-X_0)C\theta C\beta+(Y-Y_0)C\theta S\alpha S\beta+C\alpha S\theta+(Z-Z_0)(-C\alpha S\beta C\theta+S\theta S\alpha)}{(X-X_0)S\beta+(Y-Y_0)(-S\alpha C\beta)+(Z-Z_0)C\alpha C\beta-f}$$

$$y = -f\frac{(X-X_0)(-S\theta C\beta)+(Y-Y_0)(S\theta S\alpha S\beta+C\alpha C\theta)+(Z-Z_0)(C\alpha S\beta S\theta+C\theta S\alpha)}{(X-X_0)S\beta+(Y-Y_0)(-S\alpha C\beta)+(Z-Z_0)C\alpha C\beta-f}$$

(A6)

References

1. Slaughter, D.C.; Giles, D.K.; Downey, D. Autonomous robotic weed control systems: A review. *Comput. Electron. Agric.* **2008**, *61*, 63–78. [CrossRef]
2. Shalal, N.; Low, T.; McCarthy, C.; Hancock, N. A review of autonomous navigation systems in agricultural environments. In Proceedings of the SEAg 2013: Innovative Agricultural Technologies for a Sustainable Future, Barton, Australia, 22–25 September 2013; Available online: http://eprints.usq.edu.au/24779/ (accessed on 20 July 2015).
3. Mousazadeh, H. A technical review on navigation systems of agricultural autonomous off-road vehicles. *J. Terramech.* **2013**, *50*, 211–23. [CrossRef]
4. López-Granados, F. Weed detection for site-specific weed management: Mapping and real-time approaches. *Weed Res.* **2011**, *51*, 1–11. [CrossRef]
5. Romeo, J.; Pajares, G.; Montalvo, M.; Guerrero, J.M.; Guijarro, M.; Ribeiro, A. Crop row detection in maize fields inspired on the human visual perception. *Sci. World J.* **2012**, *2012*, 484390. [CrossRef] [PubMed]
6. Romeo, J.; Pajares, G.; Montalvo, M.; Guerrero, J.M.; Guijarro, M.; de la Cruz, J.M. A new expert system for greenness identification in agricultural images. *Exp. Syst. Appl.* **2013**, *40*, 2275–2286. [CrossRef]
7. Guerrero, J.M.; Pajares, G.; Montalvo, M.; Romeo, J.; Guijarro, M. Support vector machines for crop/weeds identification in maize fields. *Exp. Syst. Appl.* **2012**, *39*, 11149–11155. [CrossRef]
8. Gée, Ch.; Bossu, J.; Jones, G.; Truchetet, F. Crop/weed discrimination in perspective agronomic images. *Comput. Electron. Agric.* **2008**, *60*, 49–59. [CrossRef]
9. Zheng, L.; Zhang, J.; Wang, Q. Mean-shift-based color segmentation of images containing green vegetation. *Comput. Electron. Agric.* **2009**, *65*, 93–98. [CrossRef]
10. Montalvo, M.; Pajares, G.; Guerrero, J.M.; Romeo, J.; Guijarro, M.; Ribeiro, A.; Ruz, J.J.; de la Cruz, J.M. Automatic detection of crop rows in maize fields with high weeds pressure. *Exp. Syst. Appl.* **2012**, *39*, 11889–11897. [CrossRef]
11. Guijarro, M.; Pajares, G.; Riomoros, I.; Herrera, P.J.; Burgos-Artizzu, X.P.; Ribeiro, A. Automatic segmentation of relevant textures in agricultural images. *Comput. Electron. Agric.* **2011**, *75*, 75–83. [CrossRef]
12. Burgos-Artizzu, X.P.; Ribeiro, A.; Tellaeche, A.; Pajares, G.; Fernández-Quintanilla, C. Improving weed pressure assessment using digital images from an experience-based reasoning approach. *Comput. Electron. Agric.* **2009**, *65*, 176–185. [CrossRef]
13. Sainz-Costa, N.; Ribeiro, A.; Burgos-Artizzu, X.P.; Guijarro, M.; Pajares, G. Mapping wide row crops with video sequences acquired from a tractor moving at treatment speed. *Sensors* **2011**, *11*, 7095–7109. [CrossRef] [PubMed]
14. Tellaeche, A.; Burgos-Artizzu, X.P.; Pajares, G.; Ribeiro, A. A new vision-based approach to differential spraying in precision agriculture. *Comput. Electron. Agric.* **2008**, *60*, 144–155. [CrossRef]
15. Jones, G.; Gée, Ch.; Truchetet, F. Assessment of an inter-row weed infestation rate on simulated agronomic images. *Comput. Electron. Agric.* **2009**, *67*, 43–50. [CrossRef]
16. Tellaeche, A.; Burgos-Artizzu, X.P.; Pajares, G.; Ribeiro, A. A vision-based method for weeds identification through the Bayesian decision theory. *Pattern Recognit.* **2008**, *41*, 521–530. [CrossRef]
17. Li, M.; Imou, K.; Wakabayashi, K.; Yokoyama, S. Review of research on agricultural vehicle autonomous guidance. *Int. J. Agric. Biol. Eng.* **2009**, *2*, 1–26.
18. Reid, J.F.; Searcy, S.W. Vision-based guidance of an agricultural tractor. *IEEE Control. Syst.* **1997**, *7*, 39–43. [CrossRef]
19. Billingsley, J.; Schoenfisch, M. Vision-guidance of agricultural vehicles. *Auton. Robots* **1995**, *2*, 65–76. [CrossRef]

20. Rovira-Más, F.; Zhang, Q.; Reid, J.F.; Will, J.D. Machine vision based automated tractor guidance. *Int. J. Smart Eng. Syst. Des.* **2003**, *5*, 467–480. [CrossRef]
21. Kise, M.; Zhang, Q. Development of a stereovision sensing system for 3D crop row structure mapping and tractor guidance. *Biosyst. Eng.* **2008**, *101*, 191–198. [CrossRef]
22. Xue, J.; Zhang, L.; Grift, T.E. Variable field-of-view machine vision based row guidance of an agricultural robot. *Comput. Electron. Agric.* **2012**, *84*, 85–91. [CrossRef]
23. Wei, J.; Rovira-Mas, F.; Reid, J.F.; Han, S. Obstacle detection using stereo vision to enhance safety autonomous machines. *Trans. ASABE* **2005**, *48*, 2389–2397. [CrossRef]
24. Nissimov, S.; Goldberger, J.; Alchanatis, V. Obstacle detection in a greenhouse environment using the Kinect sensor. *Comput. Electron. Agric.* **2015**, *113*, 104–115. [CrossRef]
25. Campos, Y.; Sossa, H.; Pajares, G. Spatio-temporal analysis for obstacle detection in agricultural videos. *Appl. Soft Comput.* **2016**, *45*, 86–97. [CrossRef]
26. Cheein, F.A.; Steiner, G.; Paina, G.P.; Carelli, R. Optimized EIF-SLAM algorithm for precision agriculture mapping based on stems detection. *Comput. Electron. Agric.* **2011**, *78*, 195–207. [CrossRef]
27. Pajares, G. Overview and Current Status of Remote Sensing Applications Based on Unmanned Aerial Vehicles (UAVs). *Photogramm. Eng. Remote Sens.* **2015**, *81*, 281–329. [CrossRef]
28. RHEA. Robot Fleets for Highly Effective Agriculture and Forestry Management. Available online: http://www.rhea-project.eu/ (accessed on 19 August 2016).
29. Exelis Visual Information Solutions. Available online: http://www.exelisvis.com/docs/VegetationIndices. html (accessed on 19 August 2016).
30. Meyer, G.E.; Camargo-Neto, J. Verification of color vegetation indices for automated crop imaging applications. *Comput. Electron. Agric.* **2008**, *63*, 282–293. [CrossRef]
31. Point Grey Innovation and Imaging. How to Evaluate Camera Sensitivity. Available online: https://www. ptgrey.com/white-paper/id/10912 (accessed on 19 August 2016).
32. Schneider Kreuznach. Tips and Tricks. Available online: http://www.schneiderkreuznach.com/en/photo-imaging/product-field/b-w-fotofilter/products/filtertypes/special-filters/486-uvir-cut/ (accessed on 22 August 2016).
33. Tucker, C.J. Red and photographic infrared linear combinations for monitoring vegetation. *Remote Sens. Environ.* **1979**, *8*, 127–150. [CrossRef]
34. Ollinger, S.V. Sources of variability in canopy reflectance and the convergent properties of plants. *New Phytol.* **2011**, *189*, 375–394. [CrossRef] [PubMed]
35. Rabatel, G.; Gorretta, N.; Labbé, S. Getting NDVI Spectral Bands from a Single Standard RGB Digital Camera: A Methodological Approach. In Proceedings of the 14th Conference of the Spanish Association for Artificial Intelligence, CAEPIA 2011, La Laguna, Spain, 7–11 November 2011; Volume 7023, pp. 333–342.
36. Xenics Infrared Solutions. Bobcat-640-GigE High Resolution Small form Factor InGaAs Camera. Available online: http://www.applied-infrared.com.au/images/pdf/Bobcat-640-GigE_Industrial_LowRes. pdf (accessed on 22 August 2016).
37. Kiani, S.; Kamgar, S.; Raoufat, M.H. Machine Vision and Soil Trace-based Guidance-Assistance System for Farm Tractors in Soil Preparation Operations. *J. Agric. Sci.* **2012**, *4*, 1–5. [CrossRef]
38. Hague, T.; Tillet, N.; Wheeler, H. Automated crop and weed monitoring in widely spaced cereals. *Precis. Agric.* **2006**, *1*, 95–113. [CrossRef]
39. JAI 2CCD Cameras. Available online: http://www.jai.com/en/products/ad-080ge (accessed on 22 August 2016).
40. 3CCD Color cameras. Image acquisition. Resource Mapping. Remote Sensing and GIS for Conservation. Available online: http://www.resourcemappinggis.com/image_technical.html (accessed on 22 August 2016).
41. Kise, M.; Zhang, Q.; Rovira-Más, F. A Stereovision-based Crop Row Detection Method for Tractor-automated Guidance. *Biosyst. Eng.* **2005**, *90*, 357–367. [CrossRef]
42. Rovira-Más, F.; Zhang, Q.; Reid, J.F. Stereo vision three-dimensional terrain maps for precision agriculture. *Comput. Electron. Agric.* **2008**, *60*, 133–143. [CrossRef]
43. Svensgaard, J.; Roitsch, T.; Christensen, S. Development of a Mobile Multispectral Imaging Platform for Precise Field Phenotyping. *Agronomy* **2014**, *4*, 322–336. [CrossRef]

44. Li, L.; Zhang, Q.; Huang, D. A Review of Imaging Techniques for Plant Phenotyping. *Sensors* **2014**, *14*, 20078–20111. [CrossRef] [PubMed]
45. Rasmussen, J.; Ntakos, G.; Nielsen, J.; Svensgaard, J.; Poulsen, R.N.; Christensen, S. Are vegetation indices derived from consumer-grade camerasmounted on UAVs sufficiently reliable for assessing experimentalplots? *Eur. J. Agron.* **2016**, *74*, 75–92. [CrossRef]
46. Bockaert, V. Sensor sizes. Digital Photography Review. Available online: http://www.dpreview.com/glossary/camera-system/sensor-sizes (accessed on 22 August 2016).
47. Emmi, L.; Gonzalez-de-Soto, M.; Pajares, G.; Gonzalez-de-Santos, P. Integrating Sensory/Actuation Systems in Agricultural Vehicles. *Sensors* **2014**, *14*, 4014–4049. [CrossRef] [PubMed]
48. Choosing the Right Camera Bus. Available online: http://www.ni.com/white-paper/5386/en/ (accessed on 22 August 2016).
49. Cambridge in Colour. Available online: http://www.cambridgeincolour.com/tutorials/camera-exposure.htm (accessed on 22 August 2016).
50. Montalvo, M.; Guerrero, J.M.; Romeo, J.; Guijarro, M.; de la Cruz, J.M.; Pajares, G. Acquisition of Agronomic Images with Sufficient Quality by Automatic Exposure Time Control and Histogram MatchingLecture Notes in Computer Science. In Proceedings of the Advanced Concepts for Intelligent Vision Systems (ACIVS'13), Poznan, Poland, 28–31 October 2013; Volume 8192, pp. 37–48.
51. Cinegon 1.9/10 Ruggedized Lens. Available online: http://www.schneiderkreuznach.com/fileadmin/user_upload/bu_industrial_solutions/industrieoptik/16mm_Lenses/Compact_Lenses/Cinegon_1.9--10_ruggedized.pdf (accessed on 27 March 2015).
52. Optical Filters. Edmund Optics. Available online: http://www.edmundoptics.com/technical-resources-center/optics/optical-filters/?&#guide (accessed on 22 August 2016).
53. Point Grey Innovation and Imaging. Selecting a lens for Your Camera. Available online: https://www.ptgrey.com/KB/10694 (accessed on 22 August 2016).
54. Jeon, H.Y.; Tian, L.F.; Zhu, H. Robust Crop and Weed Segmentation under Uncontrolled Outdoor Illumination. *Sensors* **2011**, *11*, 6270–6283. [CrossRef] [PubMed]
55. Linker, R.; Cohen, O.; Naor, A. Determination of the number of green apples in RGB images recorded in orchard. *Comput. Electron. Agric.* **2012**, *81*, 45–57. [CrossRef]
56. Moshou, D.; Bravo, D.; Oberti, R.; West, J.S.; Ramon, H.; Vougioukas, S.; Bochtis, D. Intelligent multi-sensor system for the detection and treatment of fungal diseases in arable crops. *Biosyst. Eng.* **2011**, *108*, 311–321. [CrossRef]
57. Oberti, R.; Marchi, M.; Tirelli, P.; Calcante, A.; Iriti, M.; Borghese, A.N. Automatic detection of powdery mildew on grapevine leaves by image analysis: Optimal view-angle range to increase the sensitivity. *Comput. Electron. Agric.* **2014**, *104*, 1–8. [CrossRef]
58. Blas, M.R.; Blanke, M. Stereo vision with texture learning for fault-tolerant automatic baling. *Comput. Electron. Agric.* **2011**, *75*, 159–168. [CrossRef]
59. Farooque, A.A.; Chang, Y.K.; Zaman, Q.U.; Groulx, D.; Schumann, A.W.; Esau, T.J. Performance evaluation of multiple ground based sensors mounted on a commercial wild blueberry harvester to sense plant height, fruit yield and topographic features in real-time. *Comput. Electron. Agric.* **2012**, *84*, 85–91. [CrossRef]
60. Dworak, V.; Huebner, M.; Selbeck, J. Precise navigation of small agricultural robots in sensitive areas with a smart plant camera. *J. Imaging* **2015**, *1*, 115–133. [CrossRef]
61. Fu, K.S.; Gonzalez, R.C.; Lee, C.S.G. *Robótica: Control, Detección, Visión e Inteligencia*; McGraw-Hill: Madrid, Spain, 1988.
62. Herrera, P.J.; Dorado, J.; Ribeiro, A. A Novel Approach for Weed Type Classification Based on Shape Descriptors and a Fuzzy Decision-Making Method. *Sensors* **2014**, *14*, 15304–15324. [CrossRef] [PubMed]
63. Li, P.; Lee, S.H.; Hsu, H.Y. Review on fruit harvesting method for potential use of automatic fruit harvesting systems. *Procedia Eng.* **2011**, *23*, 351–366. [CrossRef]
64. Nguyen, T.T.; Slaughter, D.C.; Hanson, B.D.; Barber, A.; Freitas, A.; Robles, D.; Whelan, E. Automated mobile system for accurate outdoor tree crop enumeration using an uncalibrated camera. *Sensors* **2015**, *15*, 18427–18442. [CrossRef] [PubMed]
65. Vázquez-Arellano, M.; Griepentrog, H.W.; Reiser, D.; Paraforos, D.S. 3-D imaging systems for agricultural applications-a review. *Sensors* **2016**, *16*, 618. [CrossRef] [PubMed]

66. Rong, X.; Huanyu, J.; Yibin, Y. Recognition of clustered tomatoes based on binocular stereo vision. *Comput. Electron. Agric.* **2014**, *106*, 75–90.

67. Steen, K.A.; Christiansen, P.; Karstoft, H.; Jørgensen, R.N. Using deep learning to challenge safety standard for highly autonomous machines in agriculture. *J. Imaging* **2016**, *2*, 6. [CrossRef]

68. Barnard, S.; Fishler, M. Computational stereo. *ACM Comput. Surv.* **1982**, *14*, 553–572. [CrossRef]

69. Cochran, S.D.; Medioni, G. 3-D Surface Description from binocular stereo. *IEEE Trans. Pattern Anal. Mach. Intell.* **1992**, *14*, 981–994. [CrossRef]

70. Pajares, G.; de la Cruz, J.M. On combining support vector machines and simulated annealing in stereovision matching. *IEEE Trans. Syst. Man Cybern. Part B* **2004**, *34*, 1646–1657. [CrossRef]

71. Correal, R.; Pajares, G.; Ruz, J.J. Automatic expert system for 3D terrain reconstruction based on stereo vision and histogram matching. *Expert Syst. Appl.* **2014**, *41*, 2043–2051. [CrossRef]

72. Rovira-Más, F.; Wang, Q.; Zhang, Q. Design parameters for adjusting the visual field of binocular stereo cameras. *Biosyst. Eng.* **2010**, *105*, 59–70.

73. Pajares, G.; de la Cruz, J.M. *Visión por Computador: Imágenes Digitales y Aplicacione*; RA-MA: Madrid, Spain, 2007. (In Spanish)

74. MicroStrain Sensing Systems. Available online: http://www.microstrain.com/inertial/3dm-gx3--35 (accessed on 22 August 2016).

75. SVS-VISTEK. Available online: https://www.svs-vistek.com/en/svcam-cameras/svs-svcam-search-result.php (accessed on 22 August 2016).

76. National Instruments. CompactRIO. Available online: http://sine.ni.com/nips/cds/view/p/lang/es/nid/210001 (accessed on 22 August 2016).

77. National Instruments. LabView. Available online: http://www.ni.com/labview/esa/ (accessed on 22 August 2016).

78. Cyberbotics. Webots Robot Simulator. Available online: https://www.cyberbotics.com/ (accessed on 24 August 2016).

79. Gazebo. Available online: http://gazebosim.org/ (accessed on 24 August 2016).

80. Guerrero, J.M.; Guijarro, M.; Montalvo, M.; Romeo, J.; Emmi, L.; Ribeiro, A.; Pajares, G. Automatic expert system based on images for accuracy crop row detection in maize fields. *Exp. Syst. Appl.* **2013**, *40*, 656–664. [CrossRef]

81. Gonzalez-de-Santos, P.; Ribeiro, A.; Fernandez-Quintanilla, C.; López-Granados, F.; Brandstoetter, M.; Tomic, S.; Pedrazzi, S.; Peruzzi, A.; Pajares, G.; Kaplanis, G.; et al. Fleets of robots for environmentally-safe pest control in agriculture. *Precis. Agric.* **2016**, 1–41. [CrossRef]

82. Conesa-Muñoz, J.; Pajares, G.; Ribeiro, A. Mix-opt: A new route operator for optimal coverage path planning for a fleet in an agricultural environment. *Exp. Syst. Appl.* **2016**, *54*, 364–378. [CrossRef]

Journal of
Imaging

MDPI

Article

Precise Navigation of Small Agricultural Robots in Sensitive Areas with a Smart Plant Camera

Volker Dworak [1,*], Michael Huebner [2,†] and Joern Selbeck [1,†]

1   Leibniz Institute for Agricultural Engineering Potsdam-Bornim e.V., Max-Eyth-Allee 100, D-14469 Potsdam, Germany; jselbeck@atb-potsdam.de

2   Embedded Systems for Information Technology, Ruhr-University of Bochum, Universitätsstraße 150, D-44801 Bochum, Germany; michael.huebner@rub.de

*   Author to whom correspondence should be addressed; vdworak@atb-potsdam.de; Tel.: +49-331-5699-420; Fax: +49-331-5699-849.

†   These authors contributed equally to this work.

Academic Editors: Gonzalo Pajares Martinsanz and Francisco Rovira-Más

Received: 4 September 2015; Accepted: 30 September 2015; Published: 13 October 2015

Abstract: Most of the relevant technology related to precision agriculture is currently controlled by Global Positioning Systems (GPS) and uploaded map data; however, in sensitive areas with young or expensive plants, small robots are becoming more widely used in exclusive work. These robots must follow the plant lines with centimeter precision to protect plant growth. For cases in which GPS fails, a camera-based solution is often used for navigation because of the system cost and simplicity. The low-cost plant camera presented here generates images in which plants are contrasted against the soil, thus enabling the use of simple cross-correlation functions to establish high-resolution navigation control in the centimeter range. Based on the foresight provided by images from in front of the vehicle, robust vehicle control can be established without any dead time; as a result, off-loading the main robot control and overshooting can be avoided.

Keywords: vision-based navigation; infield navigation; steering control; plant camera

---

1. Introduction

In the field of agricultural research, precision farming and bio-farming agricultural robots are becoming more important because of the growing availability of robots as well as new and alternative applications that robots can provide or will provide in the near future. These alternative applications include testing and measurement applications, and research has focused on the placement of small chemical or insect bombs at precise infield positions for pest control. When wind from helicopter propellers prevents precise application, the payload is too small for deployment using helicopters, or views close to the ground [1] or mechanical manipulation are required, infield robots are the right choice. However, only robots with small wheels and low weight can be used when young or expensive plants need to be protected. Therefore, the vehicle cannot drive over the plant lines, and a method for inter-row weeder guidance is required. The typical drilling distance for wheat is 16 cm in Germany; thus, navigation precision must be in the centimeter range, which can be accomplished with a vision-based system [2–4]. Standard Global Positioning System (GPS) approaches often fail to produce high resolution over the entire field because when a satellite is hidden by an obstacle, such as a tree, hill, building, or the horizon, the resolution of the calculated position jumps to the meter range. Even a high-resolution real-time kinematic GPS that uses two GPS devices and a radio connection to transfer the correction data faces similar problems when the radio connection is lost. For a ground-based radio connection, such a loss of connection can occur without obstacles because ground-reflected radio waves experience a 180° phase shift of directly transmitted waves, which results in the attenuation of the transmitted wave [5]. Under good satellite and radio conditions, real-time kinematic GPS can reach centimeter-scale resolution for slow vehicle speeds [6] or at larger time steps of one second [7]. In addition, typical maps uploaded to agricultural machineries have a grid size in the meter range [8]. A meter-scale grid is sufficient for large agricultural machines but not for small field robots. Accordingly, alternative or complementary techniques are required for precise navigation control. The best practice is to use an actual view of the plant lines to determine the correct direction to navigate [2,9–11]. For an automobile, navigational laser scanners and camera systems are common [12]; however, laser scanners are expensive and optimized for automobile applications; thus, they have large and overlapping spots to ensure the safe detection of all potential obstacles instead of centimeter-scale resolution [13,14]. Low-cost laser scanners are produced for indoor use and are not designed to operate in conditions with high amounts of water, dust and vibration, and they fail when exposed to direct sunlight, which is the greatest disadvantage for infield applications. In research applications, laser scanners are versatile under good weather conditions for a range of applications, including measurements of tree-row crops [15] or corn crops [16]. Although the cost of camera systems is decreasing, they lack robust plant detection software, which must be implemented in the system by the user. However, a normalized difference vegetation index (NDVI) processing system may be implemented along with plant detection software. NDVI is commonly used in agriculture to detect chlorophyll activity and separate plants from the soil. Although different formulas can be used to process NDVI signals and images, they always result in a grayscale or binary image with an adequate threshold. Both types of images can be used for high-resolution navigation control, and most high-resolution navigation applications use Hough transformations to detect plant rows [2,9–11,17], the cross-correlation is a simpler approach in terms of processing power for a small embedded system. The combination of these images with a mask representing the plant line, for cross-correlation result in a precise position signal, which can be used to (lock-in) follow along the plant line. The cross-correlation function has a high filtering effect and is thereby good for noise reduction and outlier suppression. Using the binary image and mask results in a dramatic reduction of calculation power for microcontrollers/processors and field programmable gate arrays (FPGA), because they can use the logical "AND" and a counter for the whole math. Additionally, the horizontal use of the images has the grade advantage for parallelizing the calculation process, and thereby accelerating the result determination. The foresight provided by the image in front of the vehicle provides sufficient calculation time and timely decision support. Accordingly, vehicle control can be established without any dead time, and overshooting of the control output can be avoided. The calculating power costs in a small embedded system are low and will continue to decrease in the future, thereby enabling effective image-based navigation control systems for infield robot applications.

2. Methods

The plant-based navigation described in this article is a combination of two robust concepts: the image of a known scene with a known plant camera system and the cross-correlation mathematical operation that determines the degree of similarity of two functions. In this case, the cross-correlation determines the degree of similarity of a pixel line from both an image and mask, which corresponds to the periodic plant line structure.

2.1. Plant Camera and Imaging

The plant camera should be mounted as high as possible on the field robot. The mounting angle should provide a good compromise between foresight and high-resolution views close to the robot, which also depends on the viewing angle of the objective. A typical arrangement is shown in Figure 1.

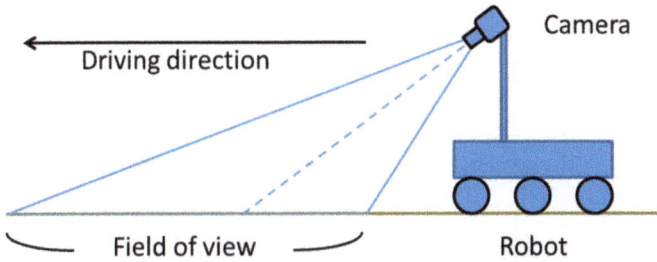

Figure 1. Mounting position of the plant camera on top of the field robot.

With respect to costs, any color complementary metal-oxide-semiconductor (CMOS) charge-coupled device (CCD) cameras with near infrared (NIR) sensitivity (~800 nm to 900 nm) can be used. The IR-cutoff filter must be removed, and a low-pass filter from approximately 645 nm to 950 nm (RG645, SCHOTT AG, 55122 Mainz, Germany) should be used. Better results can be achieved with an adapted double band-pass filter [18], although this configuration has a higher cost. An adequate CMOS chip is the Aptina MT9V032STC (Aptina Imaging Corporation, San Jose, CA, USA), which has high NIR sensitivity at 850 nm (Figure 2) and provides an image with 752 × 480 pixels.

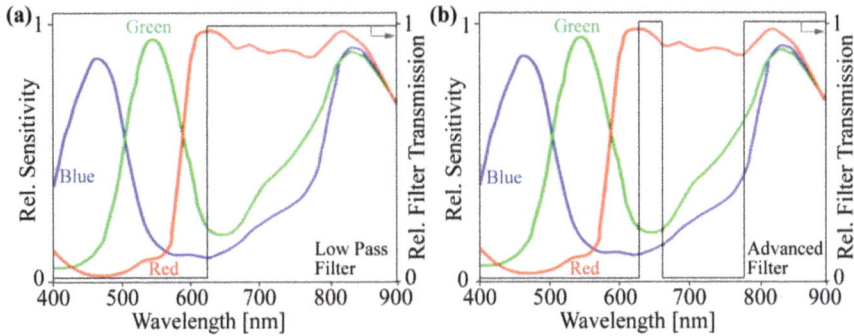

Figure 2. Spectral distribution of the RGB color chip MT9V032STC. Additional spectral range of the low pass filter (**a**) and optimized double band-pass filter (**b**) [18].

Figure 2 shows the usable spectral range of the camera chip after implementing an optical filter. The green and blue channels are only sensitive in the NIR range, and the red channel is sensitive to red and NIR light. Figure 3 shows the spectral characteristics of a plant, and the highest amplitude of spectral response is between the red and NIR range. Therefore, these spectral components are often used for the NDVI.

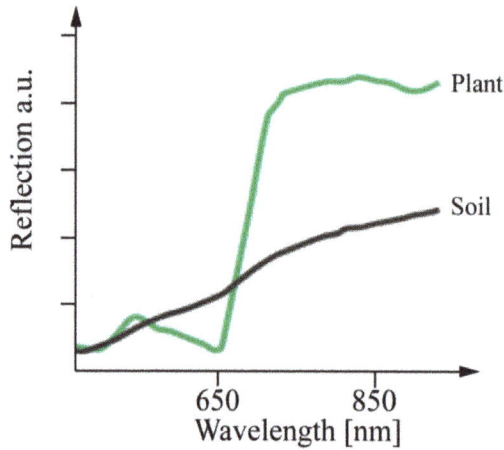

Figure 3. Characteristic spectral distribution of plants and soil. Measured 2012 at a campaign by Gebbers *et al.* with a spectrophotometer (400 to 1000 nm, build of MMS1 NIR enhanced optical modules (Zeiss, Jena Germany) and LOE-USB controller (tec5, Oberursel, Germany)) [19].

The formula for the NDVI must be adapted to the spectral composition of the new "RGB" channels of the color chip.

$$(\text{NIR} - \text{R}) / (\text{NIR} + \text{R}) \rightarrow ((\text{B}_{channel} + \text{G}_{channel}) - \text{R}_{channel}) / \text{R}_{channel} \qquad (1)$$

Several optimizations can be used improve the contrast between plants and soil in the images [18], but the most important are debayering control and white balancing termination in custom cameras. These algorithms are usually optimized for RGB images and do not operate properly for new channel configurations. With optimized NDVI images, enhanced binary images can be produced [18], and they simplify the subsequent image processing applications [20].

Images of winter wheat were taken at the early growth state, and they were not optimal in terms of quality and viewing direction compared with images from fixed-mounted camera on a field robot platform. Although the images were not optimal, they demonstrate the robustness of the cross-correlation algorithm.

2.2. Cross-Correlation

The discrete form of the cross-correlation is shown in Equation (2). The calculation width from $-\text{N}$ to N corresponds to the search window width of the cross-correlation, which will be subsequently used for tracking the position of the resulting maximum.

$$CC(k) = \sum_{n=-N}^{N} B(n) \cdot A(n+k) \qquad (2)$$

The cross-correlation algorithm describes the identity of two functions, with one discrete function moved over the other and each data point (pixel) then multiplied. Finally, all of the results are summed, and the position of this sum is then stored or displayed (Figure 4).

J. Imaging **2015**, *1*, 115–133

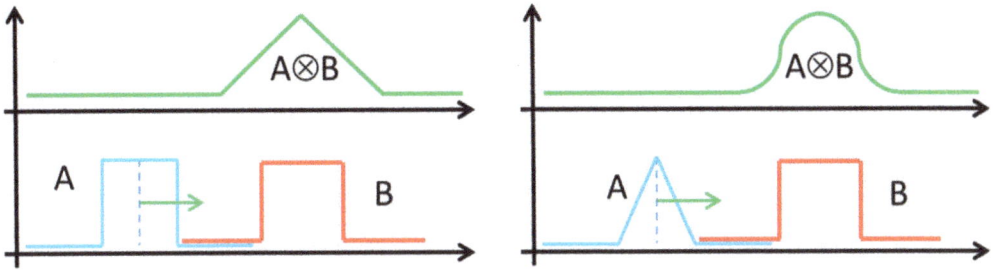

Figure 4. Correlation between simplified signals A and B.

The maximum of the correlation result provides the highest identity position of the two functions. For a periodic function, the result will have a periodic maximum. With respect to the round shape of the maxima peak, the center of the peak can be calculated by the distance between the -3 dB $(1/\sqrt{2})$ and -6 dB (0.5) points or displayed in a small search window for peak tracking of the median position width. Navigation control must follow the peak signal; therefore, the mask function does not have to be moved over the entire pixel line of the image, which reduces the amount of calculation cycles to the number of pixels from the mask function ($-N$ to N). During perfect navigation control, the peak signal of the correlation is always in the middle of the results (Figure 5), and the input signals for the control loop to follow the specific plant line represent deviations from optimal conditions. Therefore, the search window for tracking can be even smaller than the mask window. For example, the window size for the cross-correlation is four or five periods wide, and the window size for the tracking search is one period.

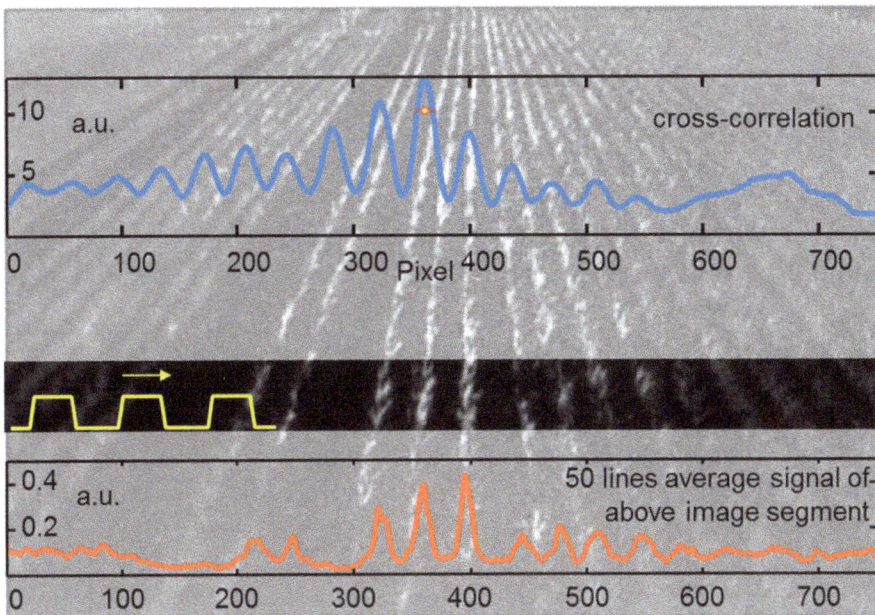

Figure 5. Application of the discrete cross-correlation between a 50-line averaged image signal and three-pulse rectangular shape mask. The center point of the correlation result is marked with a yellow and red point.

Because the middle range of the image is optimal for this application, an inexpensive objective for the camera system can be used. Optical distortions caused by the objective affect the outer part of the image, although they can be ignored for this application. The image will also be used for additional image processing applications [21], with a higher quality lens used in practice.

3. Strategies

Several strategies can be used to obtain adequate results for this navigation approach, including mask design, average number of image line determination, error correction establishment, and embedded system design.

3.1. Mask

The mask function can be estimated or calculated by different functions or methods. With a fixed camera mount and known drilling distance, an empirical mask can be used in most cases. At the starting point in front of the field, the robot system can calculate the actual mask. These calculations do not affect the control loop because the mask will be calculated only once at the initializing/starting phase after each turnaround at field ends. After stretching out the perspective in the image, a fast Fourier transform (FFT) function can be used to determine the basic frequency of the plant line distance, which should be the frequency for the periodic mask function. Based on the stretched image, the maximum peak intensities of the cross-correlation can be used by varying the periodicity of the rectangular-shaped mask from the expected nearby lower periodicity to higher periodicity. Alternatively, a low-pass filter, such as a Gaussian-shaped filter, can be applied to the image lines in the x-direction. The threshold based binary result can be used to obtain a rectangular signal; in this case, a median duty cycle is a good choice for the mask function.

In addition, the mask function length, or the number of periods to be used for the correlation, must be set. A short mask produces a small number of calculations, whereas a large mask provides a better filtering effect and is more robust against outliers. For this application, a very large mask must be adapted at the border side because the perspective image causes changes in both the periodicity and duty cycle (Figure 5). For the same camera mounting position, a correction function or lookup table can be used to adapt the mask limb. However, with respect to calculation power, a medium-size mask is a good compromise. One rectangle is too sensitive and can lead to frequent missing plants in the line. Four rectangles are robust as long as four plants are not missed; this exception will be discussed later. Image stretching evens the linear mask correction function for the y position, with the mask decreasing in size from the bottom to the top y lines. In addition, small variations of mask scaling can be used to calculate results with the highest minimum-to-maximum distance or the best fit. Increasing differences from the best mask scale to expected scale provide additional terrain information. Larger scales indicate that a hill is coming, whereas smaller scales indicate that a hill has passed. For a field robot with an integrated hybrid power system, this information can be used to direct the robot to provide a higher amount of power, such as by increasing the generator turns per minutes.

3.2. Number of Tracking Points and Averaged Image Lines

In natural scenes, the field arrangement is never perfect. Therefore, an individual image line in the x direction can appear as noise, which is useless for a tracking result or point and may be caused by drilling errors or animal interference at the specific area. Significant effort is required to write a program that can manage all of the existing or possible exceptions, and a more effective solution is to reduce the number of tracking points and calculate them with an averaged image line. Due to the perspective distortion in the image, a limited number of lines is available for averaging in the y direction because the result is approaching an increasingly flat line. Depending on the mounting position and camera resolution, each setup will have its own optimal compromise for the number of tracking points and averaged image lines, which can be performed by averaging 10 or 50 lines or using a certain percentage of lines above and below the actual y position, such as 20 lines combined with an additional 20 lines above and below. This process results in a high degree of filtering, but the average of each 20-line package must be calculated only once. These packages can be weighted by 0.25, 0.5 and 0.25, which results in a Gaussian filter response. Thousands of combinations of filter types and lengths are possible. With respect to calculation power, simple algorithms are preferable because the cross-correlation can filter as well.

Figure 6 demonstrates that the differences at the center region are minimal as long as plants are not missing in the line.

The number of calculated tracking points is not constrained because the averaged regions can overlap. A larger overlap results in smaller potential movements of the tracking points, although it requires additional calculation power to perform averaging and has an additional disadvantage of reduced tracking at small curve radius.

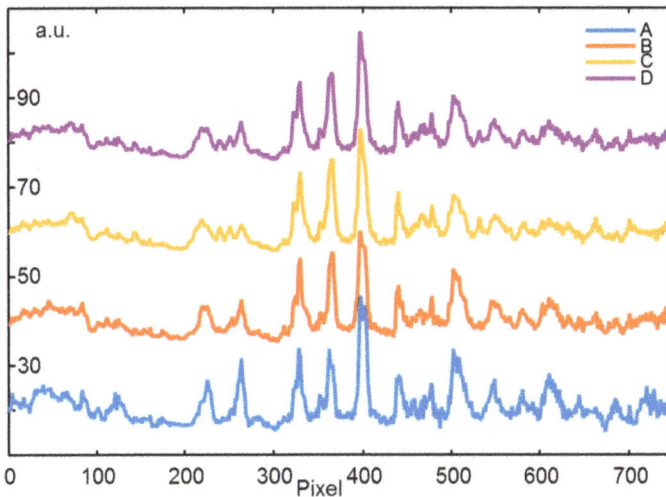

Figure 6. Four plotted lines showing the normalized average result at the same y position. Line A is an average of 20 lines, line B is an average of 40 lines, line C is an average of 60 lines, and line D is the Gaussian result ($20 \times 0.25 + 20 \times 0.5 + 20 \times 0.25$).

3.3. Error Correction

Because of the imperfect field situation, the cross-correlation can produce higher maxima to the left and right of our tracking maximum. Therefore, the search window size should be reduced to one maximum. Determining the field conditions can reduce the window size [22]. If the maximum jump is towards the window corner or the distance between minimal and maximal values is too small, then a warning or error signal is indicated. If the error signal is missing for results at higher y positions, then certain errors can be ignored and compensated by a linear regression as described below. Figure 7 shows a typical search window result with the recused width during a normal operation cycle, and the gray areas indicate the warning region for the center point. The warning region is an example and can be adapted to the camera resolution and view field.

Larger field areas may not be in the correct order, which might be caused by animals or drilling errors. Drilling errors are not frequent in gardening, but such errors occur more frequently during grain cultivation. At the drilling machine stopping position for grain refilling, islands of excessive or missing plants can occur, and the cross-correlation cannot find a tracking point in such areas; as a result, the entire solution will be adapted with a spline or linear regression calculation for the expected driving direction. This function helps to find outliers and can bridge gaps in an image. Possible tracking points in line and close to the spline position will be used in the algorithm, and the robot can follow the spline interpolation over the gap areas. This process is the correct procedure for use in an area with excessive numbers of plants, which is determined by the NDIV information. For missing plants, this strategy is justified if the gap is smaller than the robot. For larger gaps, the robot should ask the supervisor to drive around the gap or use additional image analysis techniques to ensure that holes in the ground are not present. Due to the reduced curve-driving capability of drilling machines, the interpolation function requires only several terms for fitting the curve shape. The main driving direction is the y direction in the image; therefore, the function depends on the y coordinate:

$$f(y) = a + by + cy^2 + dy^3 \qquad\qquad (3)$$

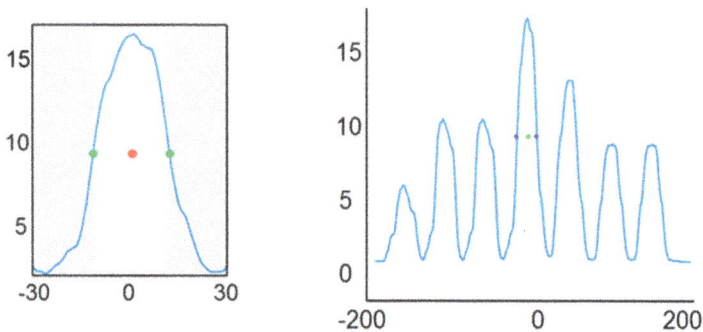

Figure 7. Search window at the center position of the cross-correlation result. The median is used to detect the nearest thickness points of the peak function, and the middle position is marked. Both plots were obtained from the results of a binary image. The gray areas indicate the warning regions for tracking point quality.

If the regression coefficient is inadequate, then historical tracking points from previous images can be used to strengthen the interpolation function. Therefore, the movement should be stored in memory, with at least the last image saved.

Additional error corrections can be implemented by using alternative information sources [17], such as gyroscopes, accelerators, magnetic compasses, barometers and GPS. This process is called "sensor fusion", and in combination with calculated information from the sensor signals, it helps to resolve critical issues.

All needed parameters and functions for the cross correlation are summarized in the workflow diagram in Figure 8.

3.4. Embedded System

To offload the main computer of the robot system, the direction correction data should be calculated by an independent system. Many semiconductor chips can be used to perform such calculations at low cost and low power consumption. For simple solutions, it is impractical for the robot to transport a large computer workstation because of the increased power consumption and payload, although the advantage of this solution is the high degree of potential parallelization. XMOS has developed an XS1-L16A-128-QF124 microprocessor with 16 processor cores for $20, and FPGAs are available from multiple companies at prices ranging from $2 to $10,000. The NVIDIA Tegra K1 chip has quad ARM processors and 192 graphics processing units (GPUs), and a computer-like evaluation kit (Jetson TK1) costs €170. However, a more optimized solution is to use modern software to design and evaluate application specific optimized processors.

Get NDVI image from camera system	
Gray scale image	Binary image

Select mask type and periodicity	
Gray scale mask	Binary mask
A. By image analysis like: FFT, cross correlation, averaging, … B. Use stored mask (typical by known plant type and fixed camera mounting)	

Select starting point
Bottom line or center of the image

Select averaging or filter strategy and length (n= number of pixel lines)
A. n and overlapping area (→resulting number of loops per image) B. Filter type like: averaging, Gaussian, … C. Strategy: First filtering and then cross correlation; First cross correlation and then filtering D. Thresholding for binary result

Select tracking point parameters
A. Select search window size (typical width of one period) B. Use min-max criteria for the quality of the cross correlation result C. Use linear regression for interpolating the expected window position

Run cross correlation
A. Plot tracking points B. Output steering command

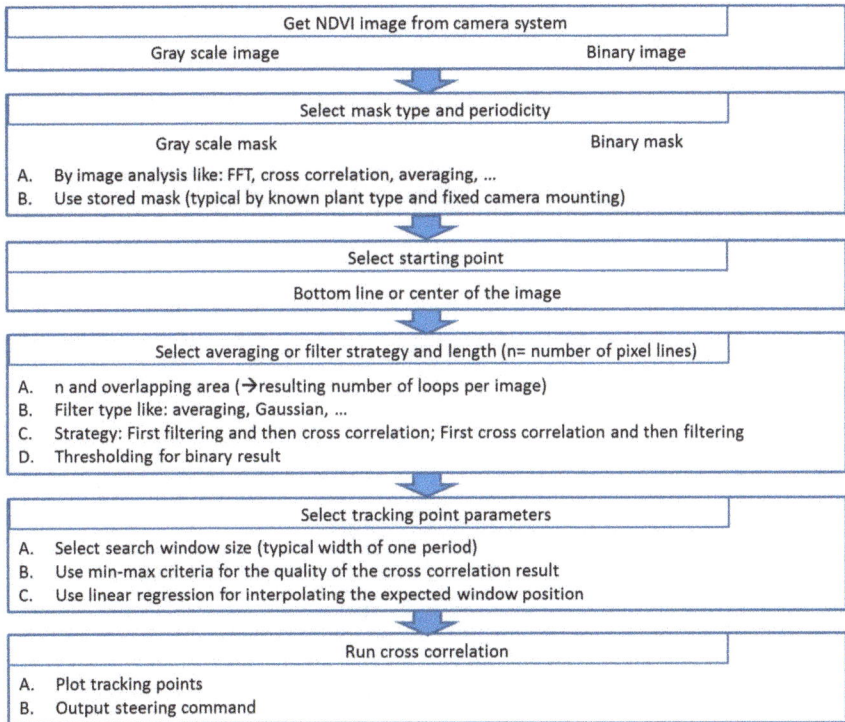

Figure 8. Simplified workflow diagram for the main path.

The required processing architecture for robot applications must be heterogeneous because both robotics control applications as well as dataflow-oriented applications are used. Robot control is clearly part of the application, which is more control-flow oriented. Therefore, the most suitable target architecture is a standard central processing unit (CPU) and varies according to the algorithm used for image processing. Here, a clear data-flow orientation is most suitable; therefore, hardware with maximal parallelization would be most beneficial to host this part of the application and specific processor architectures, such as very long instruction word (VLIW) processors, GPUs and FPGAs, are most suitable. Because the FPGA architecture can combine control- and data-flow oriented processors, this hardware is most suitable for the target application described in this paper. Promising FPGA architecture is found in the Zynq platform by Xilinx, which combines a dual core ARM 9 processor with reconfigurable hardware and a number of standard interfaces, such as the controller area network (CAN), Peripheral Component Interconnect Express (PCI Express), Serial Peripheral Interface (SPI), analogue, and a high number of digital I/Os [23]. More suitable architecture, Zynq Ultrascale, will be available in future and will include an ARM Mali GPU, quad core ARM A53, dual core R5 and many more features.

The aforementioned architectures exhibits the trends of current and future embedded system platforms, which clearly follow the trend of heterogeneity in terms of their processing units because of applications that have different requirements, particularly for embedded applications. These requirements can be either functional or non-functional. Functional requirements include specific algorithms that deliver results according to a quality of service request (e.g., image resolution, frame rate), whereas nonfunctional requirements include real-time operation, high throughput, high reliability, and availability; these functions are crucial because real-time requirements are essential for the proper and safe operation of a system. Most of the recent achievements have enabled embedded platforms to measure and control themselves to adapt to these requirements, even during run-time operations. [24]. Here, a specific adaptation of the processing element is used to self-tune the complete architecture according to the current status of the processor and environment [25] to resolve the issue of a static computing architecture that cannot be optimized for a specific and dynamic application. These architectures are able to self-tune the processor, accelerator and specific interface cores using the reconfigurable portion of the chip by exploiting dynamic and partial reconfigurations. Here, a chip component is updated during run-time operations, whereas the rest of the chip remains in operation. This feature enables chip configurations according to the changing requirements of an application, thus increasing the flexibility of an embedded system tremendously.

4. Results

Figures 9–11 illustrate the cross-correlation algorithms implemented on plant images with real in-field conditions. Figures 9 and 10 use the average of 20 lines and were calculated every 10 lines. Figure 9 shows two scenes with small angles and middle position errors, and the mounting height and viewing direction provide an adequate foresight that is typically 20 m.

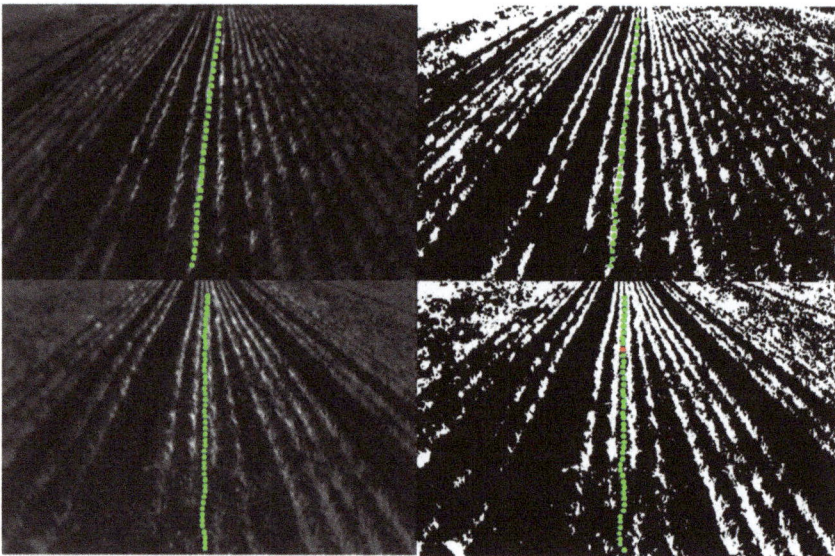

Figure 9. Two winter wheat scenes photographed using the low-cost plant camera. The left side shows a gray-scale NDVI image, and the right side shows a binary image. Both sides are overlain with the individually calculated tracking points.

Figure 9 shows that the tracking points exhibit small variations from a straight line. The differences from a linear regression line and between grayscale and binary images are discussed later. For this camera perspective, the tracking point presents excellent following of the plant lines, and variations from the linear regression are small (see Table 1). Figure 10 shows a similar field scene but with different mounting angles. The resulting images include the horizon and have a maximum field of view, although with higher restrictions for pixel resolutions in the upper 30% of the images.

Figure 10 demonstrates the enormous potential foresight of this plant-based navigation solution, with the algorithm losing tracking at several pixel lines before the horizon. Nevertheless, tracking is possible for over 50 m using a low-resolution camera. Figure 11 shows two field scenes with curves in the plant line.

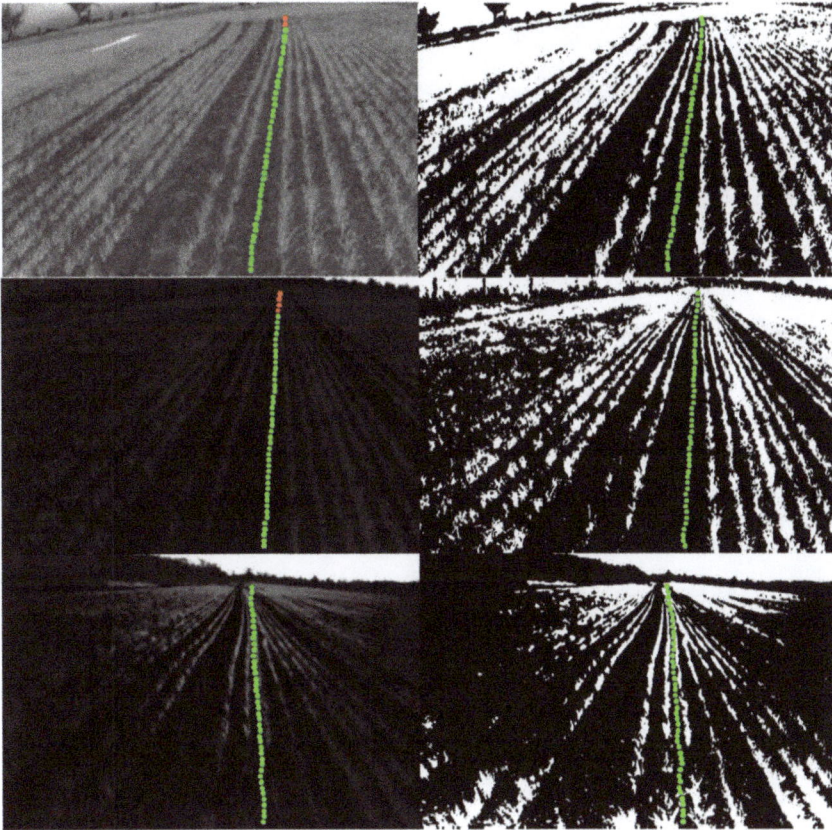

Figure 10. Three winter wheat scenes photographed using the low-cost plant camera. The left side shows a gray-scaled NDVI image, and the right side shows a binary image. Both sides are overlain with the individually calculated tracking points. Scenes one and two in the grayscale images show two and four red tracking points, respectively, which indicate exceedance of the warning level and an excessively small difference between the maximum and minimum.

The results in Figure 11 demonstrate that the algorithm can find tracking points without requiring straight lines in the image. The fitting curves demonstrate the good predictive potential of future directions, even with poor quality images in which the bright-sky pixel intensities reduce the dynamic range for infield pixels. Figure 12 shows two extreme situations for which the field of view is inadequate for this application.

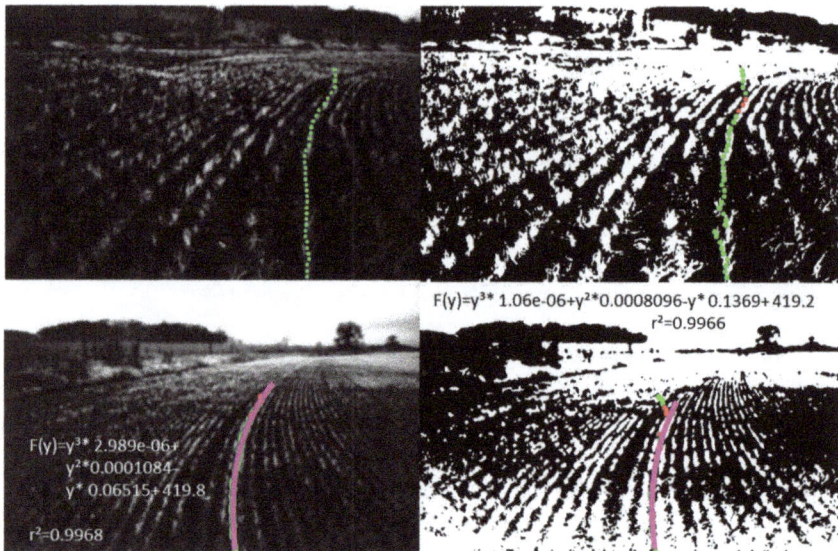

$$F(y)=y^{3*}\,1.06e\text{-}06+y^{2*}0.0008096\text{-}y^*\,0.1369+419.2$$
$$r^2=0.9966$$

$$F(y)=y^{3*}\,2.989e\text{-}06+$$
$$y^{2*}0.0001084+$$
$$y^*\,0.06515+419.8$$

$$r^2=0.9968$$

Figure 11. Two winter wheat scenes with a curved path. The left side shows the gray-scaled NDVI image, and the right side shows the binary image. Both sides are overlain with the individual calculated tracking points. Scene 1 uses an average of over 11 lines for the gray-scale image, over 15 lines for the binary image and a three period mask. Scene 2 uses an average of over seven lines for both the gray-scale and binary image and a four period mask. Scene 2 is overlain with a third-order fit.

Figure 12 demonstrates the robustness of the algorithm, even under the worst image quality conditions. Here, the viewing angle is too flat and the image contrast is reduced by the bright-sky pixels. In addition, the size of the plants is highly variable, and larger gaps with missing plants are observed. Under these extreme conditions, it is important to determine if the mask is appropriate for the actual scene or image. The width of the mask can be analyzed with additional loops that vary the width. The median level in Figure 7 is caused by the maximum and minimum values in the search window, and this difference also indicates the quality of the tracking point, is used as a warning signal and can be used to manipulate the mask size for the cross-correlation calculation. Figure 13 shows the position of the maximum difference for the calculations using masks varying from −2 to +2 pixel widths per period. Multiple points at the +2 level indicate that the mask should be wider, and multiple points at the −2 level indicate that the mask should be narrower.

As shown in Figure 11, a third-order fit is a simple but adequate function for interpolating or evaluating individual tracking points. Table 1 presents the R^2 and root mean square error (RMSE) statistics for the third-order regression for the demonstrated scenes.

A stability index with values greater than 0.9 demonstrates accuracy in the tracking points determined with the cross-correlation application. The binary results exhibit nearly the same stability as long as the image quality is adequate. For low-quality images from Scenes 6 to 9, the binary results differ from the gray-scale results. Therefore, the faster binary approach requires an adequate camera mounting and good plant camera system with efficient binary results independent of lighting [18]. The first scene shows an orthogonal tracking line, for which the R^2 value is useless for typical straight driving directions; as a result, the RMSE value should be used as the quality indicator. With respect to the different viewing angles and view fields, the RMSE values at the starting point can be compared after normalization to the given plant line distance of 160 mm. This line distance is equivalent to the mask periodicity.

Figure 12. Two winter wheat scenes with both curved and flat viewing directions. The left side shows the gray-scaled NDVI images, and the right side shows the binary images. Both sides are overlain with the individual calculated tracking points.

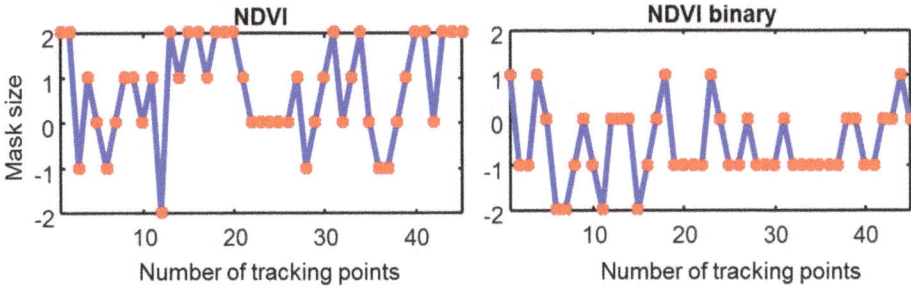

Figure 13. The quality of the mask size is indicated by the maximum difference between the maximum and minimum values in the search window for each tracking point. The mask size varies from +2 to −2 additional pixels for the periodic structure of the mask. This example plot was produced from the first image in Figure 10.

Table 1. Results for the third-order regressions.

	R^2		RMSE			Normalized RMSE in mm	
Scene	Gray	Binary	Gray	Binary	Pixel/Row	Gray	Binary
1	0.3362	0.1477	1.508	2.299	64	3.77	5.75
2	0.9962	0.9909	0.9015	0.9328	47	3.07	3.18
3	0.9984	0.9935	0.8452	1.345	64	2.11	3.36
4	0.9904	0.9753	0.870	1.071	59	2.36	2.90
5	0.9728	0.9205	1.758	1.210	77	3.65	2.51
6	0.9303	0.5164	2.844	4.820	80	5.69	9.64
7	0.9968	0.9866	0.8468	1.029	38	3.57	4.33
8	0.9463	0.9744	2.726	1.957	58	7.52	5.40
9	0.9644	0.6476	2.365	4.021	114	3.32	5.64

Binary Results

The binary images should be processed by appropriate algorithms in the plant camera system; however, this processing is beyond the scope of this paper. Regardless of how the binary images are obtained, proper binary image processing improves the performance of the cross-correlation because the resulting binary images exhibit sharper peaks as shown in Figure 14.

Figure 15 shows a small terrain effect in the image. Variations from the perspective function used to shrink the masks width indicate changes in the terrain.

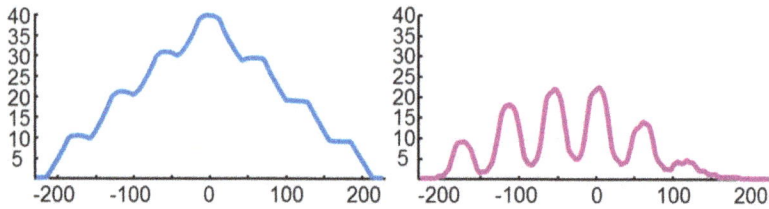

Figure 14. Cross-correlation results from one averaged pixel line over a window size with three periods. The left side shows the gray-scale results, and the right side shows the binary results.

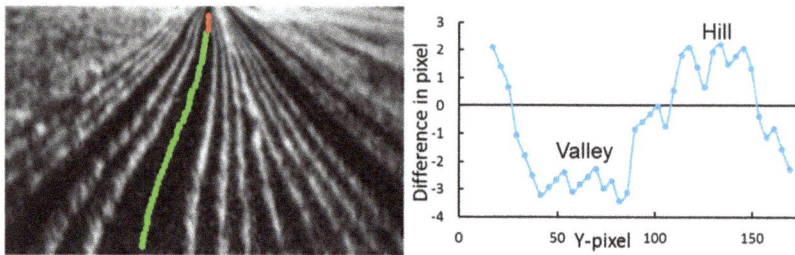

Figure 15. Terrain effect is indicated by a difference between the linear mask shrinking by the perspective in the image, and the mask size with highest maximum. Below zero indicates a valley and above zero indicates a hill.

The diagram in Figure 15 demonstrates the possibility for detecting terrain features. For a more difficult situation with a curve, the mask shrinking factor caused by the curve must be additionally considered.

5. Conclusions

The combination of a plant camera and cross-correlation algorithm results in a robust in-field navigation solution for robots working with sensitive plants. This solution uses the plant lines themselves to follow precise tramlines. The results of this study demonstrate that the proposed approach avoids driving over plants and provides accurate navigation control in the centimeter range. In addition, the proposed approach overcomes the issue of jumps that hinder GPS-driven solutions. Because plant lines were drilled with large agricultural machinery, the minimum curve radius is restricted, which provides a number of possibilities for reducing the power required to calculate the algorithms.

1.   Image lines in the x direction can be concentrated by averaging in the y direction.
2.   The cross-correlation function does not have to move over the entire pixel line.
3.   The moving mask of the cross-correlation can be reduced to a few periodic replications with a rectangular shape, thereby reducing the length of the used pixels and number of multiplications.

4. The reduced mask must only move over a length smaller than one period to have only one maximum peak of the cross-correlation in the inspection window.
5. Missing tracking points can be interpolated using a simple linear regression function.

In addition, the required calculation power can be reduced using the binary result from the plant camera. The multiplied binary image line and a binary mask can be replaced by the logical conjunction "AND", which can be performed in parallel during one clock cycle in a FPGA.

All of the algorithms required for this application are reduced to multiplications and summations, which is an important point for implementing algorithms in a small embedded system with restricted resources. All line averaging can be performed in parallel, and for reduced packages, even the cross-correlation can be performed in parallel. The inspection window for the cross-correlation must follow shifts in the tracking points and is therefore a cascaded operation. After two or three parallel correlations, the shift must be added and then the new run starts with the shifted window position. Using this window provides a substantial advantage in potential error detection. If the calculated tracking points fall in the edge region of the window or if the difference between the maximum and minimum value in the window is too small, then these tracking points can be ignored. A simple linear regression can be used to calculate a tracking function, and outlier tracking points can be overbridged, which also illustrates the robustness of this navigation solution. The detailed description of this application and its resulting simplicity are significant advantages in the establishment of steering control in small embedded systems that offload the main system of the field robot. The steering command is not the only output of the solution, and the foresight provides additional information on field conditions, such as direction, hills and valleys, assumed obstacles, and the field ends. Such additional information is important for providing correct overall field driving plans or management. When this information is combined with sensor signals, such as gyros and accelerators, the main system can determine if the approaching hill has too great of an ascending slope for the robot.

The next step in this line of study will be changing the correlation direction from the x axis to the orthogonal axes of the fitted line. This modification could enable the application to follow sharper curves in the image because the averaging of lines in the y direction can filter out the plant lines, while the averaging in the expected direction cannot perform this filtering. Figure 16 shows the difference between the averaging directions. In addition, the cross-correlation will output higher peaks in the expected direction, thus improving the quality of the tracking points.

J. Imaging **2015**, *1*, 115–133

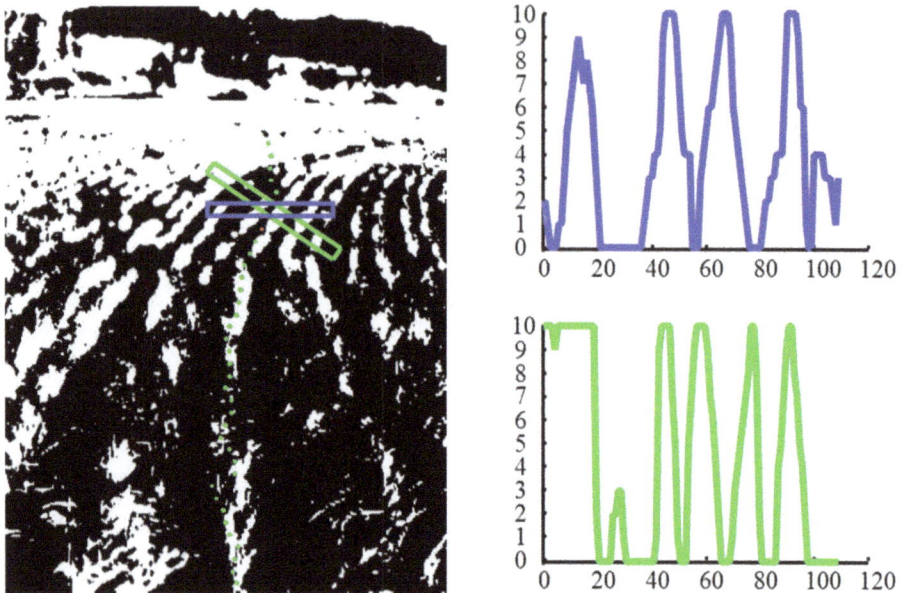

Figure 16. Different averaging directions determining different resulting input functions for the cross correlation. Upper plot is the result of the averaging in y-direction. Lower plot is the result of the 33° direction.

Acknowledgments: We thank Mathias Hoffmann for his work during the measurement campaign and for his support for the first MATLAB implementation.

Author Contributions: Volker Dworak conceived, designed, and performed the experiments. Volker Dworak, Michael Hübner, and Jörn Selbeck analyzed the data and programmed the solution in a MATLAB environment. Volker Dworak and Michael Hübner wrote the paper.

Conflicts of Interest: The authors declare no conflict of interest.

References

1. Shi, Y.Y.; Wang, N.; Taylor, R.K.; Raun, W.R.; Hardin, J.A. Automatic corn plant location and spacing measurement using laser line-scan technique. *Precis. Agric.* **2013**, *14*, 478–494. [CrossRef]
2. Astrand, B.; Baerveldt, A.J. A vision based row-following system for agricultural field machinery. *Mechatronics* **2005**, *15*, 251–269. [CrossRef]
3. Olsen, H.J. Determination of row position in small-grain crops by analysis of video images. *Comput. Electron. Agric.* **1995**, *12*, 147–162. [CrossRef]
4. Máthé, K.; Buşoniu, L. Vision and control for UAVs: A survey of general methods and of inexpensive platforms for infrastructure inspection. *Sensors* **2015**, *15*, 14887–14916. [CrossRef] [PubMed]
5. Wallace, R. *Achieving Optimum Radio Range*; Application Report from Texas Instruments Incorporated SWRA479: Dallas, TX, USA, 2015.
6. Norremark, M.; Griepentrog, H.W.; Nielsen, J.; Sogaard, H.T. The development and assessment of the accuracy of an autonomous GPS-based system for intra-row mechanical weed control in row crops. *Biosyst. Eng.* **2008**, *101*, 396–410. [CrossRef]
7. Mathanker, S.K.; Maughan, J.D.; Hansen, A.C.; Grift, T.E.; Ting, K.C. Sensing miscanthus swath volume for maximizing baler throughput rate. *Trans. ASABE* **2014**, *57*, 355–362.
8. Molin, J.P.; Colacco, A.F.; Carlos, E.F.; de Mattos, D. Yield mapping, soil fertility and tree gaps in an orange orchard. *Revista Brasileira de Fruticultura* **2012**, *34*, 1256–1265. [CrossRef]

9. Jiang, G.Q.; Wang, Z.H.; Liu, H.M. Automatic detection of crop rows based on multi-ROIs. *Expert Syst. Appl.* **2015**, *42*, 2429–2441. [CrossRef]
10. Bakker, T.; Wouters, H.; van Asselt, K.; Bontsema, J.; Tang, L.; Muller, J.; van Straten, G. A vision based row detection system for sugar beet. *Comput. Electron. Agric.* **2008**, *60*, 87–95. [CrossRef]
11. Torres-Sospedra, J.; Nebot, P. A new approach to visual-based sensory system for navigation into orange groves. *Sensors* **2011**, *11*, 4086–4103. [CrossRef] [PubMed]
12. Fernandez, J.; Calavia, L.; Baladron, C.; Aguiar, J.M.; Carro, B.; Sanchez-Esguevillas, A.; Alonso-Lopez, J.A.; Smilansky, Z. An intelligent surveillance platform for large metropolitan areas with dense sensor deployment. *Sensors* **2013**, *13*, 7414–7442. [CrossRef] [PubMed]
13. Dworak, V.; Selbeck, J.; Ehlert, D. Ranging sensors for vehicle-based measurement of crop stand and orchard parameters: A review. *Trans. ASABE* **2011**, *54*, 1497–1510. [CrossRef]
14. Martínez, M.A.; Martínez, J.L.; Morales, J. Motion detection from mobile robots with fuzzy threshold selection in consecutive 2D Laser scans. *Electronics* **2015**, *4*, 82–93. [CrossRef]
15. Sanz, R.; Rosell, J.R.; Llorens, J.; Gil, E.; Planas, S. Relationship between tree row LIDAR-volume and leaf area density for fruit orchards and vineyards obtained with a LIDAR 3D Dynamic Measurement System. *Agric. For. Meteorol.* **2013**, *171*, 153–162. [CrossRef]
16. Hoefle, B. Radiometric correction of terrestrial LiDAR point cloud data for individual maize plant detection. *IEEE Geosci. Remote Sens. Lett.* **2014**, *11*, 94–98. [CrossRef]
17. Hague, T.; Marchant, J.A.; Tillett, N.D. Ground based sensing systems for autonomous agricultural vehicles. *Comput. Electron. Agric.* **2000**, *25*, 11–28. [CrossRef]
18. Dworak, V.; Selbeck, J.; Dammer, K.H.; Hoffmann, M.; Zarezadeh, A.A.; Bobda, C. Strategy for the development of a smart NDVI camera system for outdoor plant detection and agricultural embedded systems. *Sensors* **2013**, *13*, 1523–1538. [CrossRef] [PubMed]
19. Gebbers, R.; Tavakoli, H.; Herbst, R. Crop sensor readings in winter wheat as affected by nitrogen and water supply. In *Precision Agriculture'13*; Stafford, J.V., Ed.; Wageningen Academic Publishers: Gelderland, The Netherlands, 2013; pp. 79–86.
20. Tillett, R.D. Image analysis for agricultural processes: A review of potential opportunities. *J. Agric. Eng. Res.* **1991**, *50*, 247–258. [CrossRef]
21. Peteinatos, G.G.; Weis, M.; Andujar, D.; Ayala, V.R.; Gerhards, R. Potential use of ground-based sensor technologies for weed detection. *Pest Manag. Sci.* **2014**, *70*, 190–199. [CrossRef] [PubMed]
22. Billingsley, J.; Schoenfisch, M. Vision-guidance of agricultural vehicles. *Auton. Robot.* **1995**, *2*, 65–76. [CrossRef]
23. Zynq-7000 AP SoC Technical Reference Manual, UG585 (V1.10). Available online: www.xilinx.com (accessed on 23 February 2015).
24. Janßen, B.; Schwiegelshohn, F.; Hübner, M. Adaptive computing in real-time applications. In Proceedings of the 13th IEEE International NEW Circuits And Systems (NEWCAS) Conference, Grenoble, France, 7–10 June 2015; pp. 166–173.
25. Janssen, B.; Mori, J.Y.; Navarro, O.; Gohringer, D.; Hubner, M. Future trends on adaptive processing systems. In Proceedings of the 12th IEEE International Symposium on Parallel and Distributed Processing with Applications (ISPA 2014), Milan, Italy, 26–28 August 2014.

Journal of
Imaging

MDPI

Article

Using Deep Learning to Challenge Safety Standard for Highly Autonomous Machines in Agriculture

Kim Arild Steen *,†, Peter Christiansen †, Henrik Karstoft and Rasmus Nyholm Jørgensen

Department of Engineering, Aarhus University, Finlandsgade 22 8200 Aarhus N, Denmark;
pech@eng.au.dk (P.C.); hka@eng.au.dk (H.K.); rnj@eng.au.dk (R.N.J.)
* Correspondence: kim.steen@eng.au.dk; Tel.: +45-3116-8628
† These authors contributed equally to this work.

Academic Editors: Francisco Rovira-Más and Gonzalo Pajares Martinsanz
Received: 18 December 2015; Accepted: 2 February 2016; Published: 15 February 2016

Abstract: In this paper, an algorithm for obstacle detection in agricultural fields is presented. The algorithm is based on an existing deep convolutional neural net, which is fine-tuned for detection of a specific obstacle. In ISO/DIS 18497, which is an emerging standard for safety of highly automated machinery in agriculture, a barrel-shaped obstacle is defined as the obstacle which should be robustly detected to comply with the standard. We show that our fine-tuned deep convolutional net is capable of detecting this obstacle with a precision of 99.9% in row crops and 90.8% in grass mowing, while simultaneously not detecting people and other very distinct obstacles in the image frame. As such, this short note argues that the obstacle defined in the emerging standard is not capable of ensuring safe operations when imaging sensors are part of the safety system.

Keywords: deep learning; obstacle detection; autonomous; ISO

1. Introduction

In order for an autonomous vehicle to operate safely and be accepted for unsupervised operation, it must perform automatic real-time risk detection and avoidance in the field with high reliability [1]. This property is currently being described in an ISO/DIS standard [2], which contains a short description of how to meet requirements for obstacle detection performance. The requirements for tests and the test object are described in Section 5 in the standard. The standard uses a standardized object, shown in Figure 1 which is meant to mimic a human seated (torso and head). This standardized object makes sense for non-imaging sensors such as ultrasonic sensors, LiDARs or Time-of-Flight cameras, which measures the geometrical properties of the objects or distance to the objects. However, for an imaging sensor such as an RGB camera, the definition of this standardized object does not guarantee safety. In this short note, we will present how deep learning methods can be used to design an algorithm that robustly detects the standardized object in various situations, including high levels of occlusion. Based on this, the algorithm is able to comply with the standard, but at the same time, it is not detecting people and animals, as they are not part of the trained model.

Deep convolutional neural networks have demonstrated outstanding performance in various vision tasks such as image classification [3], object classification [4,5], and object detection [4–6]. LeCun et al. formalized the idea of the deep convolutional architecture [7], and Krizhevsky *et al.* introduced a paradigm shift in image classification with the AlexNet [3]. In recent years the AlexNet has been used in various deep learning related publications.

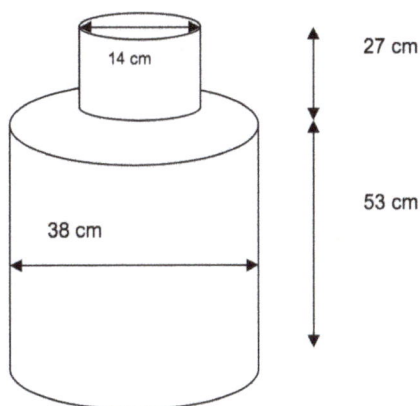

Figure 1. Standardized obstacle.

2. Materials and Methods

In this section, we present the data used for this paper, together with a short description of how the deep learning algorithm was trained and implemented.

2.1. Data Collection

The results in this paper are based on recordings performed in both row-crop fields and grass fields. The recordings were part of a research project aimed at improving safety for semi-autonomous and autonomous agricultural machines [1]. The recordings contain various kinds of obstacles, such as people, animals, well covers and the ISO standardized obstacle. All obstacles are stationary and the data is recorded while driving towards and past them. The sensor kit (seen in Figure 2a) used for the experiments includes a number of imaging devices: a thermal camera, a 3D LiDAR, an HD-webcam and a stereo camera. In this paper, we only focus on the RGB-images. Images from the experiments and recordings are seen in Figure 2.

(a)   (b)   (c)

Figure 2. Images from experiments and recordings. (**a**) The sensor kit mounted on a mower; (**b**) An example image from the recordings in grass; (**c**) An example image from the recordings in row crops.

2.1.1. Training Data

An iPhone 6 was used to record five short videos of the ISO standardized obstacle. The recordings include various rotations, scales, and intensity of the object. A total of 437 frames from the videos were extracted and bounding boxes of the object were created. Example frames from the videos can be seen in Figure 3.

Figure 3. Examples of training data.

2.2. Training of Deep Convolutional Network

Deep learning is utilized for barrel detection using a sliding window approach similar to [6]. We start by fine-tuning AlexNet (Available at the Caffe model zoo [8]), which is pre-trained on data from the image-net competition [9] for ISO standardized obstacle detection. By fine-tuning the neural network to images of the ISO obstacle, which has a specific shape, texture and color, the algorithm will be very good at detecting the occurance of this specific object in the image. At the same time, the algorithm will also be very good at rejecting other objects in the image (animals, people, etc.), thereby meeting the standard with respect to performance, but not with respect to safety.

To increase the number of training examples (both positive and negative), we randomly sampled sub-windows of the extracted training images. A sub-window was labeled as positive (containing an object) if it had over 50% intersection over union with the labeled bounding boxes. To include additional negatives, non-face sub-windows are collected from *Annotated Facial Landmarks in the Wild* database [10] in a similar approach. A total of 1925 positive and 11,550 negative samples have been used in this paper. These examples were then resized to 114 × 114 pixels and used to fine-tune a pre-trained AlexNet model. The original AlexNet model outputs a vector of 1000 units, each representing the probability of the 1000 different classes. In our case, we only want to detect if an image patch contains an ISO object or not. Hence, the last layer is changed to output a two-dimensional vector. For fine-tuning, we used 14K iterations and batch size of 100 images, where each batch contained 67 positive and 33 negative examples. During fine-tuning, the learning rate for the convolutional layers was 10 times smaller than the learning rate of the fully connected layers.

After fine-tuning, the fully-connected layers of the AlexNet model can be converted into convolutional layers by reshaping layer parameters [11]. This makes it possible to efficiently run the network on images of any size and obtain a heatmap of the ISO obstacle classifier. Each pixel in a heatmap shows the network response, which is the likelihood of having an ISO obstacle, corresponding to the 114 pixel ×114 pixel region in the original image. In order to detect ISO obstacles of different sizes, the input images can be scaled up or down. The chosen training image resolution is half the resolution used in the original AlexNet. Reducing the resolution of the training images allows us to reduce the input image by half, thus reducing processing time, while maintaining the resolution of the resulting heatmaps. An example of a resulting heatmap is illustrated in Figure 4.

(a)

(b)

Figure 4. Illustration of ISO obstacle and resulting heatmap. (a) RGB image from the row crop field; (b) Resulting heatmap.

The model was trained using the Caffe framework [12] using a single GPU (4 GB Quadro K2100M). The training time was approximately 1–2 h.

2.3. Detection of ISO Obstacle using Deep Convolutional Network

When the deep convolutional network has been trained, it can be used to detect the ISO obstacle in color images. In order to detect the obstacle at multiple distances, the input image needs to be scaled accordingly. In this paper, we use 13 scales, which are all processed by the same network structure. We use 13 scales to be able to detect the barrel when it is far away (57 pixel × 57 pixel in the original image) and up close (908 pixel × 908 pixel). As described in the previous section, the output is a heatmap, where the intensity reflects the likelihood of an ISO obstacle.

Based on the heatmap, one can detect the obstacle and draw a bounding box. Each pixel in the heatmap corresponds to a 114 pixel × 114 pixel sub-window in the input image. To remove redundant overlapping detection boxes, we use non-maximum suppression [6] with 50% overlap threshold.

3. Results

The results are based on data collected in two different cases, at three different dates. The data has been collected at different times during the days to ensure different lighting conditions. In both cases, we use the model trained on the data presented in Section 2.1.1. Despite this, the results in this paper are of a preliminary nature, as various weather conditions and scenarios are not included in the data.

3.1. Row Crops

Based on the presented algorithm, a total of 7 recordings have been evaluated with respect to detection of the ISO obstacle in row crops. The recordings also contain other kinds of obstacles such as people and animals. The recordings contain a total of 14, 153 frames with 20, 414 annotated obstacles (8126 of those are the ISO obstacle).

In the ISO standard, the obstacles needs to be detected within a defined safety distance. The safety distance is a product of the expected working speed and machine type. Hence, there is no fixed value for this. In Figure 5, a histogram of the achieved detection distances is shown. It is seen that the algorithm is able to detect the obstacle both at close range (3–6 m) and also at far range (over 15 m). The ISO obstacle is present in front of the machine a total of 14 times and the algorithm is able to detect the obstacle everytime. The detection distance, which is the distance of the first positive detection, for these 14 times, ranges from 10 m to 20 m, with an average of 14.56 m.

Evaluating all frames, at frame level with all annotated objects, we achieve TP (true positive) = 2851, FP (false positive) = 1, TN (true negative) = 7105 and FN (false negative) = 4919. The high number of false negatives is a result of annotations, where the ISO obstacle is located more than 20 m away, which is more than the achieved detection range. Based on this, we achieve a hit rate of 36.7% Equation (1), which is the ratio between positive detections and all annotations of the ISO obstacle, and a precision of 99.9% Equation (2) (As there are less than 10, 000 datapoints, we are not able to present the last decimal as a result). In the standard, it is stated that the system must achieve a success rate of 99.99%, however, it is unclear how this should be tested. As stated above, the algorithm is able to detect the ISO obstacle when the ISO obstacle is within 10 m of the machine at a precision of almost 100%.

$$hit\ rate = \frac{TP}{TP + FN} = \frac{2851}{2851 + 4919} = 0.3669 \tag{1}$$

$$precision = \frac{TP}{TP + FP} = \frac{2851}{2851 + 1} = 0.9996 \tag{2}$$

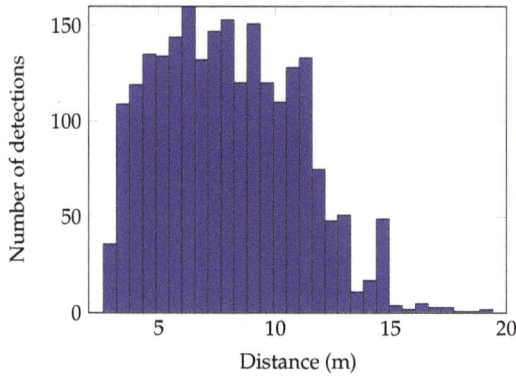

Figure 5. Histogram of detection distances.

In Figure 6, the achieved hit rate for different distance ranges is shown. It is seen that the algorithm is not able to detect the obstacle in all frames (the hit rate is below 1). However, with a precision close to 100%, a single detection is reliable enough to be considered a detection of the ISO obstacle present in the recordings. Hence, the success rate is 99.9%, estimated at frame level, within an average safety distance of 14.56 m.

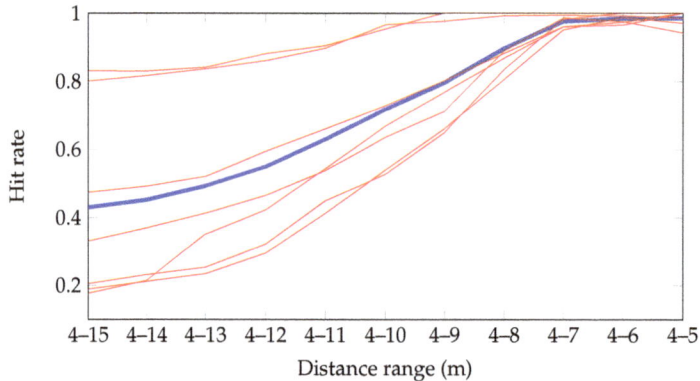

Figure 6. Hit rate, evaluated at frame level, for different distance ranges (e.g. 0.97 is the mean hit rate for all frames in range 4–7 m). Red: hit rate for the 7 recordings, and blue: average hit rate.

3.2. Grass Mowing

We also evaluated the algorithm in a much more difficult case; grass mowing. In grass mowing, the obstacles are often highly occluded (a scenario that is not described in detail in the standard). The recording contains the ISO obstacle and people. The recordings contain a total of 19,390 frames with 936 annotated ISO obstacles.

As with the row crops case, we evaluate the achieved detection distance. In the recording, the ISO obstacle is detected when the ISO obstacle is within approximately 6 m of the machine. The ISO obstacle is present in front of the machine a total of 8 times and the algorithm is able to detect the obstacle everytime.

Evaluating all frames at frame level, we achieve TP = 307, FP = 31, TN = 18,337 and FN = 787. The high number of false negatives is a result of annotations, where the ISO obstacle is located more than 6 m away, which is more than the achieved detection range. Based on this, we achieve a hit rate of 28.1% Equation (3)

$$hit\ rate = \frac{TP}{TP + FN} = \frac{307}{307 + 787} = 0.281 \tag{3}$$

$$precision = \frac{TP}{TP + FP} = \frac{307}{307 + 31} = 0.908 \tag{4}$$

and a precision of 90.8% Equation (4). Again, the ISO obstacle is successfully detected everytime we are driving towards it, however, the achieved precision is lower than 99.99% which is stated in the standard. As seen in Figure 7, the ISO obstacle is highly occluded. In the standard, it is noted that the obstacle must appear unobscured to ensure high levels of system confidence and if the obstacle is obscured, the manufacturer must understand how this affects performance. In the grass case, we show how the system is affected by this. The precision drops from 99.9% to 90.8% and the achieved detection distance drops from an average of 14.56 m to 6 m. The lower precision is due to a higher number of false positives. Most of these false positives are the tractor in the image. The tractor will always be present in the same position in the image and further developments could remove this in post-processing or by including tractor images as negatives in the training data.

Figure 7. Detections in grass mowing case. Notice that people are not detected.

4. Discussion

The results show that we are able to robustly detect the presence of the ISO obstacle in various conditions including different lighting and heavy occlusion. The results also show that the algorithm is not able to detect the presence of the other obstacles within the image frame—even people. This is not a surprise, as the model has been trained on the ISO obstacle and not on other types of obstacles.

We are using a large deep learning model to detect a simple object and it might seem like we are overdoing it. However, by using deep learning and pre-trained networks, we have been able to design a robust classifier for detecting the ISO obstacle in various scenarios, using only a few minutes of training video. This shows the power of these models and how they can be exploited.

In this paper, we have implemented the algorithm using Caffe and the corresponding MATLAB interface, which means that it does not run in real-time. However, CNN implementations ready for real-time applications exist in literature [13]. Furthermore, deep learning algorithms are also being utilized in detection systems for the car industry, where deep learning models are able to classify between a number of different obstacles. This is powered by high-end embedded boards from NVIDIA, which has recently launched a 300 USD board [14], enabling real-time deep neural nets on UAVs and UGVs. These observations make it fair to assume that agricultural machines could also benefit from that computing power.

The ISO standard states that the system must have a success rate of 99.99% in various weather conditions, however, it is unclear how this success rate should be measured. We are not able to detect the ISO obstacle in all frames, however, we achieve a precision of 99.9%, and are able to detect the ISO obstacle at an average distance of 14.56 m in row crops. In the grass mowing case, the obstacle was highly occluded which affected the achieved detection distance. Furthermore, the presence of the tractor within the image frame resulted in more false positives. These should be removed to ensure higher precision. We have not been able to test the performance in various weather conditions, hence, the results obtained in this paper is of a preliminary nature.

The most important result in this paper is that we are able to show that an imaging system can be designed to comply with the ISO standard and completely miss important obstacles such as people—both kids and adults. We argue that the standardized obstacle presented in the standard is not fully adequate to ensure safe operations with highly autonomous machines in agriculture. The design of the obstacle is based on a human head and torso (human sitting down), but we show that we can detect this type of obstacle and at the same time completely miss the people in front of the machines. As the ISO standard aims to represent a human sitting down, we suggest that the standardized obstacle should resemble a more life-like person, if a standardized object is required. In our experiments, we have used mannequins for this task. It is important that the life-like obstacles have the same properties as the current ISO obstacles with respect to color, material, and temperature (hot water). The life-like obstacle could be used to test different possible postures, such as standing, lying and sitting. However, other kinds of obstacles could also be present in the fields. Including other types of obstacles, such as animals, in the requirements, could potentially increase safety overall. However, even doing this, there is no guarantee that the methods will be able to detect real obstacles in the fields, as they might not be perfectly described by the trained algorithm. Deep learning methods have achieved very good performance in very difficult image and object recognition tasks. This is accomplished through access to a vast amount of image training data, where objects like people, animals and cars are depicted in thousands of different postures, sizes, colors and situations. The deep learning framework is able to encapsulate this great amount of variability and thereby produce beyond-human performance in object recognition tasks. This is being exploited in the work towards autonomous cars and could also be done in agriculture.

5. Conclusions

In an emerging standard for safety of highly automated machinery in agriculture, a barrel-shaped obstacle is defined as the obstacle which should be robustly detected to comply with the standard. In this paper, we show that by using deep learning techniques, we are able to achieve a high level of precision in two different cases in agricultural operations, with one of the cases concerning highly occluded obstacles. The algorithm detects the ISO specified obstacle in every test run, but it completely misses important obstacles such as people.

Therefore, we argue that the standardized obstacle presented in the standard is not fully adequate to ensure safe operations with highly autonomous machines in agriculture and further work should be conducted to describe an adequate procedure for testing the obstacle detection performance of highly autonomous machines in agriculture.

Acknowledgments: This research is sponsored by the Innovation Fund Denmark as part of the project "SAFE - Safer Autonomous Farming Equipment" (Project No. 16-2014-0) and "Multi-sensor system for ensuring ethical and efficient crop production" (Project No. 155-2013-6). We would also like to thank Ole Green from Agro Intelligence ApS for his contributions to the projects.

J. Imaging **2016**, *2*, 6

Author Contributions: Kim Arild Steen and Peter Christiansen contributed to the data aquisition, algorithm development, algorithm testing and writing of the manuscript. Henrik Karstoft has contributed with internal review of the manuscript, and Rasmus Nyholm Jørgensen has contributed with formalizing the hypothesis of this contribution, field experiment planning, algorithm testing and internal review of the manuscript.

Conflicts of Interest: The authors declare no conflict of interest.

References

1. Christiansen, P.; Hansen, M.; Steen, K.; Karstoft, H.; Jørgensen, R. Advanced sensor platform for human detection and protection in autonomous farming. In *Precision Agriculture'15*; Wageningen Academic Publishers: Wageningen, The Netherlands, 2015; pp. 1330–1334.
2. ISO/DIS 18497. Available online: http://www.iso.org/iso/home/store/catalogue_tc/catalogue_detail.htm?csnumber=62659 (accessed on 17 December 2015).
3. Krizhevsky, A.; Sutskever, I.; Hinton, G.E. Imagenet classification with deep convolutional neural networks. In *Advances in Neural Information Processing Systems*; MIT Press: Cambridge, MA, USA, 2012; pp. 1097–1105.
4. Szegedy, C.; Liu, W.; Jia, Y.; Sermanet, P.; Reed, S.; Anguelov, D.; Erhan, D.; Vanhoucke, V.; Rabinovich, A. Going deeper with convolutions. Available online: http://arxiv.org/pdf/1409.4842v1.pdf (accessed on 18 December 2015).
5. Simonyan, K.; Zisserman, A. Very deep convolutional networks for large-scale image recognition. Available online: http://arxiv.org/pdf/1409.1556v6.pdf (accessed on 18 December 2015).
6. Farfade, S.S.; Saberian, M.; Li, L.J. Multi-view Face Detection Using Deep Convolutional Neural Networks. Available online: http://arxiv.org/pdf/1502.02766v3.pdf (accessed on 18 December 2015).
7. LeCun, Y.; Bottou, L.; Bengio, Y.; Haffner, P. Gradient-based learning applied to document recognition. *Proc. IEEE* **1998**, *86*, 2278–2324.
8. AlexNet. Available online: http://caffe.berkeleyvision. org/model_zoo.html (accessed on 10 October 2015).
9. Russakovsky, O.; Deng, J.; Su, H.; Krause, J.; Satheesh, S.; Ma, S.; Huang, Z.; Karpathy, A.; Khosla, A.; Bernstein, M.; *et al.* Imagenet large scale visual recognition challenge. *Int. J. Comput. Vis.* **2014**, *115*, 1–42.
10. Koestinger, M.; Wohlhart, P.; Roth, P.M.; Bischof, H. Annotated facial landmarks in the wild: A large-scale, real-world database for facial landmark localization. In Proceedings of the First IEEE International Workshop on Benchmarking Facial Image Analysis Technologies, Barcelona, Spain, 6–13 Novenber 2011.
11. Long, J.; Shelhamer, E.; Darrell, T. Fully convolutional networks for semantic segmentation. Available online: http://arxiv.org/pdf/1411.4038v2.pdf (accessed on 18 December 2015).
12. Jia, Y.; Shelhamer, E.; Donahue, J.; Karayev, S.; Long, J.; Girshick, R.; Guadarrama, S.; Darrell, T. Caffe: Convolutional architecture for fast feature embedding. In Proceedings of the 22nd ACM international conference on Multimedia, Orlando, FL, USA, 3–7 November 2014; pp. 675–678.
13. Redmon, J.; Divvala, S.; Girshick, R.; Farhadi, A. You only look once: Unified, real-time object detection. Available online: http://arxiv.org/pdf/1506.02640v4.pdf (accessed on 18 December 2015).
14. Nvidia Jetson TX1. Available online: http://www.nvidia.com/object/jetson-tx1-dev-kit.html (accessed on 1 December 2015).

Journal of
Imaging

MDPI

Article

3D Reconstruction of Plant/Tree Canopy Using Monocular and Binocular Vision

Zhijiang Ni [†], Thomas F. Burks * and Won Suk Lee [†]

Department of Agricultural & Biological Engineering, University of Florida, Gainesville, FL 32611, USA;
zhijiangnicau@gmail.com (Z.N.); wslee@ufl.edu (W.S.L.)
* Correspondence: tburks@ufl.edu; Tel.: +1-352-392-1864
† These authors contributed equally to this work.

Academic Editors: Gonzalo Pajares Martinsanz and Francisco Rovira-Más
Received: 29 August 2016; Accepted: 19 September 2016; Published: 29 September 2016

Abstract: Three-dimensional (3D) reconstruction of a tree canopy is an important step in order to measure canopy geometry, such as height, width, volume, and leaf cover area. In this research, binocular stereo vision was used to recover the 3D information of the canopy. Multiple images were taken from different views around the target. The Structure-from-motion (SfM) method was employed to recover the camera calibration matrix for each image, and the corresponding 3D coordinates of the feature points were calculated and used to recover the camera calibration matrix. Through this method, a sparse projective reconstruction of the target was realized. Subsequently, a ball pivoting algorithm was used to do surface modeling to realize dense reconstruction. Finally, this dense reconstruction was transformed to metric reconstruction through ground truth points which were obtained from camera calibration of binocular stereo cameras. Four experiments were completed, one for a known geometric box, and the other three were: a croton plant with big leaves and salient features, a jalapeno pepper plant with median leaves, and a lemon tree with small leaves. A whole-view reconstruction of each target was realized. The comparison of the reconstructed box's size with the real box's size shows that the 3D reconstruction is in metric reconstruction.

Keywords: 3D images; multiple view reconstruction; metric reconstruction; plant reconstruction; machine vision; stereo vision

1. Introduction

Three-dimensional (3D) reconstruction of a plant/tree canopy can not only be used to measure the height, width, volume, area, and biomass of the target, but also can be used to visualize the object in virtual 3D space. 3D reconstruction is also called 3D digitizing or 3D modeling. Plant/tree 3D reconstruction could be cataloged into two types: (1) depth-based 3D modeling; and (2) image-based 3D modeling. Depth-based 3D modeling involves using sensors, such as, ultrasonic sensors, lasers, Time-of-Flight (ToF) cameras, and Microsoft red, green, and blue depth (RGB-D) cameras.

Using ultrasonic sensors, Sinoquet et al. [1] created a 3D model of corn plant profiles and canopy structure. The 3D results were used to calculate the leaf area and its distribution in the plant. Tumbo et al. [2] used ultrasonics in the field to measure citrus canopy volume. Twenty ultrasonic transducers were arranged on vertical boards (10 sensors per side). The ultrasonic sensors were installed behind a tractor, which was assumed to travel at an approximate speed of 0.5 km/h. A formula was provided to calculate the volume. To study the accuracy of this calculation, Zaman and Salyani [3] conducted research on the effect of ground speed and foliage density on canopy volume measurement. The experimental results showed that there was a 17.37% to 28.71% difference between the estimated and manually measured volumes.

Also using laser sensors, Tumbo et al. [2] described how to measure citrus canopy volume. Comparisons were made between the estimated volume and manually measured volume. The results showed high correlation. Wei and Salyani [4] employed a laser scanner and developed a laser scanning system, data acquisition system, and corresponding algorithm to calculate tree height, width, and canopy volume. To evaluate the accuracy of their system, a rectangular box was used as a target. Five repeated experiments were conducted to measure the box's height, length, and volume. However, no direct comparison between estimated volume and manually measured volume of citrus trees was made. Wei and Salyani [5] extended the same laser scanning system to calculate foliage density. They defined foliage density as the ratio of foliage volume to tree canopy volume, where foliage volume was defined as the space contained within the laser incident points and the tree row plane, while canopy volume was defined as the space enclosed between outer canopy boundary and the tree row plane. Lee and Ehsani [6] developed a laser scanner-based system to measure citrus geometric characteristics. After the experimental trees were trimmed to an ellipsoid shape, whose volumes were easy to manually measure, the surface area and volume were estimated by using a laser scanner. Rosell et al. [7] reported the use of a 2D light detection and ranging (LIDAR) scanner to obtain the 3D structures of plants. Sanz-Cortiella et al. [8] assumed that there was a linear relationship between the tree leaf area and the number of impacts of laser beam on the target. The point clouds generated by the laser scanner were used to calculate the total leaf area. Both indoor and outdoor experiments were conducted to validate this assumption. Zhu et al. [9] reconstructed the shape of a tree crown from scanned data-based on alpha shape modeling. A boundary mesh model was extracted from the boundary point cloud. This method resulted in a rough shape reconstruction of a big (20-meter high) tree.

Studying the application of a ToF camera, Cui et al. [10] described a 3D reconstruction method initiated by scanning the object using the ToF camera, and the reconstruction was realized through the combination of 3D super-resolution and a probabilistic multiple scan alignment algorithm. In 3D reconstruction, a ToF camera was usually used in combination with a red, green, and blue (RGB) camera. The ToF camera provided depth information, and the RGB camera would give color information. Shim et al. [11] presented a method to calibrate a multiple view acquisition system composed of ToF cameras and RGB color cameras. This system has the ability to calibrate multi-modal sensors in real time. Song et al. [12] combined a ToF image with images taken from stereo cameras to estimate a depth map for plant phenotyping. The experiments were conducted in a glasshouse using green pepper plants as targets. The canopy characteristics such as stem length, leaf area, and fruit size were estimated. This estimation was a challenging task since occlusion was occuring. The depth information from the ToF image was used to assist the determination of the disparity between left and right images. A global optimization method, using graph cuts developed by Boykov and Kolmogorov [13], was also used to find the disparity. The result using the graph cuts (GC) method was compared with the one resulting from combing graph cuts and ToF depth information. A quality evaluation was conducted, and GC + ToF gave the highest score. A smooth surface reconstruction of a pepper leaf was obtained using this method. Adhikari and Karkee [14] developed a 3D vision system to automatically prune apple trees. The vision system was composed of a ToF 3D camera and a RGB color camera. Experimental results showed that this system had about 90% accuracy in identifying pruning points.

The RGB-D camera is a Microsoft [15] product called Kinect that is designed for Xbox360. Kinect is composed of a RGB camera, a depth camera, and an infrared laser projector. Kinect was mostly used indoors for video game and view reconstruction. Izadi et al. [16] and Newcombe et al. [17] used a moving Kinect to reconstruct a dense indoor view. Kinect Fusion was employed to realize the reconstruction in real time because there was a special requirement on their hardware, specifically the GPU, to use it. Chene et al. [18] applied Kinect on 3D phenotyping of plants. An algorithm was developed to segment the depth image from the top view of the plant. The 3D view of the plant was then reconstructed from the segmented depth image. Azzari et al. [19] used Kinect to characterize vegetation structure. The measurements calculated from their depth image matched well with the results of a plant size measured manually. Different experiments were conducted in the lab, and in an outdoor field under different light conditions—such as early afternoon, late afternoon, and night. Experimental results showed that the Kinect had a limitation under direct sunlight. Wang and Zhang [20] used two Kinect devices to make a 3D reconstruction of a dormant cherry tree that was moved into a laboratory environment. During the experiment, some parts of the branches were missed due to occlusion and a long distance between camera and tree. The reconstructed results could be used for automatic pruning.

Image-based 3D modeling involved reconstructing the 3D properties from 2D images by using single camera or stereo cameras. Zhang et al. [21] used stereo vision to reconstruct a 3D corn model. The boundaries of the corn leaves were extracted and matched. The 3D leaves were modeled using a space intersection algorithm from 2D boundaries. This was a two-image reconstruction. Song [22] used stereo vision to model crops in horticulture. The cameras were installed on the top of the crops, and a top view of the crop was reconstructed. Han and Burks [23] did work on 3D reconstruction of a citrus canopy. Multiple images were used, and consecutive images were stitched together through image mosaic techniques. The canopy was reconstructed from the stitched image. The results did not realize real-size reconstruction.

The estimation of camera matrices is the first step in 3D reconstruction. The method of self-calibration described by Pollefeys et al. [24,25] is usually used. Fitzgibbon and Zisserman [26] described a method to automatically recover camera matrices and 3D scene points from a sequence of images. These images were sequentially acquired through an uncalibrated camera, and image triplets were used to estimate camera matrices and 3D points. Then the consecutive image triplets were formed into a sequence through one-view overlapping or two-view overlapping. Snavely et al. [27] developed a novel method to recover camera matrices and 3D points from unordered images. All these technologies were known as Structure from Motion (SfM). The sparse feature points were used to match the images. The most often used features were called Scale Invariant Feature Transform (SIFT) as described by Lowe [28].

Quan et al. [29] did research on plant modeling based on multiple images. SfM was used to estimate camera motion from multiple images. Here, instead of using sparse feature points, quasi-dense feature points as described by Lhuillier and Quan [30] were used to estimate camera matrices and 3D points of the plant. The leaves of the plant were modeled by segmenting the 2D images and computing the depths using the computed 3D points, and the branches were drawn through an interactive procedure. This modeling method was suitable for a plant with distinguishable leaves. To model a tree, which has small leaves, Tan et al. [31] did research on image-based 3D reconstruction. SfM was also employed to recover camera matrices and 3D quasi-dense points. To make a full 3D reconstruction of the tree, the visible branches were first reconstructed, followed by the occluded branches. The occluded branches were reconstructed through an unconstrained growth and constrained growth method. Subsequently, the leaves were added to the branches. Some of the leaves were from segmented images, while others were derived from the synthesizing methodology. Teng et al. [32] used machine vision to recover the sparse and unoccluded leaves in three dimensions. The method used was similar to the work of Quan et al. [29]. The results of the 3D reconstruction were used to classify the leaves and to identify the plant's type.

Furukawa and Ponce [33] provided a patch-based multiple view stereo (PMVS) algorithm to produce dense points to model the target. Small rectangular patches, called surfel, were used as feature points. The cameras' matrices were pre-calibrated using the method provided by Snavely et al. [27]. Features in each image were detected, then matched across multiple images. An expansion procedure, similar to the method provided by Lhuillier and Quan [30], was used to produce a denser set of patches.

Santos and Oliveira [34] applied the PMVS method to agricultural crops, such as basil and ixora. Plants with big and unoccluded leaves were well reconstructed. The reported processing time for 143 basil images was approximately 110 min, and almost 40 min for 77 ixora images. The image numbers will increase with the plant's size, consequently the processing time required will also increase with the increased number of images. Most of the processing was spent on feature detection and matching. The matching procedure was conducted through serial computation; however, if it could be conducted in parallel computation, the processing time would be significantly reduced.

Currently, a Graphics processing unit (GPU)-based SIFT, which is known as SiftGPU described by Wu [35], is available to do key points detection and matching via parallel computing. The bundler package described by Snavely [36] and the PMVS package developed by Furukawa and Ponce [33] were combined into a single package called VisualSFM by Wu [37], which involved using parallel computing technology. This would significantly decrease the running time.

The objectives of our study were to:

- Provide a new method to calibrate camera calibration matrix in metric level.
- Apply the fast software 'VisualSFM' on complicate objects, e.g., plant/tree, to generate a full-view 3D reconstruction.
- Generate the metric 3D reconstruction from projective reconstruction and achieve real-size 3D reconstruction for complicate agricultural plant scenes.

2. Materials and Methods

2.1. Hardware

In this paper, two Microsoft LifeCam Studio web high definition (HD) cameras (1080 p) were assembled inside a wooden box, and mounted approximately in parallel, with the baseline at 30 mm, as shown in Figure 1. To acquire images, they were connected to a Lenovo IdeaPad Y500 laptop with a NVIDIA GeForce GT650M GPU, which can be used in parallel computation to accelerate the computing time in feature points detection and matching.

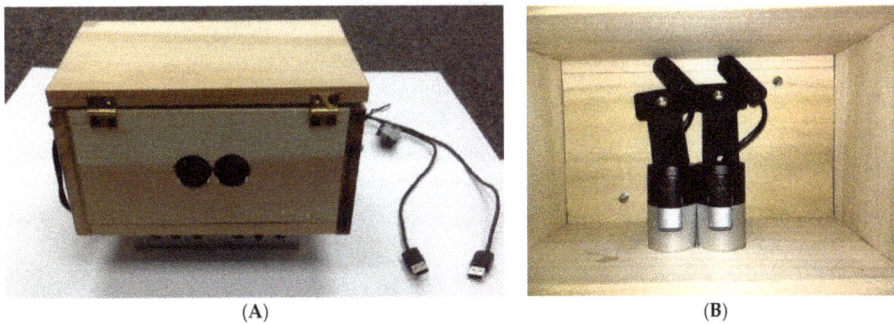

(A) (B)

Figure 1. Stereo cameras which are used to acquire images: (**A**) whole view; (**B**) inside view.

2.2. Stereo Camera Calibration

A 3D point ($\overrightarrow{X}$) and its projection ($\overrightarrow{x}$) in 2D image is related through camera calibration matrix P. The relationship is expressed as $s\overrightarrow{x} = P\overrightarrow{X}$, where $\overrightarrow{x} = (x, y, 1)^T$ is in homogenous form in 2D, $\overrightarrow{X} = (X, Y, Z, 1)^T$ is in homogenous form in 3D, P is a 3 × 4 matrix, and s is a scale. The objective of camera calibration is to determine camera calibration matrix P, which includes both intrinsic parameters and extrinsic parameters. Zhang [38] provided a flexible technique for camera calibration using only five images taken from different angles. A checkerboard was used as calibration pattern. For each image, the plane of the checkerboard was assumed as z-plane, so the Z coordinates for all the 3D points were zero. X and Y coordinates could be obtained from the actual checkerboard size. All these provided ground truth. A Matlab toolbox, developed by Bouguet [39], was used to solve camera calibration matrix using Zhang's algorithm. This toolbox is not only suitable for a single camera, but is also suitable for stereo cameras.

The external camera parameters provided by Zhang's [38] method were built on each checkerboard's own coordinate system, not on the same world coordinate system. In order to build the same world coordinate system, a large 2D x-z coordinate system was plotted on an A0-size paper, together with a vertical checkerboard (Figure 2), and all of these provided the 3D ground truth on the same coordinate system. The detailed 2D *x-z* coordinate system is shown in Figure 3. Each line in the *x-z* plane was at 50 mm spacing. The middle line (oz) was rotated −10° around O_{orig} to get the left line, and rotated +10° around O_{orig} to get the right line. The checkerboard was then placed at different locations on the left, middle, and right line (marked as 1 through 45 in Figure 3). The 3D coordinates of each corner on the checkerboard, at each location, could be solved as the ground truth. Two images were taken at each location from the left and right cameras. From these 2D images, the 2D projection of these corners could also be solved.

Based on these 2D and 3D coordinates, the gold standard algorithm of Hartley and Zisserman [40] was used to calculate the camera matrices for both left and right cameras.

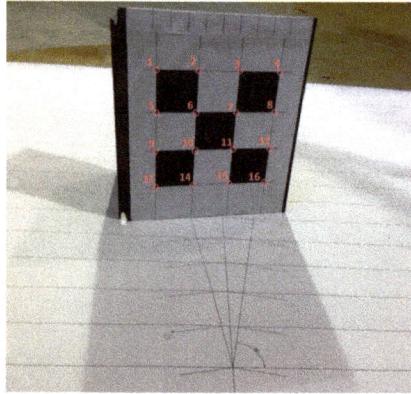

Figure 2. World coordinate system (2D *x-z* system plus vertical checkerboard).

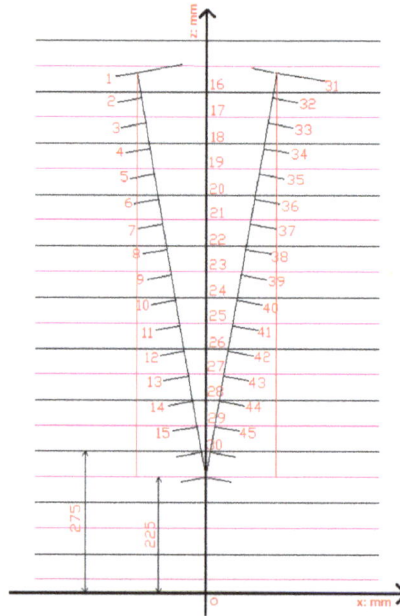

Figure 3. 2D *x-z* coordinate system (each line is 50 mm separated).

Based on camera calibration matrix and 2D image coordinates, we can get estimated 3D points. When compared to the actual 3D points, we can estimate the error in *X*, *Y*, and *Z* directions (Figure 4). These experimental results showed that this stereo camera set had good accuracy when the distance between cameras and the target was less than 800 mm. The statistical analysis for errors in the *X*, *Y*, and *Z* directions are shown in Table 1. The mean error in *x* direction is 0.42 mm, the mean error in *y* direction is 0.36 mm, and the mean error in *z* direction is 2.78 mm.

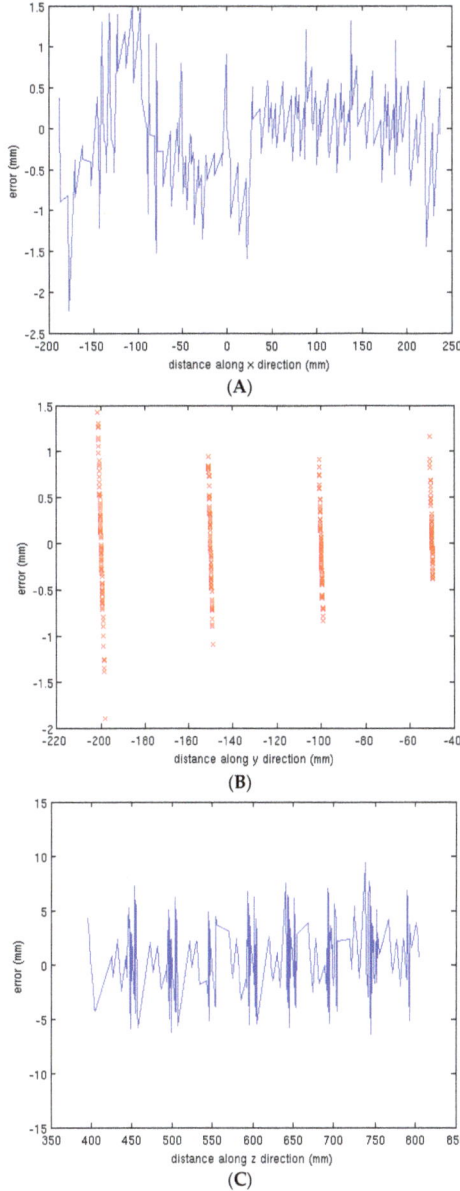

Figure 4. Error plots in *X*, *Y*, and *Z* direction. (**A**) errors in *x* direction; (**B**) errors in *y* direction; (**C**) errors in *z* direction.

Table 1. Statistical analysis of errors between estimated and actual corners.

Axis	Mean Absolute Error (mm)	Standard Deviation (mm)
X	0.42	0.35
Y	0.36	0.31
Z	2.78	1.74

2.3. Image Acquisition

To make a full view reconstruction of the plant or tree, multiple images from different view angles had to be taken over the target. The stereo camera (shown in Figure 1) and a laptop with image acquisition software were used to acquire the images. One setup of the experiment is shown in Figure 5, where the target plant was in the center, and the stereo cameras positions are shown around it. The images taken from the adjacent locations should have an overlapping region.

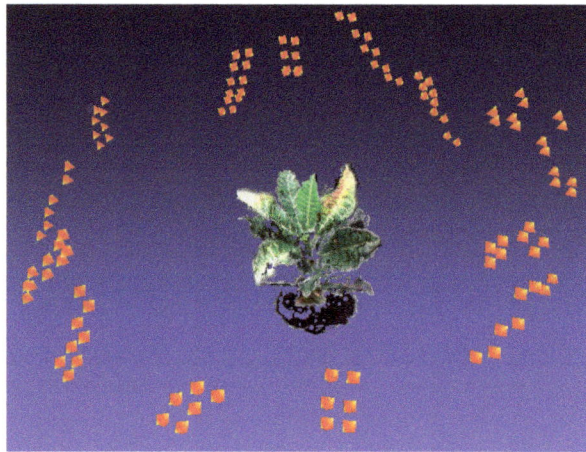

Figure 5. An example of stereo camera setup for image acquisition with the 3D reconstruction results (56 stereo pairs were used).

2.4. Feature Points Detection and Matching

At the beginning, feature points were detected as Harris corners [41]. A pixel was selected as a salient pixel if its response was an eight-way local maximum. Normalized cross correlation (NCC) and normalized sum of squared differences (NSSD) described by Richard [42] could be used to match the features. Harris corner features were not invariant to affine and scale transform. Mikolajczyk and Schmid [43] provided different scale and affine invariant feature point detectors, such as Harris-Laplace and Harris-Affine. Mikolajczyk and Schmid [44] did a performance evaluation for four different local feature detectors (Harris-Laplace, Hessian-Laplace, Harris-Affine, and Hessian-Affine) and 10 different feature descriptors. Lowe [28] provided a Scale Invariant Feature Transform (SIFT) descriptor to describe the detected keypoints. Using Lowe's SIFT research, Yan and Sukthankar [45] derived a PCA-based SIFT (PCA-SIFT), and Morel and Yu [46] provided an affine SIFT (ASIFT). To enhance the computing speed of SIFT, a speeded up robust features (SURF) was provided by Bay et al. [47]. To further improve the computation speed of SIFT, a parallel algorithm called SiftGPU was provided by Wu [35].

Snavely et al. [27] applied SIFT on multiple-view reconstruction from unordered images. Snavely [36] provided the software called Bundler to realize this method. In Snavely's research, the SIFT feature points for each image were detected. Each pair of two images were then matched using ANN algorithm from Arya et al. [48]. This process was conducted in serial computation. The computation time required increased significantly as the number of input images and the number of feature points per image increased. Santos and Oliveira [34] applied Bundler on their plant phenotyping, and they reported that almost one hour would be needed to match the features for each two images of the total 143 images, and almost 30 min for 77 images.

Wu [37] provided a fast method called visual structure from motion (SfM) method to accelerate the feature points' detection, matching, bundle adjustment, and 3D reconstruction. Wu's method was applied in this paper.

2.5. Sparse Bundle Adjustment

Given a set of images, the matched feature points, also known as 2D projections, could be found through the feature matching algorithm introduced in the previous section. Each matched feature point had a corresponding 3D point in the scene. The camera matrices and 3D points could be estimated through bundle adjustment method [40]. The j-th 3D point $\hat{X}_j$ will be projected on the i-th image as $\hat{x}_j^i$ through the i-th camera calibration matrix $\hat{P}^i$, where $\hat{x}_j^i = \hat{P}^i \hat{X}_j$ [40]. By minimizing the errors between re-projected projection $\hat{x}_j^i$ and the actual projection x_j^i, the camera calibration matrix $\hat{P}^i$, and sparse 3D points $\hat{X}_j$ could be estimated. A software package called sparse bundle adjustment (SBA) was provided by Lourakis and Argyros [49] to realize this minimization.

2.6. Dense 3D Reconstruction Using CMVS and PMVS

The patch model developed by Furukawa and Ponce [33,50] to produce 3D dense reconstruction from multiple view stereo (MVS) was used in this research. Patch was reconstructed through three steps: feature matching, patch expansion, and patch filtering. Feature matching was used to generate an initial bundle of patches. Then the patches were made denser. The outliers were removed by filtering. Finally, the patches were used to build a polygonal mesh. Furukawa and Ponce [51] developed software called PMVS to implement this method. PMVS used the output (camera matrices) from Bundler as the input. Other inputs for PMVS were from another software called CMVS [52].

2.7. Stereo Reconstruction Using VisualSFM

VisualSFM, which was proposed by Wu [37], integrated three technologies together: feature points detection and matching [35], multicore bundle adjustment [53], and dense 3D reconstruction [33]. Multiple images from a full view of the plant/tree would be imported into this software. A fully reconstructed result would be generated through the previously mentioned three steps.

2.8. Metric Reconstruction

The result from bundle adjustment was not metric reconstruction, which means that the reconstructed result did not show the actual size of the target.

A direct reconstruction method using ground truth was provided by Hartley and Zisserman [40] to realize the metric reconstruction. Using pre-calibrated stereo cameras, the Euclidean ground truth of a set of 3D points X_{euc}^i could be solved from the 2D correspondence $x_1^i \leftrightarrow x_2^i$, and the estimated 3D points X_{est}^i could be obtained from bundle adjustment. The estimated 3D points and the Euclidean 3D points were related through a homography transformation (H). Then we have $X_{euc}^i = H \cdot X_{est}^i$. The first two images from the stereo camera were used to solve the Euclidean ground truth from the 2D projection. From our stereo camera calibration we knew that this stereo camera pair had good accuracy only when the distance between camera and the target was less than 580 mm. Therefore, those 3D points whose Z coordinates were bigger than 580 mm would be filtered as outliers.

To minimize the homography fitting error, these two sets of 3D points had to be normalized. After normalization, using the method described by Hartley and Zisserman [40], the centroid of the new points was at the origin, and the average distance from the origin is $\sqrt{3}$. After applying normalization,

$X^i_{newpts1} = T_1 \cdot X^i_{euc}$ and $X^i_{newpts2} = T_2 \cdot X^i_{est}$ the homography between $\{X^i_{newpts1}\}$ and $\{X^i_{newpts2}\}$ was estimated using rigid transformation Forsyth and Ponce [54]. By fitting rigid transformation, we get $X^i_{newpts1} = H_{est} \cdot X^i_{newpts2}$. To de-normalize it, we have $X^i_{euc} = H \cdot X^i_{est}$, where $H = T_1^{-1} \cdot H_{est} \cdot T_2$. Applying H on all the 3D points from bundle adjustment, we can transfer them back to metric scale. The new camera calibration matrix was $P^i_{euc} = P^i_{est} \cdot H^{-1}$.

3. Experimental Results and Discussion

Four test experiments were conducted, one was a box with known geometry, and the other three were a croton plant with salient features, a jalapeno pepper plant with medium-size leaves, and a lemon tree with small leaves.

Test 1: A hexagon box with a given geometry was used to verify the reconstruction result. The box was placed on the top of a table. The stereo camera was manually moved around the box to take the images. Images taken at the adjacent locations should have some overlap, which is good for feature matching. The side length of the hexagon is 64 mm, and the height is 70 mm. To give the box texture, paper with printed citrus leaf images was wrapped around the box, as shown in Figure 6. Approximately 86 images were taken from various positions around this box using the stereo camera. The box was first reconstructed by using VisualSFM [37]. The result is shown in Figure 7A. The box was then was reconstructed by applying the metric reconstruction method (mentioned in step 2.8). The result is shown in Figure 7B, which shows the real size of the target.

Figure 6. Hexagon box with texture.

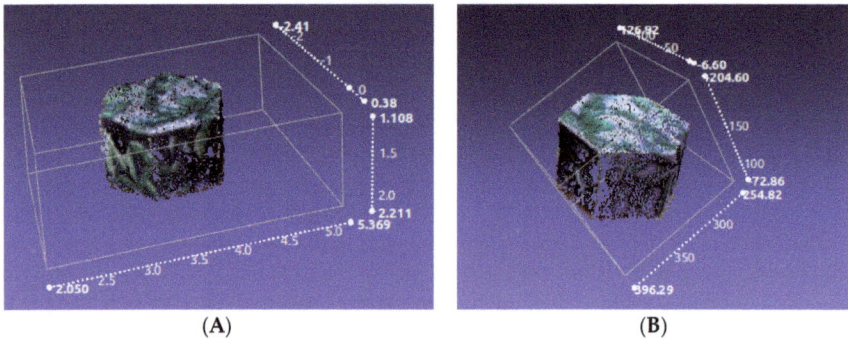

Figure 7. 3D reconstruction. (**A**) projective reconstruction; (**B**) metric reconstruction.

The reconstructed length for each side of the above hexagon and the reconstructed height of each side face are shown in Tables 2 and 3.

Table 2. Estimated length vs. actual length.

Length	L1	L2	L3	L4	L5	L6
Estimated length (mm)	64.19	63.47	68.82	65.59	63.00	61.99
Actual length (mm)	64.00	64.00	64.00	64.00	64.00	64.00
error (mm)	0.19	−0.53	4.82	1.59	−1.00	−2.01

Table 3. Estimated height vs. actual height.

Height	H1	H2	H3	H4	H5	H6
Estimated height (mm)	70.45	68.53	71.10	68.13	70.68	69.03
Actual height (mm)	70.00	70.00	70.00	70.00	70.00	70.00
error (mm)	0.45	−1.47	1.10	−1.83	0.68	−0.97

From this verifying test, we can see that the hexagon box is well reconstructed. The estimated length and height of the box is very close to the actual size. This method was then applied to complicated objects, such as a plant and a small tree.

Test 2: Three kinds of plants with different leaf sizes were reconstructed using the method introduced. The croton plant has big and sparse leaves with salient features. The jalapeno pepper has medium and sparse leaves. The lemon tree has small and dense leaves. They are shown in Figure 8.

| (A) | (B) | (C) |

Figure 8. Experimental plants/tree: (**A**) Croton plant; (**B**) Jalapeno pepper plant; (**C**) Lemon tree.

Firstly, the objects were reconstructed in projective views by using VisualSFM from Wu [37]. Then the metric reconstruction algorithm (mentioned in step 2.8) was applied to get the 3D reconstruction in Euclidean space. For croton plants, the first pair of images were used as the ground truth. The feature points for these two images were extracted and matched. Together with the camera matrices of the stereo cameras, the actual 3D points could be calculated by using triangulation method of Hartley and Zisserman [40]. The estimated 3D points for the same 2D correspondences could be found from the reconstructed results of VisulaSFM. By applying rigid transform, the transformation between actual 3D points and estimated 3D points could be achieved. Applying this transformation to all the estimated 3D points for all the images, the final metric 3D reconstruction could be obtained, which is shown in Figure 9A. A similar process was applied to the other two plants. For the pepper plant, the first pair of images was used, and for the lemon tree, the seventh pair of images was used. The reconstructed view of the target was displayed in a bounding box, which was shown in Figure 9B,C respectively.

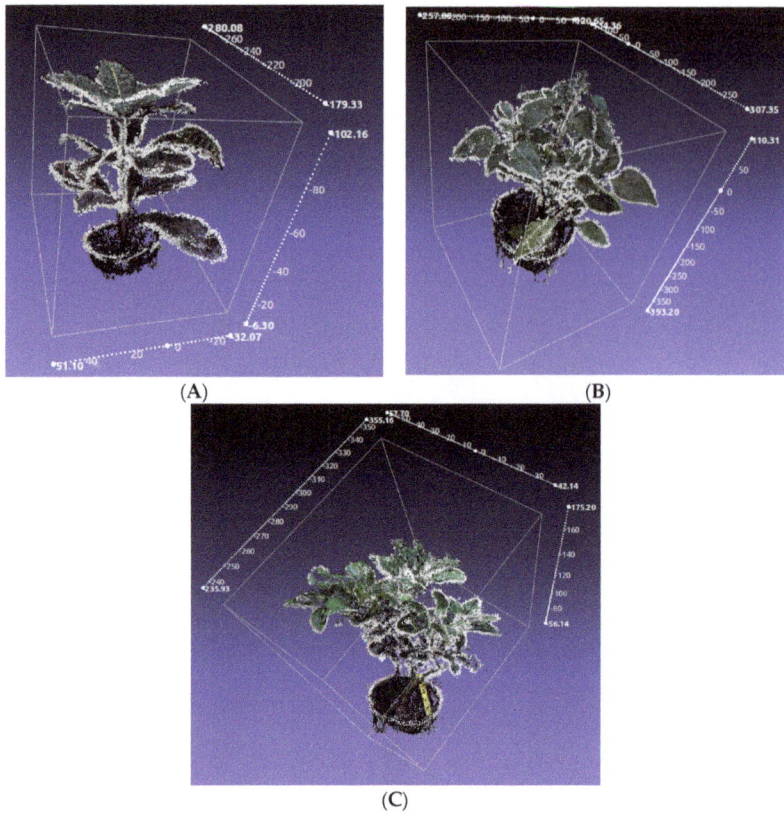

Figure 9. Reconstructed plants. (**A**) croton; (**B**) pepper; (**C**) lemon.

To roughly calculate the volume of the reconstructed plant canopy, the bounding box was divided into voxels. If the 3D point is inside the voxel, then that voxel will be marked as used. Unused voxels will be removed, as shown in Figure 10. All the 3D reconstructed points reside inside some voxels. The summation of the volume of all these voxels will be the canopy volume. There is a tradeoff between the size of the voxel and the volume of canopy. This tradeoff was not analyzed in this research since it is not the primary task. The estimated volume for these three plants are shown in Table 4.

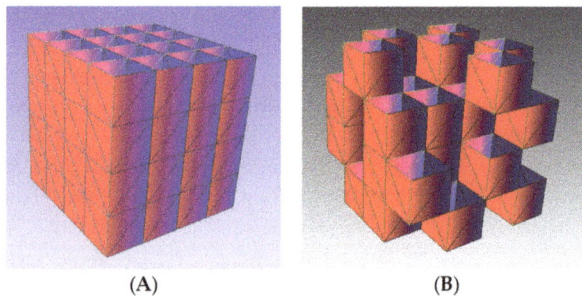

Figure 10. Demo of volume calculation. (**A**) Bounding box was divided into voxels; (**B**) Voxels left after removing unused voxels.

Table 4. Volumes of these three plants.

Experimental Targets	# of Voxel Hits/# of Total 3D Points	Voxel Size (mm^3)	Volume (cm^3)
Croton	16,156/19,579	28.46	1.23×10^3
Jalapeno pepper	28,591/38,773	12.61	3.61×10^2
Lemon tree	48,609/96,680	3.76	1.83×10^2

4. Conclusions

This paper demonstrated a new approach to calibrate the camera calibration matrix on a metric level and then implemented the VisualSFM method to make a projective reconstruction of a plant/tree canopy. Stereo cameras were employed to estimate the actual 3D points for image pairs. The projective reconstructed view was then transformed to metric reconstruction by applying rigid transformation. A verifying experiment was performed by reconstructing a hexagon box. The result showed that this method can reach the true size of the target object. The same method was applied on three kinds of plant/tree with different leaf sizes. The reconstructed results presented a good visual view in 3D with reconstructed leaf features retaining their defining characteristics. This approach provides a metric reconstruction method that can achieve real-size reconstruction, which is a significant accomplishment in practical applications such as, 3D visualization, plant phenotypeing, robotic harvesting, and precision spraying, where real-size characteristics of plants are important for successful production practices.

Acknowledgments: The authors would like to acknowledge the contribution to this research provided by sponsors GeoSpider, Inc. and their funding source through the USDA SBIR program. This research is supported in part by a grant from the United States Department of Agriculture Small Business Innovation Research (USDA-SBIR) award contract #2012-33610-19499 and the USDA NIFA AFRI National Robotics Initiative #2013-67021-21074.

Author Contributions: Zhijiang Ni was the principle researcher and author of the paper. Thomas F. Burks was the graduate advisor/mentor and conceiver of the project who oversaw research and writing and provided editorial support. Won Suk Lee was an advisor to the project and also provided editorial support.

Conflicts of Interest: The authors declare no conflict of interest.

References

1. Sinoquet, H.; Moulia, B.; Bonhomme, R. Estimating the three-dimensional geometry of a maize crop as an input of radiation models: Comparison between three-dimensional digitizing and plant profiles. *Agric. For. Meteorol.* **1991**, *55*, 233–249. [CrossRef]
2. Tumbo, S.D.; Salyani, M.; Whitney, J.D.; Wheaton, T.A.; Miller, W.M. Investigation of laser and ultrasonic ranging sensors for measurements of citrus canopy volume. *Appl. Eng. Agric.* **2002**, *18*, 367–372. [CrossRef]
3. Zaman, Q.U.; Salyani, M. Effects of foliage density and ground speed on ultrasonic measurement of citrus tree volume. *Appl. Eng. Agric.* **2004**, *20*, 173–178. [CrossRef]
4. Wei, J.; Salyani, M. Development of a laser scanner for measuring tree canopy characteristics: Phase 1. Prototype development. *Trans. ASAE* **2004**, *47*, 2101–2107. [CrossRef]
5. Wei, J.; Salyani, M. Development of a laser scanner for measuring tree canopy characteristics: Phase 2. Foliage density measurement. *Trans. ASAE* **2005**, *48*, 1595–1601. [CrossRef]
6. Lee, K.H.; Ehsani, R. A laser scanner based measurement system for quantification of citrus tree geometric characteristics. *Appl. Eng. Agric.* **2009**, *25*, 777–788. [CrossRef]
7. Rosell, J.R.; Llorens, J.; Sanz, R.; Arnó, J.; Ribes-Dasi, M.; Masip, J.; Escolà, A.; Camp, F.; Solanelles, F.; Gràcia, F.; et al. Obtaining the three-dimensional structure of tree orchards from remote 2D terrestrial LIDAR scanning. *Agric. For. Meteorol.* **2009**, *149*, 1505–1515. [CrossRef]
8. Sanz-Cortiella, R.; Llorens-Calveras, J.; Escola, A.; Arno-Satorra, J.; Ribes-Dasi, M.; Masip-Vilalta, J.; Camp, F.; Gracia-Aguila, F.; Solanelles-Batlle, F.; Planas-DeMarti, S.; et al. Innovative LIDAR 3D dynamic measurement system to estimate fruit-tree leaf area. *Sensors* **2011**, *11*, 5769–5791. [CrossRef] [PubMed]
9. Zhu, C.; Zhang, X.; Hu, B.; Jaeger, M. Reconstruction of tree crown shape from scanned data. In *Technologies for E-Learning and Digital Entertainment*; Pan, Z., Zhang, X., Rhalibi, A., Woo, W., Li, Y., Eds.; Springer: Berlin, Germany, 2008; pp. 745–756.

10. Cui, Y.; Schuon, S.; Chan, D.; Thrun, S.; Theobalt, C. 3D shape scanning with a time-of-flight camera. In Proceedings of the IEEE Conference on Computer Vision and Pattern Recognition, San Francisco, CA, USA, 13–18 June 2010; pp. 1173–1180.
11. Shim, H.; Adelsberger, R.; Kim, J.; Rhee, S.M.; Rhee, T.; Sim, J.Y.; Gross, M.; Kim, C. Time-of-flight sensor and color camera calibration for multi-view acquisition. *Vis. Comput.* **2012**, *28*, 1139–1151. [CrossRef]
12. Song, Y.; Glasbey, C.A.; Heijden, G.A.M.; Polder, G.; Dieleman, J.A. Combining stereo and time-of-flight images with application to automatic plant phenotyping. In *Image Analysis*; Heyden, A., Kahl, F., Eds.; Springer: Berlin, Germany, 2011; pp. 467–478.
13. Boykov, Y.; Kolmogorov, V. An experimental comparison of min-cut/max- flow algorithms for energy minimization in vision. *IEEE Trans. Pattern Anal. Mach. Intell.* **2004**, *26*, 1124–1137. [CrossRef] [PubMed]
14. Adhikari, B.; Karkee, M. 3D reconstruction of apple trees for mechanical pruning. In Proceedings of the ASABE Annual International Meeting, Louisville, KY, USA, 7–10 August 2011.
15. Microsoft, Kinect for Xbox 360. Available online: http://www.xbox.com/en-US/KINECT (accessed on 10 March 2012).
16. Izadi, S.; Kim, D.; Hilliges, O.; Molyneaux, D.; Newcombe, R.; Kohli, P.; Shotton, J.; Hodges, S.; Freeman, D.; Davison, A. KinectFusion: Real-time 3D reconstruction and interaction using a moving depth camera. In Proceedings of the 24th Annual ACM Symposium on User Interface Software and Technology, Santa Barbara, CA, USA, 16–19 October 2011; pp. 559–568.
17. Newcombe, R.A.; Davison, A.J.; Izadi, S.; Kohli, P.; Hilliges, O.; Shotton, J.; Molyneaux, D.; Hodges, S.; Kim, D.; Fitzgibbon, A. KinectFusion: Real-time dense surface mapping and tracking. In Proceedings of the 10th IEEE International Symposium on Mixed and Augmented reality (ISMAR), Basel, Switzerland, 26–29 October 2011; pp. 127–136.
18. Chene, Y.; Rousseau, D.; Lucidarme, P.; Bertheloot, J.; Caffier, V.; Morel, P.; Belin, E.; Chapeau-Blondeau, F. On the use of depth camera for 3D phenotyping of entire plants. *Comput. Electron. Agric.* **2012**, *82*, 122–127. [CrossRef]
19. Azzari, G.; Goulden, M.; Rusu, R. Rapid characterization of vegetation structure with a Microsoft Kinect sensor. *Sensors* **2013**, *13*, 2384–2398. [CrossRef] [PubMed]
20. Wang, Q.; Zhang, Q. Three-dimensional reconstruction of a dormant tree using RGB-D cameras. In Proceedings of the Annual International Meeting, Kansas City, MI, USA, 21–24 July 2013; p. 1.
21. Zhang, W.; Wang, H.; Zhou, G.; Yan, G. Corn 3D reconstruction with photogrammetry. *Int. Arch. Photogramm. Remote Sens. Spat. Inf. Sci.* **2008**, *37*, 967–970.
22. Song, Y. Modelling and Analysis of Plant Image Data for Crop Growth Monitoring in Horticulture. Ph.D. Thesis, University of Warwick, Coventry, UK, 2008.
23. Han, S.; Burks, T.F. 3D reconstruction of a citrus canopy. In Proceedings of the 2009 ASABE Annual International Meeting, Reno, NV, USA, 21–24 June 2009.
24. Pollefeys, M.; Koch, R.; van Gool, L. Self-calibration and metric reconstruction in spite of varying and unknown internal camera parameters. In Proceedings of the 6th International Conference on Computer Vision, Bombay, India, 4–7 January 1998; pp. 90–95.
25. Pollefeys, M.; Koch, R.; Vergauwen, M.; van Gool, L. Automated reconstruction of 3D scenes from sequences of images. *ISPRS J. Photogramm. Remote Sens.* **2000**, *55*, 251–267. [CrossRef]
26. Fitzgibbon, A.W.; Zisserman, A. Automatic camera recovery for closed or open image sequences. In Proceedings of the 5th European Conference on Computer Vision-Volume I, Freiburg, Germany, 2–6 June 1998; pp. 311–326.
27. Snavely, N.; Seitz, S.M.; Szeliski, R. Photo tourism: Exploring photo collections in 3D. *ACM Trans. Graph.* **2006**, *25*, 835–846. [CrossRef]
28. Lowe, D.G. Distinctive image features from scale-invariant keypoints. *Int. J. Comput. Vis.* **2004**, *60*, 91–110. [CrossRef]
29. Quan, L.; Tan, P.; Zeng, G.; Yuan, L.; Wang, J.; Kang, S.B. Image-based plant modeling. *ACM Trans. Graph.* **2006**, *25*, 599–604. [CrossRef]
30. Lhuillier, M.; Quan, L. A quasi-dense approach to surface reconstruction from uncalibrated images. *IEEE Trans. Pattern Anal. Mach. Intell.* **2005**, *27*, 418–433. [CrossRef] [PubMed]
31. Tan, P.; Zeng, G.; Wang, J.; Kang, S.B.; Quan, L. Image-based tree modeling. *ACM Trans. Graph.* **2007**, *26*, 87. [CrossRef]

32. Teng, C.H.; Kuo, Y.T.; Chen, Y.S. Leaf segmentation, classification, and three-dimensional recovery from a few images with close viewpoints. *Opt. Eng.* **2011**, *50*, 037003.

33. Furukawa, Y.; Ponce, J. Accurate, dense, and robust multiview stereopsis. *IEEE Trans. Pattern Anal. Mach. Intell.* **2010**, *32*, 1362–1376. [CrossRef] [PubMed]

34. Santos, T.T.; Oliveira, A.A. Image-based 3D digitizing for plant architecture analysis and phenotyping. In Proceedings of the Workshop on Industry Applications (WGARI) in SIBGRAPI 2012 (XXV Conference on Graphics, Patterns and Images), Ouro Preto, Brazil, 22–25 August 2012.

35. Wu, C. SiftGPU: A GPU Implementation of Scale Invariant Feature Transform (SIFT). 2007. Available online: http://www.cs.unc.edu/~ccwu/siftgpu/ (accessed on 20 April 2013).

36. Snavely, N. Bundler: Structure from Motion (SfM) for Unordered Image Collections. 2010. Available online: http://phototour.cs.washington.edu/bundler/ (accessed on 8 August 2011).

37. Wu, C. VisualSFM: A Visual Structure from Motion System. 2011. Available online: http://homes.cs.washington.edu/~ccwu/vsfm/ (accessed on 20 April 2013).

38. Zhang, Z. A flexible new technique for camera calibration. *IEEE Trans. Pattern Anal. Mach. Intell.* **2000**, *22*, 1330–1334. [CrossRef]

39. Bouguet, J.Y. Camera Calibration ToolBox for Matlab. 2008. Available online: http://www.vision.caltech.edu/bouguetj/calib_doc/ (accessed on 27 October 2011).

40. Hartley, R.; Zisserman, A. *Multiple View Geometry in Computer Vision*; Cambridge University Press: Cambridge, UK, 2003.

41. Harris, C.; Stephens, M. A combined corner and edge detector. In Proceedings of the 4th Alvey Vision Conference, Manchester, UK, 31 August–2 September 1988; pp. 147–151.

42. Richard, S. *Computer Vision: Algorithms and Applications*; Springer: Berlin, Germany, 2011.

43. Mikolajczyk, K.; Schmid, C. Scale & affine invariant interest point detectors. *Int. J. Comput. Vis.* **2004**, *60*, 63–86.

44. Mikolajczyk, K.; Schmid, C. A performance evaluation of local descriptors. *IEEE Trans. Pattern Anal. Mach. Intell.* **2005**, *27*, 1615–1630. [CrossRef] [PubMed]

45. Yan, K.; Sukthankar, R. PCA-SIFT: A more distinctive representation for local image descriptors. In Proceedings of the IEEE Computer Society Conference on Computer Vision and Pattern Recognition, Washington, DC, USA, 27 June–2 July 2004; Volume 502, pp. II-506–II-513.

46. Morel, J.M.; Yu, G. ASIFT: A new framework for fully affine invariant image comparison. *SIAM J. Imaging Sci.* **2009**, *2*, 438–469. [CrossRef]

47. Bay, H.; Tuytelaars, T.; van Gool, L. Surf: Speeded up robust features. In Proceedings of the 9th European Conference on Computer Vision, Graz, Austria, 7–13 May 2006.

48. Arya, S.; Mount, D.M.; Netanyahu, N.S.; Silverman, R.; Wu, A.Y. An optimal algorithm for approximate nearest neighbor searching fixed dimensions. *J. ACM* **1998**, *45*, 891–923. [CrossRef]

49. Lourakis, M.I.A.; Argyros, A.A. SBA: A software package for generic sparse bundle adjustment. *ACM Trans. Math. Softw.* **2009**, *36*, 1–30. [CrossRef]

50. Furukawa, Y.; Ponce, J. Accurate, dense, and robust multi-view stereopsis. In Proceedings of the IEEE Conference on Computer Vision and Pattern Recognition, Minneapolis, MN, USA, 17–22 June 2007; pp. 1–8.

51. Furukawa, Y.; Ponce, J. Patch-Based Multi-View Stereo Software (PMVS—Version 2). 2010. Available online: http://www.di.ens.fr/pmvs/ (accessed on 9 September 2012).

52. Furukawa, Y. Clustering Views for Multi-View Stereo (CMVS). 2010. Available online: http://www.di.ens.fr/cmvs/ (accessed on 9 September 2012).

53. Wu, C.C.; Agarwal, S.; Curless, B.; Seitz, S.M. Multicore bundle adjustment. In Proceedings of the IEEE Conference on Computer Vision and Pattern Recognition, Colorado Springs, CO, USA, 20–25 June 2011; pp. 3057–3064.

54. Forsyth, D.A.; Ponce, J. *Computer Vision: A Modern Approach*; Prentice Hall: Upper Saddle River, NJ, USA, 2003.

Journal of
Imaging

MDPI

Article

Peach Flower Monitoring Using Aerial Multispectral Imaging

Ryan Horton [1], Esteban Cano [1], Duke Bulanon [1,*] and Esmaeil Fallahi [2]

[1] Department of Physics and Engineering, Northwest Nazarene University, Nampa, ID 83686, USA; ryanhorton@nnu.edu (R.H.); ecano@nnu.edu (E.C.)
[2] Parma Research and Extension Center, University of Idaho, Parma, ID 83660, USA; efallahi@uidaho.edu
* Correspondence: dbulanon@nnu.edu; Tel.: +1-208-467-8047

Academic Editors: Gonzalo Pajares Martinsanz and Francisco Rovira-Más
Received: 25 October 2016; Accepted: 29 December 2016; Published: 6 January 2017

Abstract: One of the tools for optimal crop production is regular monitoring and assessment of crops. During the growing season of fruit trees, the bloom period has increased photosynthetic rates that correlate with the fruiting process. This paper presents the development of an image processing algorithm to detect peach blossoms on trees. Aerial images of peach (*Prunus persica*) trees were acquired from both experimental and commercial peach orchards in the southwestern part of Idaho using an off-the-shelf unmanned aerial system (UAS), equipped with a multispectral camera (near-infrared, green, blue). The image processing algorithm included contrast stretching of the three bands to enhance the image and thresholding segmentation method to detect the peach blossoms. Initial results showed that the image processing algorithm could detect peach blossoms with an average detection rate of 84.3% and demonstrated good potential as a monitoring tool for orchard management.

Keywords: blossoms; digital image processing; machine vision; peaches; unmanned aerial system

1. Introduction

Idaho is popularly known for potatoes, but the state grows other specialty crops which include peaches. Numerous types of peaches are grown in the southwestern part of Idaho, which is warmer as compared to other regions. The state produces about 5300 tons of peaches [1]. In addition to peaches, Idaho agriculture produces apples, pears, cherries, apricots, nectarines, plums and grapes. The specialty crop industry in Idaho is thriving. However, the industry is currently facing the challenges of labor shortage, increasing labor cost, and the pressure of a growing market. Because of these challenges, fruit growers need to adopt new technologies that can aid in optimizing crop production.

One of these new technologies, known as precision agriculture, is an agricultural management concept based on measuring crop variability in the field and responding to field issues [2]. Crop variability has both temporal and spatial components that need to be considered. The spatial component is facilitated by the use of the global positioning system (GPS), which enables the farmer to locate the precise location in the field. In combination with advanced sensors that could measure field conditions such as moisture levels, nitrogen levels, and organic matter content, it allows the creation of maps that show the spatial variability of the field.

Although precision agriculture has been used mostly for row crops such as corn and wheat, studies have shown that the technology has been adopted for specialty crops which include fruit trees [3]. One of the precision agriculture technologies that has been reported is remote sensing. Remote sensing can be implemented using a satellite or aerial system [4]. The downsides of using satellites are the cost for real-time, high-resolution images and the frequency of data collection, which could affect the temporal aspect of crop production [5]. Another remote sensing method is using aerial systems, which can be classified as manned or unmanned. Similar to satellites, a manned aerial system is costly, and it may not be economically feasible for smaller fields. However, with the proliferation of cheap commercial unmanned aerial systems (UAS) such as the 3DR Iris and DJI Phantom series (Figure 1), remote sensing using unmanned aerial systems can be very promising for fruit growers with small acreages.

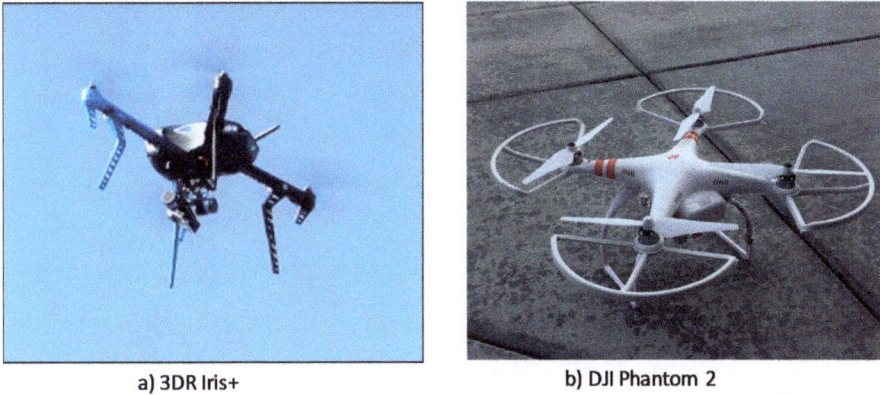

a) 3DR Iris+   b) DJI Phantom 2

Figure 1. Off-the-shelf unmanned aerial systems. (**a**) 3DR Iris+; (**b**) DJI Phantom 2.

A number of researchers have used unmanned aerial systems for civilian applications which include power line detection, roadway traffic monitoring, wetland analysis, and agriculture. Li et al. [6] developed an image processing algorithm for power line detection using Hough transform. A pulse-coupled neural filter was used to remove background noise before applying the Hough transform. Coifman et al. [7] investigated the use of UASs to monitor roadway traffic to facilitate offline planning and real-time management applications. A feasibility study by Ro et al. [8], which conducted a field experiment at a local interstate using UASs, concluded that UAS applications will become popular in the transportation area in the near future. The use of UAS photogrammetry provided a valuable and accurate enhancement to wetland delineation, classification, and health assessment [9].

Another area that has received a lot of attention for UAS application is agriculture. One of the examples of the use of unmanned aerial systems (UASs) for fruit trees is the crop monitoring and assessment platform (C-MAP) developed at Northwest Nazarene University [10]. The C-MAP is composed of an off-the-shelf UAS equipped with a multispectral camera. Figure 2 shows one of the C-MAP UASs flying over an experimental apple orchard with different watering methods, a drip and a sprinkler. An image processing algorithm was developed in this study to calculate the enhanced normalized difference vegetation index (ENDVI), which is a combination of the near-infrared band, green band, and blue band, and generated a false color image. The red color region has high ENDVI while the blue color region has the lowest ENDVI values. The false color image clearly shows the variability of the field caused by the difference in water input [11].

In this paper, the application of CMAP is extended to the detection of blossoms of peaches using a customized image processing algorithm. It has been reported that there is an increase of photosynthetic activity during the bloom period, which correlates with the fruiting process [12]. Peaches follow a linear pattern of crop development each year that allows the farmers to manage the fruit production and make sure that the crop is progressing as it should. In addition, farmers scout the orchard during the blooming season and use the observed amount of blooms with other parameters including crop density and the number of leaves on trees to predict yield. Early prediction of yield helps growers in marketing their products and in the packing operations [13]. The objectives of this study are: (1) to expand the use of CMAP to detect peach blossoms; and (2) to develop an image processing algorithm to detect peach blossoms.

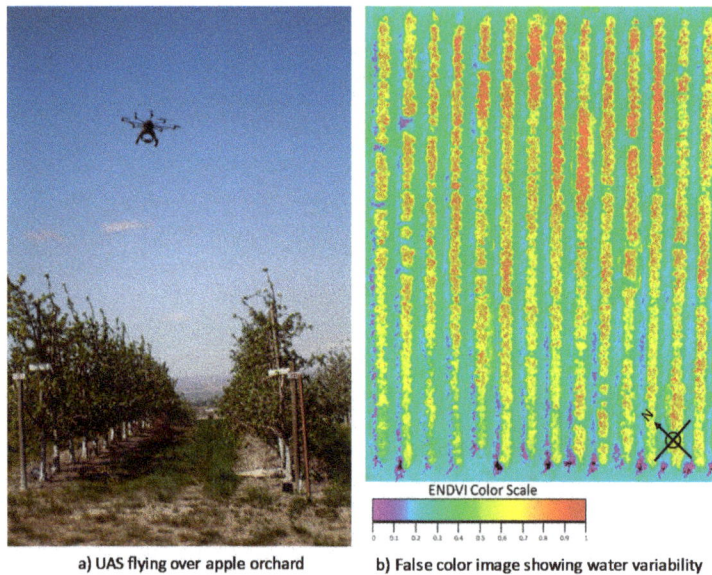

a) UAS flying over apple orchard   b) False color image showing water variability

Figure 2. Monitoring of apple orchard using C-MAP. (**a**) UAS flying over an apple orchard; (**b**) False color image showing water variability.

2. Materials and Methods

2.1. Target Field

The target fields in this study are an experimental peach orchard located north of Parma Idaho at the University of Idaho Research and Extension Center and a peach orchard located north of Marsing Idaho owned by Symms Fruit Ranch. Both orchards are located in the western part of the state of Idaho. The Parma orchard contains a variety of peach types, whereas the Symms orchard contains one type of peach (*Prunus persica*), which is the target crop in this study. The Parma orchard was approximately two acres and although the orchard at Symms was much larger, approximately only two acres were observed for the study.

2.2. Image Acquisition System

Two UASs were used in this study, both of which were DJI Phantom Quadcopters [14]. A DJI Phantom 3 Professional quadcopter was used to capture peach images in the RGB color spectrum. The camera for the DJI Phantom 3 Professional uses a 1/2.3″ complementary metal-oxide semiconductor (CMOS) sensor with 12.4 megapixels (4000 × 3000). A DJI Phantom 3 Advanced was used to capture multispectral images of the peach orchard. The camera for the DJI Phantom 3 Advanced also uses a 1/2.3″ CMOS sensor with 12.4 megapixels (4000 × 3000) but the camera was modified to capture near-infrared band centered at 750 nm, green band, and blue band. The two UASs were used to capture images in both orchards. Both DJI Phantom quadcopters utilized a navigation controller which could control the drone either manually or autonomously if interfaced with a tablet. A tablet with the DJI Go application software was used to connect and interface with the controller in order to calibrate the DJI drones and to allow for GPS and waypoint following during flights. The captured image files were written on two SD cards inside the drones and then a computer with

MATLAB software was used to access and download the image files from the cards in order to perform image processing and analysis.

2.3. DroneDeploy

The software used on the tablet to collect the images was DroneDeploy [15]. DroneDeploy is a cloud-based software compatible with DJI Phantom 3 drones which uses Google Maps and GPS to construct a flight plan. Figure 3 shows the operation of the UAS using DroneDeploy. Once a drone is calibrated with the DJI Go app, a flight plan can be created in the DroneDeploy app at any given place as long as the device has Wi-Fi or a flight be loaded without Wi-Fi if the flights were pre-synced to the device beforehand. Using the touch display of the tablet, DroneDeploy allows for the user to tap and drag the boundaries of the flight zone overlaid over the desired region shown on Google maps. The figure shows how DroneDeploy works with the DJI Phantom.

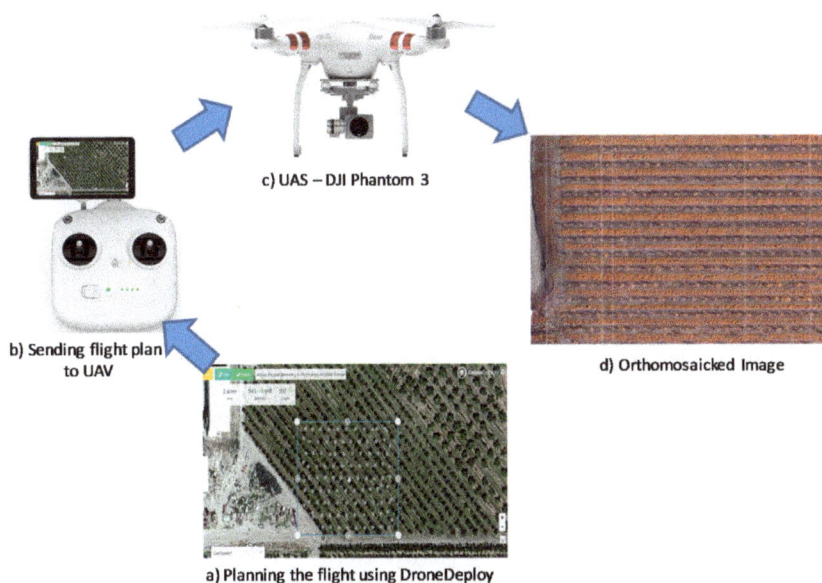

Figure 3. Operation of UAS using DroneDeploy. (**a**) Planning the flight using DroneDeploy; (**b**) Sending flight plan to UAS; (**c**) UAS – DJI Phantom 3; (**d**) Orthomosaicked image.

Figure 4 shows a screen shot of DroneDeploy, where the region enclosed by the blue rectangle is the desired field. DroneDeploy then plans the flight and calculates the position where to take the images in order to obtain pictures that cover the whole field, which is shown as grey dots on the figure. DroneDeploy also shows the coverage area, the number of images that will be taken, file size, and flight time when planning a flight. Once a flight plan is set, the DroneDeploy application allows for the user to adjust the altitude and the number of pictures the drone will take during the flight with Frontlap and Sidelap selections (Figure 5). Once the flight is initiated, DroneDeploy will automatically fly the drone along the given path and capture images at the given way points. Though the drone flies autonomously, the drone can be immediately switched back into manual flight by flipping the fight state switch on the controller. Once the images are taken, a computer accessing the DroneDeploy website can be used to upload the images and create an orthomosaic picture of the captured images.

Figure 4. Tablet screenshot of DroneDeploy.

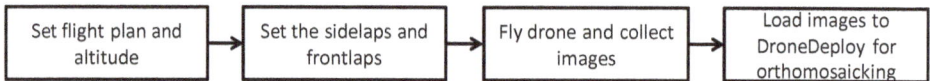

Figure 5. Image acquisition and stitching using DroneDeploy.

2.4. Image Acquisition

The images collected for this study were taken between the dates of 8 March 2016 and 20 April 2016. The pictures were taken every week except for the two weeks of bloom in which images were taken multiple days in a week. All data collection flights were dependent upon weather and solar conditions due to the impact they might have on the flight ability of the drones. All of the flights were completed between the times of 8 a.m. and 2 p.m. and were normally conducted with little to no wind. Although the conditions were clear skies, about half the images obtained were taken in cloudy weather. Sample images taken from the color camera and the modified camera are shown in Figure 6. Figure 6b shows a sample image acquired from the experimental field using the modified camera.

a) Sample RGB image of peach blossom

b) Sample multispectral image of peach blossom

Figure 6. Sample images acquired at peach orchards. (**a**) Sample RGB image of peach blossom; (**b**) Sample multispectral image of peach blossom.

2.5. Image Processing and Analysis

The acquired images were processed and analyzed using MATLAB and its Image Processing Toolbox. The focus of this paper is the processing of images from the multispectral images. The image processing involved the separation of the three bands and analyzing the color distribution. For the analysis of the color distribution, pixels of the peach blossoms and pixels of the ground (weeds) were manually selected and the pixel values of the three bands (NIR, green, and blue) were determined. The pixel values of the peach blossoms and the ground were plotted to show their distribution. Figure 7 shows the pixel value distribution of the blossom and the ground. Although we could easily draw a line and separate the peach blossom and ground, there is still some overlap between them. A contrast stretching operation was made on each band to improve the color difference between the blossom and the background [16]. Figure 7 shows the color distribution when the contrasts of each band were stretched. The near-infrared versus the blue band shows the separation between the two clusters. By using a very rudimentary thresholding process, the blossom could easily be segmented.

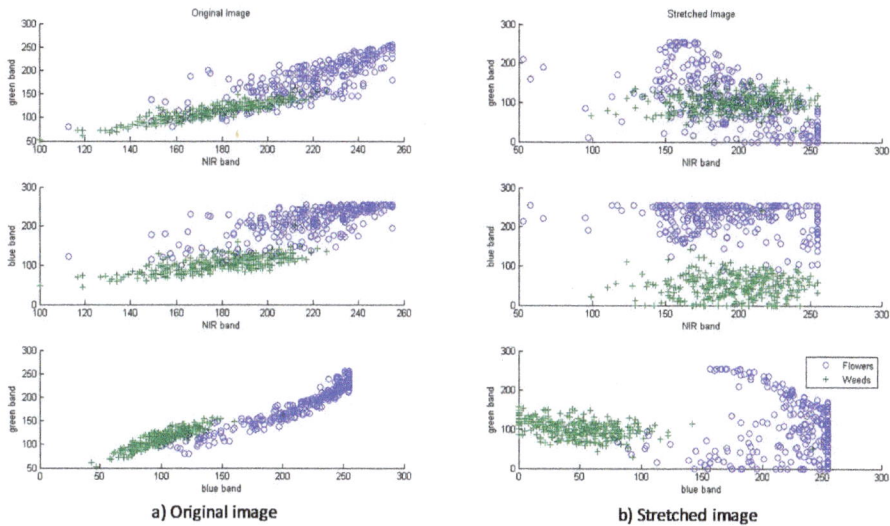

a) Original image   b) Stretched image

Figure 7. Color distribution of blossoms and weeds. (**a**) Original image; (**b**) Contrast stretched image.

2.6. Peach Blossom Detection

Figure 8 shows the image processing algorithm to detect the peach blossom. The first step is to stretch the three bands individually and then combine them. A simple thresholding operation for the near-infrared and blue bands is used for the segmentation of the blossom from the background. The thresholded image $g(x,y)$ is obtained as follows:

$$g(x,y) = \begin{cases} 1 & if \ f_{blue}(x,y) > 128 \ and \ f_{NIR}(x,y) > 128 \\ 0 & if \ f_{blue}(x,y) \leq 128 \ and \ f_{NIR}(x,y) \leq 128 \end{cases}$$

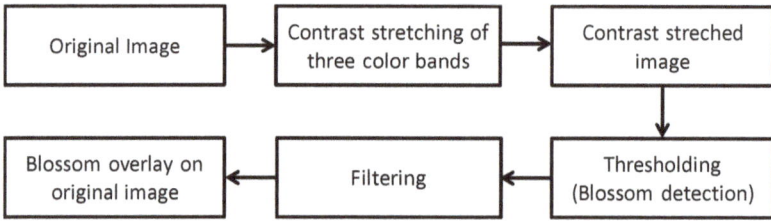

Figure 8. Image processing algorithm for blossom detection.

This segmentation process detects the blossom from the multispectral image. Figure 9 shows the image processing results. After the thresholding operation, a morphological size filtering process was used to remove "salt and pepper" noise. The overlaid image demonstrates the high success rate of detecting the blossom from the image.

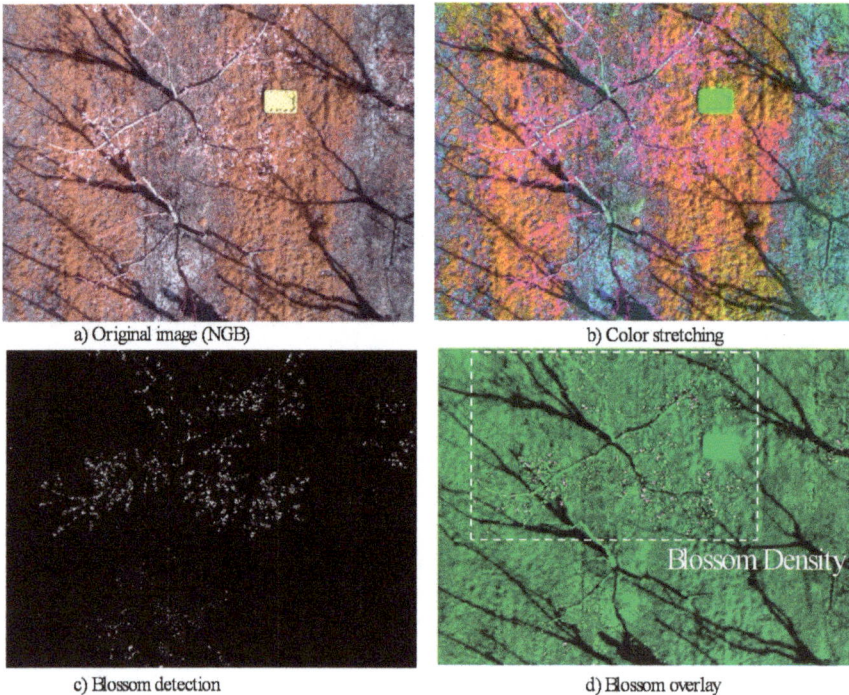

Figure 9. Image processing for blossom detection. (**a**) Original image (NGB); (**b**) Color stretching; (**c**) Blossom detection; (**d**) Blossom overlay.

The success rate of the blossom detection algorithm was evaluated using 20 randomly selected test images. The peach blossoms in the test images were manually identified and an evaluation mask was created for each test image. The success rate was calculated as follows:

$$success\ rate = \frac{number\ of\ blossom\ pixels\ from\ blossom\ detection\ algorithm}{number\ of\ blossom\ pixels\ from\ test\ image\ mask} \times 100\%$$

3. Results and Discussion

The results from the peach blossom detection algorithm showed that the blossoms were properly segmented from the raw multispectral image, with an average detection success rate of 84.3%. One of the reasons for the effective blossom detection is the use of the modified multispectral camera. With the modified filter of the camera, objects with high chlorophyll will have high reflectance in the near-infrared and green bands, but low reflectance in the blue band. In the image, the weeds have a red-brown hue because of the high chlorophyll content as compared with the other objects in the image. The colors of the peach blossoms are composed of a white and light pink hue. Some of the blossoms have a hue similar to that of the branch and some part of the ground. It can also be observed in Figure 7 that the light color of the blossom shows the high amount of near-infrared, green, blue values as compared to the weeds. Furthermore, the contrast stretching operation helped the thresholding process by increasing the separation of the pixel values between the blossoms and the ground specifically in the blue band. The contrast stretching did not affect the distribution in the near-infrared band. On the other hand, the morphological size filtering operation may have affected the detection success rate by removing small blossom pixels that were considered noise. However, the noise filtering operation was required to remove noise pixels.

Using the binary image of the blossom detection algorithm, the blossom density could be generally approximated by doing a series of calculations. Knowing the approximate height above the blossoms at which the pictures were taken, and having the images from the drone being flown over a known 2 m × 2 m square PVC pipe at that height, the density of the blossoms could be obtained. Processing this image as shown in Figure 10, the number of square meters per pixel was found for that given height, which could then be applied to the binary peach blossom detection images, yielding an approximate density of the blossoms in square units. Using the flying height of 10 m, the size of the PVC square, and the image spatial resolution of 4000 × 3000 pixels, the approximate coverage area was 600 square meters. When the density of the peach blossoms was correlated to the square units, the result would not be perfect, but as long as the height of the images was consistent across all images, a correlation to fruit yield could be attempted.

Figure 10. Image processing process for pixel density calculation.

Since the peach trees are planted at about 3 m intervals, the trees in the images were separated by creating a grid over the image and putting the trees in individual boxes. Figure 11 shows the result of this grid as well as the resulting peach segmentation over the image. The blossom density from each tree can then be estimated by doing a pixel count in each box.

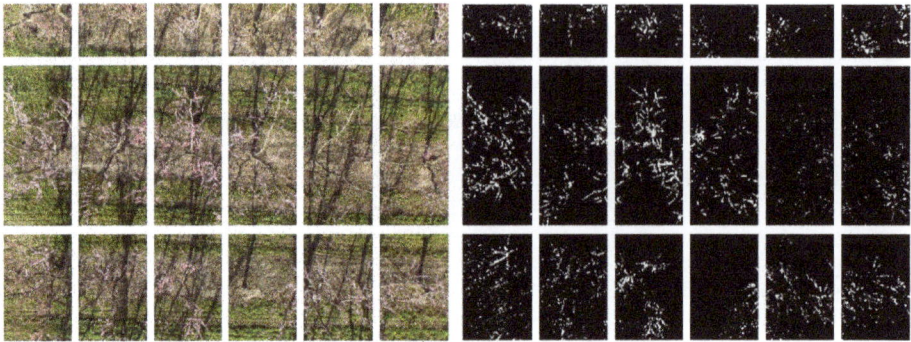

Figure 11. Image processing results for tree grid separation.

Although such a process of tree segmentation could not be done for every image and would be very inaccurate, future work of this study will involve the detection of individual trees by way of boundaries. Using the boundaries, the blossom density of each tree would then be directly and accurately calculated. A blossom density map can then be produced, which could be used to aid yield estimation and other subsequent orchard management operations. The farmer could also use the blossom density map to provide a temporal analysis of the orchard blossoms.

4. Conclusions

An image processing algorithm was developed to detect blossoms on peach trees. The image acquisition system used an on-the-shelf UAS, the DJI Phantom 3. The UAS camera was modified to allow near-infrared, green, and blue bands. Images from experimental and commercial peach orchards were used as target fields. The DroneDeploy software was used to plan the flight path, collect the images, and for image mosaicking. The image processing analysis showed that contrast stretching of the images' three bands enhanced the color of the blossoms from the background. A very basic thresholding segmentation method was used to segment the blossoms. Initial results showed that the blossoms can be detected using the thresholding operation with an average detection rate of 84.3%. Future study will involve the improvement of blossom density calculation and the development of an algorithm for exact tree segmentation.

Acknowledgments: This research was supported by the Idaho State Department of Agriculture (Idaho Specialty Crop Block Grant 2014) and Northwest Nazarene University.

Author Contributions: Duke Bulanon conceived the study, created the literature review, designed the experiments, processed the collected data and wrote the paper. Ryan Horton and Esteban Cano acquired the images, developed the image processing algorithms, and processed the data. Esmaeil Fallahi designed the target orchard and its treatment, and supervised during the data collection.

Conflicts of Interest: The authors declare no conflict of interest.

Abbreviations

The following abbreviations are used in this manuscript:

C-MAP	Crop monitoring and assessment platform
ENDVI	Enhanced Normalized Difference Vegetation Index
GPS	Global positioning system
NGB	Near infrared, Green, Blue
NIR	Near infrared
PVC	Polyvinyl chloride
RGB	Red, Green, Blue
UAS	Unmanned aerial system

J. Imaging **2017**, *3*, 2

References

1. USDA-National Agriculture Statistics Service. State Agriculture Overview (Idaho). Available online: https://www.nass.usda.gov/Quick-Stats/Ag-Overview/stateOverview.php?state=IDAHO (accessed on 30 April 2016).
2. Downey, D.R.; Ehsani, K.; Giles, S.; Haneklaus, D.; Karimi, K.; Panten, F.; Pierce, E.; Schnug, D.C.; Slaughter, S.; Upadhyaya, D. Advanced Engineering Systems for Specialty Crops: A Review of Precision Agriculture for Water, Chemical, and Nutrient Application, and Yield Monitoring. *Landbauforsch.—VTI Agric. For. Res.* **2010**, *340*, 1–88.
3. Lee, W.S.; Alchanatis, V.; Yang, C.; Hirafuji, M.; Moshou, D.; Li, C. Sensing technologies for precision specialty crop production. *Comput. Electron. Agric.* **2010**, *74*, 2–33. [CrossRef]
4. Pajares, G. Overview and current status of remote sensing applications based on Unmanned Aerial Vehicles (UAVs). *Photogramm. Eng. Remote Sens.* **2015**, *81*, 281–329. [CrossRef]
5. Johnson, A.K.L.; Kinsey-Henderson, A.E. Satellite-based remote sensing for monitoring Baath land use in sugar industry. *Proc. Aust. Soc. Sugar Cane Technol.* **1997**, *19*, 237–245.
6. Li, Z.; Liu, Y.; Walker, R.; Hayward, R.; Zhang, J. Towards automatic power line detection for a UAV surveillance system using pulse coupled neural filter and an improved Hough transform. *Mach. Vis. Appl.* **2010**, *21*, 677–686. [CrossRef]
7. Coifman, B.; McCord, M.; Mishalani, R.G.; Iswalt, M.; Ji, Y. Roadway traffic monitoring from an unmanned aerial vehicle. *IEEE Proc. Intell. Trans. Syst.* **2006**, *153*, 11–20. [CrossRef]
8. Ro, K.; Oh, J.S.; Dong, L. Lessons learned: Application of small UAV for urban highway traffic monitoring. In Proceedings of the 45th AIAA Aerospace Sciences Meeting and Exhibit, Reno, NV, USA, 8–11 January 2007; pp. 2007–2596.
9. Boon, W.A.; Greenfield, R.; Tesfamichael, S. Wetland Assessment Using Unmanned Aerial Vehicle (UAV) Photgrammetry. In Proceedings of the International Archives of the Photogrammetry, Remote Sensing and Spatial Information Sciences, XXIII ISPRS Congress, Prague, Czech Republic, 12–19 July 2016.
10. Bulanon, D.M.; Horton, M.; Salvador, P.; Fallahi, E. *Apple Orchard Monitoring Using Aerial Multispectral Imaging*; ASABE Paper No. 1913165; American Society of Agricultural and Biological Engineers (ASABE): St. Joseph, MI, USA, 2014.
11. Bulanon, D.M.; Lonai, J.; Skovgard, H.; Fallahi, E. Evaluation of different irrigation methods for an apple orchard using an aerial imaging system. *ISPRS Int. J. Geo-Inf.* **2016**, *5*, 79. [CrossRef]
12. Fujii, J.A.; Kennedy, R.A. Seasonal changes in the photosynthetic rate in apple trees—A comparison between fruiting and nonfruiting trees. *Plant Pysiol.* **1985**, *78*, 519–524. [CrossRef]
13. Aggelopoulou, K.D.; Wulfsohn, D.; Fountas, S.; Gemtos, T.A.; Nanos, G.D.; Blackmore, S. Spatial variation in yield and quality in a small apple orchard. *Precis. Agric.* **2010**, *11*, 538–556. [CrossRef]
14. Da-Jiang Innovations (DJI). Available online: http://www.dji.com/products/phantom (accessed on 30 December 2016).
15. DroneDeploy. Available online: https://www.dronedeploy.com/ (accessed on 30 December 2016).
16. Gonzalez, R.C.; Woods, R.E. *Digital Image Processing*, 3rd ed.; Pearson: New York, NY, USA, 2007.

Journal of
Imaging

MDPI

Article

Early Yield Prediction Using Image Analysis of Apple Fruit and Tree Canopy Features with Neural Networks

Hong Cheng [1,2,3], Lutz Damerow [2], Yurui Sun [3] and Michael Blanke [4,*]

[1] College of Information Science and Technology, Agricultural University of Hebei, Baoding 071001, China; chenghong@cau.edu.cn
[2] Institute of Agricultural Engineering, Faculty of Agriculture, University of Bonn, 53115 Bonn, Germany; damerow@uni-bonn.de
[3] Key Lab of Agricultural Information Acquisition Technology, China Agricultural University, Beijing 100083, China; pal@cau.edu.cn
[4] INRES-Horticultural Science, Faculty of Agriculture, University of Bonn, 53121 Bonn, Germany
* Correspondence: mmblanke@uni-bonn.de; Tel.: +49-228-735-142

Academic Editors: Gonzalo Pajares Martinsanz and Francisco Rovira-Má
Received: 30 September 2016; Accepted: 11 January 2017; Published: 19 January 2017

Abstract: (1) Background: Since early yield prediction is relevant for resource requirements of harvesting and marketing in the whole fruit industry, this paper presents a new approach of using image analysis and tree canopy features to predict early yield with artificial neural networks (ANN); (2) Methods: Two back propagation neural network (BPNN) models were developed for the early period after natural fruit drop in June and the ripening period, respectively. Within the same periods, images of apple cv. "Gala" trees were captured from an orchard near Bonn, Germany. Two sample sets were developed to train and test models; each set included 150 samples from the 2009 and 2010 growing season. For each sample (each canopy image), pixels were segmented into fruit, foliage, and background using image segmentation. The four features extracted from the data set for the canopy were: total cross-sectional area of fruits, fruit number, total cross-section area of small fruits, and cross-sectional area of foliage, and were used as inputs. With the actual weighted yield per tree as a target, BPNN was employed to learn their mutual relationship as a prerequisite to develop the prediction; (3) Results: For the developed BPNN model of the early period after June drop, correlation coefficients (R^2) between the estimated and the actual weighted yield, mean forecast error (MFE), mean absolute percentage error (MAPE), and root mean square error (RMSE) were 0.81, −0.05, 10.7%, 2.34 kg/tree, respectively. For the model of the ripening period, these measures were 0.83, −0.03, 8.9%, 2.3 kg/tree, respectively. In 2011, the two previously developed models were used to predict apple yield. The RMSE and R^2 values between the estimated and harvested apple yield were 2.6 kg/tree and 0.62 for the early period (small, green fruit) and improved near harvest (red, large fruit) to 2.5 kg/tree and 0.75 for a tree with ca. 18 kg yield per tree. For further method verification, the cv. "Pinova" apple trees were used as another variety in 2012 to develop the BPNN prediction model for the early period after June drop. The model was used in 2013, which gave similar results as those found with cv. "Gala"; (4) Conclusion: Overall, the results showed in this research that the proposed estimation models performed accurately using canopy and fruit features using image analysis algorithms.

Keywords: ANN; back propagation neural network (BPNN); image analysis; machine learning; precision horticulture; sustainability

1. Introduction

Early and accurate prediction of fruit yield is relevant for the market planning of the fruit industry, trade, supermarkets, as well as for growers and exporters to plan for the need of labour and bins, storage, packing materials, and cartons [1]. In the European Union (EU), about 12 million tons of apples are harvested every year, making it the most important fruit crop in the EU with year- to year variations of ca. two million tons. Yield prediction becomes essential after the last natural fruit abortion, i.e., June drop, when the first reliable yield estimates may be obtained and fruit are still small, green, and often occluded by leaves or other fruit, until harvest time; the accuracy of yield prediction and challenges change during fruit ontogeny. To date, yield prediction is mainly based on historic performance of an orchard in the previous years, i.e., empirical data. In order to improve the accuracy and efficiency for apple yield estimation, the automatic prediction using computer vision technology is increasingly receiving attention [2–9].

Previous apple detection studies concentrated on the late period of fruit maturation, when the number of fruits obtained from image analysis closely correlated between algorithm prediction and manually counted fruits in an image. Stajnko et al. [2] segmented cv. "Jonagold" apple fruit from the image taken on 7 September using colour features and surface texture. Fruit detection rates in the images were 89% of the apples visible on the images. Kelman and Linker [8] detected mature green apples in tree images using shape analysis and the correct detection of the apples was 85% of the apples visible in the images.

Moreover, to predict yield during the early growing period, Linker et al. [10] aimed to detect green "Golden Delicious" apple fruit from RGB images on only a part of an apple tree to facilitate the study; the algorithm accurately detected more than 85% of the apples visible in the images under natural illumination. Zhou et al. [5] proposed a recognition algorithm based on colour features to estimate the number of young "Gala" apples after June drop with a close correlation coefficient of R^2 of 0.80 between apples detected by the fruit counting algorithm and those manually counted and R^2 of 0.57 between apples detected by the fruit counting algorithm and actual harvested yield.

However, as noted in some of these studies, some regions of the apple tree are occluded, where leaves cover the apple fruit, making their assessment difficult during counting in the orchard. A certain number of fruit grow well inside the canopy close to the tree trunk, especially as trees grow older and into their high yielding phase, and pose a challenge to detect, especially, in the early stages of fruit growth. Hence, the present work is based on the hypothesis that these limitations with image processing to detect fruit early in the season may be overcome by the integration of characteristics of the canopy structure of the tree. From an apple tree canopy image, the features of the canopy structure are extracted by image processing [2,5]. There are artificial intelligence algorithms, which could be employed to model the relationship between the features and harvested yield.

An artificial neural network (ANN), as a commonly used machine learning algorithm, has the potential of solving such problems, when the relationship between the inputs and outputs is not well understood or is difficult to translate into a mathematical function. Many ANN applications dealing with this similar situation in agriculture have been reported [11]. Back Propagation Neural Network (BPNN) was used to predict maize yield from climatic data with an accuracy at least as good as polynomial regression [12]. The use of ANN for apple yield prediction was reported in [6]. They used the number of fruit at different times during fruit ontogeny as well as the actual yield for each image per tree as training parameters and reported that the application of ANN improved apple yield prediction based on image analysis.

The objective of this study was to improve the accuracy of early yield prediction by taking features of the tree-canopy structure (number of fruit F_N, area of fruits F_A, area of fruit clusters F_{CA}, and foliage leaf area (L_A)) in the canopy image into account, besides the apple fruit number. The aims of the present paper are: (1) to describe the processes of extracting canopy features and "learning" the relationship between the features and actual yield per tree by use of a back propagation neural network (BPNN) using the data from 2009–2010; (2) to evaluate the BPNN prediction models to analyse the relation between the estimated and harvested apple yield; and (3) to represent the accuracy of the BPNN models by predicting the yield for 30 samples from 2011.

2. Materials and Methods

2.1. Site Description and Image Acquisition

Apple cv. "Gala" trees, trained as slender spindles and spaced at a tree distance of 1.5 and a row distance of 3.5 m oriented North to South, were located at the University of Bonn, Campus Klein-Altendorf, Germany (50.6° N, 6.97° E, 180 m a.s.l.). The "Gala Model 1" was used to train and test 90 thirteen-year-old "Pinova" apple trees on dwarfing M9 rootstock, spaced also at 3.5 m × 1.5 m and also trained to slender spindles after June drop in 2012. In 2013, the model was used to predict the yield of 30 samples of "Pinova" apple trees, which were captured during the same period.

The soil is a rich luvisol on alluvial loess with a score of 92 on a 100 point soil fertility scale. The climate is dominated by Atlantic Western weather buffered by the mild Rhine river influence and 604 mm annual rainfall, thereby not requiring any irrigation. A total of 180 images were captured at 1.5 m height and at a constant distance of 1.4 m perpendicular to each tree row in natural daylight using a commonly available digital camera type Samsung (Seoul, South Korea) VB 2000 with German Schneider (Green Bay, WI, USA) lens with automated white calibration in "auto-focus" mode (without the use of the zoom) set to 3 Megapixels per image. White and red calibration spheres of 50 mm diameter (polystyrene) were used to determine fruit size. A 2 m × 3 m white drapery cloth was placed behind the target tree to distinguish fruits from trees in other rows. Images (Table 1) were captured twice, i.e., when fruit were light-green after June drop about three months before harvest (period 1; "Gala Model 1") and when fruit were red during the fruit ripening period half a month before harvest (period 2; "Gala Model 2"), on the preferred western side of the tree. Illumination for the first image acquisition was estimated ca. 800 µmol PAR and for the second date ca. 600 µmol PAR m$^{-2} \cdot$s^{-1} using an EGM-5 (PPSystems, Amesbury, MA, USA). Images were obtained on these apple trees, in the early afternoon (3 to 5 pm) on days with indirect light to exclude stray or blinding light, and deep shades at a time of low solar angle on the second date (period 2).

Table 1. Characteristics of samples. Tree fruit load % refers to the % of trees carrying a high (>mean + SD), low (<mean − SD), and moderate fruit load, respectively.

Season	Numbers of Trees	Number of Images (Period 1, Period 2)	Yield/Tree (mean ± SD)	Tree Fruit Load (%) High, Mod, Low
2009	60	60; 60	20.62 ± 4.90	20; 68; 12
2010	90	90; 90	16.68 ± 5.43	18; 63; 19
2009 & 2010	150	150; 150	18.26 ± 5.55	15; 67; 18
2011	30	30; 30	18.63 ± 4.17	10; 70; 20

After harvest, fruit were sorted using a commercial grading machine (type MSE2000, Greefa, Geldermalsen, The Netherlands) to provide fruit, counts for each individual tree, size for each individual fruit, and cumulative yield/tree.

Matlab (version 2011b, Mathwoks Inc., Natik, MA, USA) was used for the image processing and modeling. Typical images are shown in Figure 1. To improve the data processing speed, images were uniformly resized to 512 × 683 pixels. The parameters used in this paper, are listed in Table 2.

(a)   (b)

Figure 1. Sample apple tree at different times; left picture (**a**) was acquired in the early period after the June drop (period 1), about 3 months before harvest, right picture (**b**) was acquired during the ripening period (period 2), about 15 days before harvest.

Table 2. Description of parameters for yield prediction.

Parameter	Description	Parameter	Description
I_A	Sum of pixels of the whole images	Y_E	The estimated yield of apple tree
F_A	Sum of pixels belonging to apple fruits	F_1	F_A/I_A
F_N	Number of fruit	F_2	$F_N/200$
F_{CA}	Sum of pixels belonging to apple clusters	F_3	$(F_A - F_{CA})/I_A$
L_A	Sum of pixels belonging to foliage	F_4	L_A/I_A
Y_A	The actual yield of apple tree	F_5	$Y_A/50$
MAPE	Mean Absolute Percentage Error	SD	Standard deviation of the error
MFE	Mean Forecast Error	RMSE	Root Mean Square Error

Note: the fruit number per tree is below 200; the actual yield per tree is below 50 kg.

2.2. Apple Fruit and Leaf Feature Description

Fruit and foliage are two main components of the apple tree canopy. Based on the images of the apple tree canopy, the fruit number (F_N) and the fruit area (F_A) are the first two essential features for yield prediction. The third feature is the area of the apple clusters (F_{CA}) in the image, because apple clusters are a conspicuous characteristic of canopy structure, which can be comprised of more than two apples. Compared with the pixel proportion of the bright red calibration sphere (Figure 1a), which was of the size range for an apple fruit in period 1, if the fruit domain exceeded the size of the calibration sphere by 3-fold, it was assumed to be an apple cluster. Since the leaves can impact apple yield estimation by occluding fruit, foliage area (L_A) is the fourth one.

As we consider F_A, F_N, L_A, and F_{CA} extracted from canopy images as essential parameters for yield prediction [5], we converted them to the ratios F_1, F_2, F_3 and F_4 (Table 2). These ratios were subsequently employed for modelling and the different steps in the modelling process are visualized in a flowchart (Figure 2).

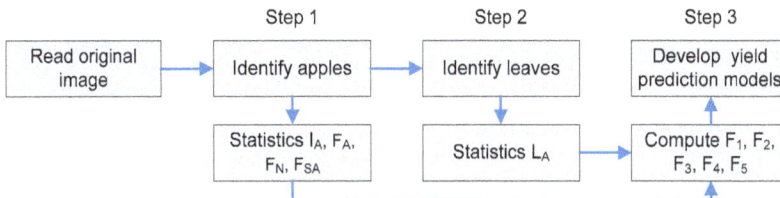

Figure 2. Outline of the processing steps.

2.3. Fruit Identification and Feature Extraction (Step 1)

The implemented fruit recognition algorithms were based on a previous study [5] (Figure 3). The RGB images were transformed to a binary image and analysed to count the apples within the picture. Pixels of each connected domain were summed as its area to computed F_A, F_N and F_{CA}.

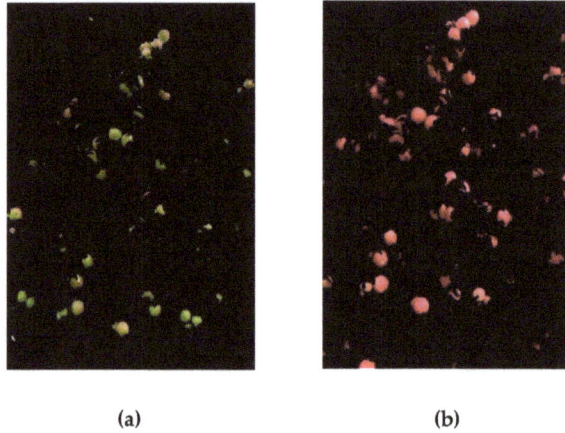

(a)   (b)

Figure 3. Image of same tree at different times, (**a**) in July; (**b**) in the beginning of September.

2.4. Leaf Identification and Feature Extraction (Step 2)

A method was developed for automated recognition of foliage within the tree image. Before foliage identification, the fruit domains were removed from the canopy image (Figure 4). Both RGB and HSI colour systems were used to segment leaves in the image automatically. Figure 4 gives an overview in order of image processing. The proposed method was employed to separate foliage from the background (white drapery, branches and sky) in the image.

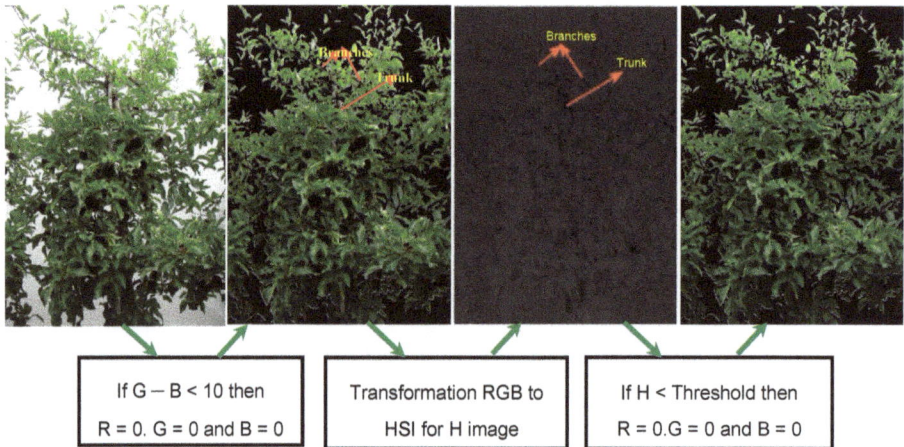

Figure 4. Proposed algorithm for leaf discrimination; an example of image processing.

The map of colour difference (green minus blue, i.e., G − B) for each pixel is shown in Figure 5. The green minus blue value was larger for the foliage compared with the background (white drapery and sky), which could be used to segment foliage from the image at a threshold colour difference G-B of 10. For each pixel, if the colour difference was below 10, then the R, G, B colour values of this pixel were set to zero (Figure 4).

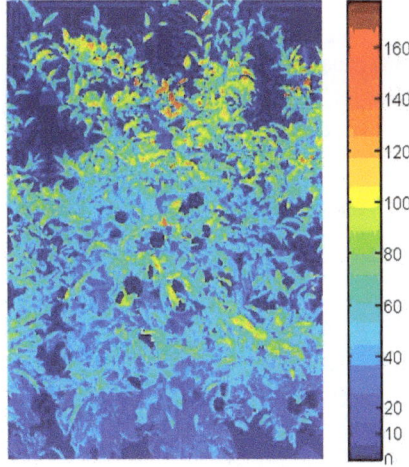

Figure 5. Example of an image of an apple tree with colour-coded mapping of colour differences between G (green) and B (blue) for each pixel, showing the leaves as bright colour dots and the background in deep blue.

However, background pixels (branches and trunk) were incorrectly assigned the same classification as foliage pixels. Specifically, in the hue (H) image, the pixels were divided into two classes (Figure 4), and one of them consisted of the deeper colour pixels represented by branches and trunk. Eventually, using the Ostu's algorithm [12], a threshold (T) was obtained. The branches and trunk were removed from the image at the threshold T. Foliage cross-sectional area (L_A) was computed by summing pixels that belonged to foliage.

2.5. Development of BPNN Yield Prediction Model (Step 3)

The ratios F_1, F_2, F_3, F_4 and F_5 (Table 3) were computed based on the parameters I_A, F_A, F_N, F_{CA}, L_A and Y_A (Table 2). One data set of 150 images acquired in period 1 was collected as set 1, and the other set of 150 images acquired in period 2 was collected as set 2. Sixty images were sampled in the summer of 2009 and ninety images were sampled in the summer of 2010 for the two sets, i.e., each set included 150 samples, respectively. Each sample consisted of five parameters. Two BPNN prediction models were built for the two periods using Sets 1 and 2, respectively.

The Back Propagation Neural Network (BPNN) was trained with an error back-propagation learning algorithm used for computing the ANN weights and biases [13]. In this present study, the inputs were the four features from the apple tree image and a vector of features in BPNN, which were different from that of Rozman [6], and the output was the forecast yield. As BPNN was trained, the weights of inputs for each processing unit were adjusted, and the network gradually "learned" the input/output relationship to minimize MSE between the actual yield of an apple tree and the estimated yield of the sample set.

Table 3. Example of 15 trees for modelling.

Sets		Set 1					Set 2			
Tree	F_1	F_2	F_3	F_4	F_5	F_1	F_2	F_3	F_4	F_5
1	0.0159	0.3100	0.0159	0.5495	0.4571	0.0552	0.5100	0.0331	0.2278	0.4571
2	0.0041	0.1800	0.0041	0.4265	0.3355	0.0279	0.3850	0.0242	0.2100	0.3355
3	0.0103	0.3100	0.0103	0.4718	0.3808	0.0348	0.4900	0.0199	0.1900	0.3808
4	0.0088	0.2550	0.0088	0.7245	0.3679	0.0347	0.6100	0.0199	0.2052	0.3679
5	0.0214	0.5250	0.0193	0.7022	0.4423	0.0590	0.5500	0.0297	0.2066	0.4423
6	0.0081	0.3850	0.0081	0.5714	0.4509	0.0318	0.5350	0.0251	0.2372	0.4509
7	0.0108	0.4000	0.0060	0.5024	0.3611	0.0320	0.4700	0.0159	0.1893	0.3611
8	0.0125	0.3750	0.0125	0.7344	0.3648	0.0440	0.5400	0.0329	0.2084	0.3648
9	0.0060	0.3100	0.0060	0.4066	0.5106	0.0549	0.5000	0.0327	0.3360	0.5106
10	0.0149	0.3900	0.0149	0.7559	0.3863	0.0537	0.5100	0.0329	0.2101	0.3863
11	0.0191	0.4550	0.0170	0.7116	0.4431	0.0596	0.4350	0.0340	0.1902	0.4431
12	0.0096	0.3350	0.0096	0.4149	0.3944	0.0445	0.5750	0.0269	0.1753	0.3944
13	0.0144	0.4150	0.0144	0.7099	0.4131	0.0546	0.3900	0.0229	0.2593	0.4131
14	0.0180	0.3150	0.0148	0.6459	0.3526	0.0585	0.3850	0.0173	0.1663	0.3526
15	0.0041	0.1800	0.0041	0.4265	0.3355	0.0279	0.3850	0.0242	0.2100	0.3355

A typical three-layer BPNN with one input layer, one hidden layer, and one output layer was employed in our present study. To determine the optimal number of hidden neurons, the number of hidden neurons was initially calculated as Equation (1), and the value of N_A was adjusted to select the best one based on the minimization of the mean squared errors (MSE), which is a statistical measure showing how well the model predicts the output value and the target value (yield).

$$N_H = (N_I + N_O)^{1/2} + N_A, \quad N_A \in [0, 10]$$ (1)

- N_I is the number of input neurons,
- N_O is the number of output neurons,
- N_H is the number of hidden neurons,
- N_A is the number of the neurons, which can be added in hidden neurons based on MSE.

Before BPNN training, the main parameters were set (Table 4). In order to reduce extra or over-fitting of the 150 samples, 90% of the samples composed the training set, which was used to develop the model, and the remaining 10% was used for the test set [14]. Both the training and test set were selected randomly, guaranteeing a balanced proportion between positive and negative outcomes. During BPNN training, the target parameter and the inputs were all inserted into the neural network. Each node in the input and hidden layers were connected to each of the nodes in the next layer (hidden or output). All connections between nodes were directed (i.e., the information flows only one way), and there are no connections between the nodes within a particular layer. Each connection between nodes had a weighting factor. These weighting factors were modified using the back-propagation algorithm based on the "error" during the training process to produce "learning".

Table 4. Relative parameter settings before the training BPNN.

Parameter	Value	Parameter	Value
Input	F_1, F_2, F_3, F_4	Hidden layer transfer function	Logarithmic sigmoid transfer function
Target	F_5	Output layer transfer function	Linear transfer function
Output	Forecast value	Learning function	Gradient descent learning function
Performance function	MSE	Training function	Levenberg-Marquardt back-propagation

2.6. The Measures for Model Evaluation

Four statistical approaches were employed to evaluate the model: R^2, MFE, RMSE, and MAPE. Mean Forecast Error (MFE) is a measure of unbiasedness of the predictions, defined as Equation (2), and MFE is closer to 0, then the model becomes less unbiased. Root mean squared error (RMSE) is an often used measure of the difference between values predicted by a model and those actually observed from the object being modeled, and is defined as Equation (3). It can rule out the possibility that large errors of opposite signs could cancel out in a MFE measure. The Mean Absolute Percentage Error (MAPE) is computed through a term-by-term comparison of the relative error in the prediction with respect to the actual value of the variable, and is defined as Equation (4). Thus, the MAPE is an unbiased statistical approach for measuring the predictive capability of a mode [14,15].

$$MFE = \frac{1}{N} \sum_{}^{N} (Y_A - Y_E) \tag{2}$$

$$RMSE = \sqrt{\frac{1}{N} \sum_{}^{N} (Y_A - Y_E)^2} \tag{3}$$

$$MAPE = \sum_{}^{N} |(Y_A - Y_E)/Y_A| \times \frac{100}{N} \tag{4}$$

3. Results

3.1. Data Analysis

The objective of the work was to evaluate algorithms for the detection of young light green apple fruit three months before harvest as accurately as possible. As shown in Table 3 for the same apple tree, the canopy feature values were developed from July (period 1) to September (period 2). In period 1, when the apple fruit was small, the computed values (fruit-related ratios) of F_1, F_2, and F_3 were much smaller compared with the foliage-related ratio F_4; in period 2, the values of F_1, F_2, and F_3 increased with increasing fruit size. Therefore, the fruit detection is strongly dependent on the amount of foliage in the canopy, which makes it almost impossible to detect apple fruit using image analysis in period 1. In period 2, the obvious colour and size changes of apple fruits make the detection easier and the influence of the foliage becomes weak. Overall, F1 appears to be the most important parameter, because it reflects the area of all apples in the tree image. Hence, it includes the overall information on the size and number of all apples.

3.2. BPNN Model Structure and Validation

The BPNN "Prediction Model 1" for the early period after June drop consists of 4 input neurons, 12 hidden neurons and 1 output neuron, and the BPNN "Prediction Model 2" for the ripening period consists of 4 input neurons, 11 hidden neurons, and 1 output neuron.

Both models performed well relative to each other, and illustrate that ANN could be employed to predict the apple yield. Table 5 shows the small differences between the early and late yield prediction model for cv. "Gala" regarding R^2 (0.02), RMSE (0.15 kg/tree), MFE (0.02), and MAPE (0.45 %), and shows that the results of Model 1 (Table 5) were similar to those of Model 2 (Table 6) with Model 1 being slightly inaccurate.

Table 5. The model structure and the evaluation of model performance for cv. "Gala".

Model	Parameter	Structure	Samples (Trees) of 2009 and 2010	RMSE in kg/Tree	MAPE (%)	MFE	R^2
Model 1	Train set	4-12-1	135	2.34	10.67	−0.05	0.81
	Test set		15	2.53	12.40	0.16	0.80
Model 2	Train set	4-11-1	135	2.27	8.9	−0.03	0.83
	Test set		15	2.31	10.36	−0.06	0.82

Table 6. Comparison between actual yield and predicted yield; $n = 150$ trees.

Model	Actual Yield (A) in kg per 150 Trees	Predicted Yield (P) in kg per 150 Trees	Difference (∣A − P∣) in kg	Mean Difference in kg per Tree
Model 1	2736	2744	8	0.05
Model 2	2736	2740	4	0.03

3.3. Yield Prediction for Subsequent Year

To evaluate their reliability and robustness, the two BPNN yield prediction models, based on combining both the 2009 and 2010 data, were used to predict the yield for next year (2011). In the growing season of 2011, 30 trees (Table 1) were selected for a wide variability of fruit load and were photographed in July and September as samples to validate the performance of "Prediction Model 1" and "Prediction Model 2", respectively. The results indicate the two models have good reliability and robustness, and the correlation increased from 0.62 to 0.75 near harvest time (Figure 6).

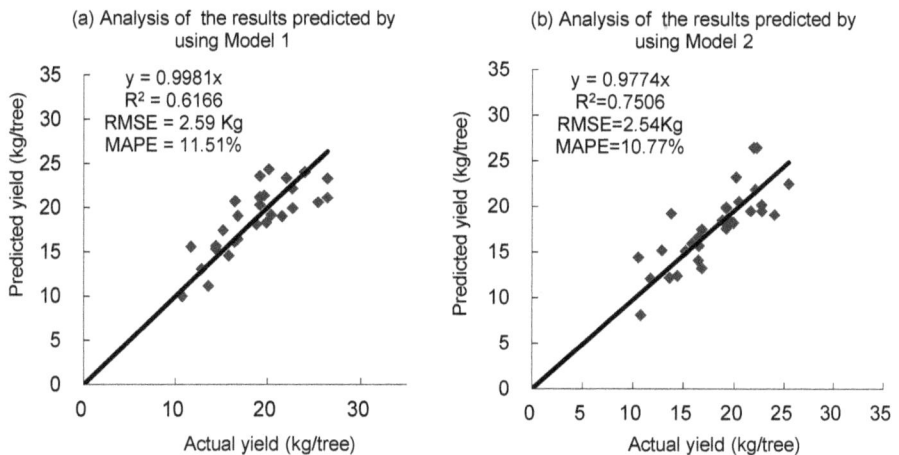

Figure 6. Yield prediction for 2011 based on (**a**) "Prediction Model 1" for young apple fruit in July and (**b**) "Prediction Model 2" for ripe apple fruit in September for the subsequent year ($n = 30$ trees).

3.4. Yield Prediction for Other Apple Varieties

The results (Tables 7 and 8) shows that the "Pinova" model performs similar to the "Gala" model 1. This "Pinova" model also could be used to predict the yield for the subsequent year (Figure 7). These again proved that a BPNN model, based on the features from the canopy, could be used for apple yield prediction at the orchard level.

Table 7. The model structure and the evaluation of model performance which was developed based on samples of 2012 for cv. "Pinova".

Model	Parameter	Structure	Samples (Trees) of 2012	RMSE in kg/Tree	MAPE (%)	MFE	R^2
"Pinova" Model	Train set	4-10-1	80	2.24	11.45	−0.14	0.89
	Test set		10	2.53	14.19	0.06	0.88

Table 8. Comparison between actual yield and predicted yield for 2012; n = 90 trees.

Model	Actual Yield (A) in kg per 100 Trees	Predicted Yield (P) in kg per 100 Trees	Difference (|A−P|) in kg	Mean Difference in kg per Tree
"Pinova" Model	1817	1822	5	0.06

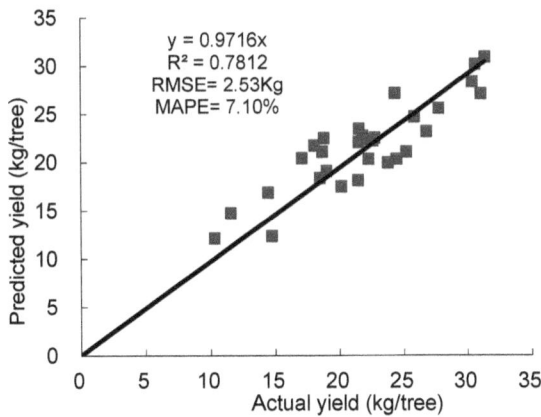

Figure 7. Yield prediction for 2013 based on the prediction "Pinova" Model for young apple fruit in July for the subsequent year (n = 34 trees).

4. Discussion

The majority of studies have used image processing algorithms to estimate total fruit number and fruit diameters to achieve yield prediction shortly before the fruit maturity and harvest [16–18]. However, counting the number and measuring the size of fruit by machine vision is based on the premise that all fruit on a tree can be seen and are not occluded by leaves. The scientific challenge is to identify each fruit in the tree image with some fruit hidden within the canopy, especially in the early period (Figure 1a). However, early prediction is essential for planning labor, bins, and harvest organization as well as transport, grading, and storage. Hence, four features were extracted from the tree image (Table 2), which were closely related to yield prediction and, moreover, changed with the growth of the fruit (Table 3). BPNN was employed for the analysis of the relationship between the four features and the actual yield to model yield prediction (Figure 6).

The small differences between the early and late yield prediction model for cv. "Gala" in Table 5 could be attributed to the fact that the proposed method with the neural network, based on the features from canopy images, can reduce the adverse influence from foliage. Hence, the method could be used for apple yield prediction in early and late growing periods after June drop, even when the apple fruit are small and green. In Table 6, the small differences between the harvested and the predicted yield for both models suggest that the BPNN model based on the features of canopy could provide accurate apple yield prediction at the individual orchard level.

In the two BPNN models for early prediction in July and pre-harvest in September, four features were extracted from the tree canopy image from the respective period as inputs into the model, and the apple yield can then be predicted. This approach with a resolution of 2.4 and 2.3 kg/tree apples (Table 5) is a further advancement to the ANN model of Rozman et al. [6], in which the numbers of fruits at different times (the time between June drop to near harvest time was divided into several periods) are input parameters resulting in a RMSE of 2.6 kg/tree for cv. "Braeburn" and 2.8 kg/tree for "Golden Delicious" both in September, possibly too short before harvest of these late ripening varieties to organize labour and bins.

In previous research, Zhou et al. [5] showed that R^2 values in the calibration data set between apple yields estimated by image processing and actual harvested yield were 0.57 for young cv. "Gala" fruit after June drop, which improved to $R^2 = 0.70$ in the fruit ripening period. By comparison, the presented combined approach (Table 5) improved the coefficient of determination (R^2) for young, small, and light-green cv. "Gala" fruit to 0.81 and for ripening fruit to 0.83. This is also an advancement of the results of Rozman et al. [6] with a correlation (r) between the forecast and actual yield of $r = 0.83$ for "Golden Delicious" and 0.78 for "Braeburn", with standard deviations (SD) of 2.83 and 2.55 kg. In our study, R^2 was 0.81 and SD was 2.28 kg for "Gala" (Table 5).

Other work concentrated on optimising recognition of green apple fruit cv. "Golden Delicious", but without yield prediction [8,18], while Cheng et al. [19] showed how yield estimation strictly depends on the crop load of the apple tree.

Aggelopoulou et al. [20] estimated the yield of apple trees based of flower counts, a method which is only suitable for med-climates such as Greece. However, the majority of apple growing countries in Northwestern Europe such as Germany, Belgium, Holland, England, and Poland encounter late frost, which can dramatically reduce the yield, accentuated by an unpredictable June drop with the same effect, making proper yield predictions before June drop impossible.

The results shown in Figure 7 validated the practicability of the models. The RMSE values between estimated apple yield and actual harvested yield was 2.6 kg/tree for the early time, and improved at near harvest time to 2.5 kg/tree. The BPNN model trained with the samples of 2009 and 2010 and with the tree images in July 2011 can be used to predict the yields of apple trees from 2011.

Further research will show where and why we under- or over-estimate the fruit yields per tree. The tree shape employed here, slender spindle, should allow similarly good results with similar tree shapes such as tall, spindle, super spindle, fruit wall, and Solaxe tree training systems.

5. Conclusions

The novelty of the approach is the combination of fruit features with (four) tree canopy features (number of fruit F_N, single fruit size F_A, area of fruit clusters F_{CA}, and foliage leaf area (L_A)) to develop two back propagation neural network (BPNN) models for early yield prediction, i.e., for young, small, green fruitlets and mature red fruits. Apple was used as a model fruit or crop and the algorithms were developed for image acquisition under natural light conditions in the orchard. The results showed that BPNN can be used for apple yield prediction and that those four selected canopy features are suitable for early yield prediction and present an elegant way for predicting fruit yield using machine vision and machine learning for apple and possibly other fruit crops.

6. Outlook

The present work was obviously conducted in our own orchard with apple trees as slender spindles, as typical for this growing region, to develop these algorithms. Hence our next project will focus to separate a site-specific model from a general model and investigate the adaptation of the proposed model to other tree forms or similar fruit such as nectarine, peach, or kaki.

Acknowledgments: We are grateful to the Chinese-German Center for Scientific Promotion (Chinesisch-Deutsches Zentrum für Wissenschaftsförderung) under the project of the Sino-German Research Group (GZ888), the German Academic Exchange Service (DAAD), and the China Scholarship Council (CSC) for supporting our bi-lateral cooperation, and MDPI for the invitation to this special edition of the J Imaging.

Author Contributions: As part of her PhD study, Hong Cheng carried out the experiments partly at the University of Bonn and partly in China under the supervision of Yurui Sun at China Agricultural University Beijing. Lutz Damerow provided the technical setup, while the image acquisition and most of the writing, publication, and horticultural input was contributed by the corresponding author Michael Blanke.

Conflicts of Interest: The authors declare no conflict of interest.

Abbreviations

The following abbreviations are used in this manuscript:

ANN   Artificial neural network
BPNN   back propagation neural network

References

1. Wulfsohn, D.; Zamora, F.A.; Téllez, C.P.; Lagos, I.Z.; Marta, G.F. Multilevel systematic sampling to estimate total fruit number for yield forecasts. *Precis. Agric.* **2012**, *13*, 256–275. [CrossRef]
2. Stajnko, D.; Rakun, J.; Blanke, M. Modelling apple fruit yield using image analysis for fruit colour, shape and texture. *Europ. J. Hortic. Sci.* **2009**, *74*, 260–267.
3. Wachs, J.P.; Sturm, H.J.; Burks, F.; Akhanais, V. Low and high-level visual feature-based apple detection from multi-modal images. *Precis. Agric.* **2010**, *11*, 717–735. [CrossRef]
4. Blanke, M.M. Prediction of apple yields in Europe—Present and new approaches in research. In Proceedings of the 106th Annual Meeting of the Washington State Horticultural Association (WSHA), Yakima, WA, USA, 8–10 December 2011; Smith, L., Ed.; WSHA Publishing: Yakima, WA, USA, 2011; pp. 68–75.
5. Zhou, R.; Damerow, L.; Sun, Y.; Blanke, M. Using colour features of cv. 'Gala' apple fruits in an orchard in image processing to predict yield. *Precis. Agric.* **2012**, *13*, 568–580. [CrossRef]
6. Rozman, C.; Cvelbar, U.; Tojnko, S.; Stajnko, D.; Karmen, P.; Pavlovie, M.; Vracko, M. Application of Neural Networks and Image Visualization for Early Predicted of Apple Yield. *Erwerbs-Obstbau* **2012**, *54*, 69–76.
7. Stajnko, D.; Rozmana, Č.; Pavloviča, M.; Beber, M.; Zadravec, P. Modeling of 'Gala' apple fruits diameter for improving the accuracy of early yield prediction. *Sci. Hortic.* **2013**, *160*, 306–312. [CrossRef]
8. Kelman, E.; Linker, R. Vision-based localisation of mature apples in trees images using convexity. *Biosyst. Eng.* **2014**, *118*, 174–185. [CrossRef]
9. Ye, X.; Sakai, K.; Garciano, L.O.; Asada, S.I.; Sasao, A. Estimation of citrus yield from airborne hyperspectral images using a neural network model. *Ecol. Model.* **2006**, *198*, 426–432. [CrossRef]
10. Linker, R.; Cohen, O.; Naor, A. Determination of the number of green apples in RGB images recorded in orchards. *Comput. Electron. Agric.* **2012**, *81*, 45–57. [CrossRef]
11. O'Neal, M.R.; Engel, B.A.; Ess, D.R.; Frankenberger, J.R. Neural network prediction of maize yield using alternative data coding algorithms. *Biosyst. Eng.* **2002**, *83*, 31–45. [CrossRef]
12. Sezgin, M.; Sankur, B. Survey over image thresholding techniques and quantitative performance evaluation. *J. Electron. Imaging* **2004**, *13*, 146–165.
13. Rumelhart, D.E.; Hinton, G.E.; Williams, R.J. Learning Representations by Back-Propagating Errors. *Nature* **1986**, *323*, 533–536. [CrossRef]
14. Esmaeili, A.; Tarazkar, M.H. Prediction of shrimp growth using an artificial neural network and regression models. *Aquac. Int.* **2010**, *19*, 705–713. [CrossRef]
15. Greene, W.H. *Econometric Analysis*, 4th ed.; Prentice-Hall: Englewood Cliffs, NJ, USA, 2000; p. 1007.
16. Annamalai, P.; Lee, W.S.; Burks, T. Color Vision Systems for Estimating Citrus Yield in Real-Time. In Proceedings of the ASAE/CSAE Annual International Meeting, Ottawa, ON, Canada, 1–4 August 2014.
17. Lee, W.S.; Chinchuluun, R.; Ehsani, R. Citrus fruit identification using machine vision for a canopy shake and catch harvester. *Acta Hortic.* **2009**, *824*, 217–222.
18. Zaman, Q.U.; Schumann, A.W.; Percival, D.C.; Gordon, R.J. Estimation of wild blueberry fruit yield using digital color photography. *Trans. ASABE* **2008**, *51*, 1539–1544. [CrossRef]

19. Cheng, H.; Damerow, L.; Sun, Y.; Blanke, M.M. Detection of apple fruit in an orchard for early yield prediction as dependent on crop load. *Acta Hortic.* **2016**, *1137*, 59–66. [CrossRef]
20. Aggelopoulou, A.D.; Bochtis, D.; Fountas, S.; Swain, K.C.; Gemtos, T.A.; Nanos, G.D. Yield prediction in apple orchards based on image processing. *Precis. Agric.* **2011**, *12*, 448–456. [CrossRef]

Journal of
Imaging

MDPI

Article

Non-Parametric Retrieval of Aboveground Biomass in Siberian Boreal Forests with ALOS PALSAR Interferometric Coherence and Backscatter Intensity

Martyna A. Stelmaszczuk-Górska [1,*], Pedro Rodriguez-Veiga [2,3], Nicolas Ackermann [4], Christian Thiel [1], Heiko Balzter [2,3] and Christiane Schmullius [1]

[1] Department of Earth Observation, Friedrich-Schiller-University Jena, Loebdergraben 32, Jena D-07743, Germany; christian.thiel@uni-jena.de (C.T.); c.schmullius@uni-jena.de (C.S.)
[2] Centre for Landscape and Climate Research, University of Leicester, Bennett Building, University Road, Leicester LE1 7RH, UK; prv4@leicester.ac.uk (P.R.-V.); hb91@leicester.ac.uk (H.B.)
[3] National Centre for Earth Observation, University of Leicester, Bennett Building, University Road, Leicester LE1 7RH, UK
[4] Gamaya AG, Bâtiment C, EPFL Innovation Park, Lausanne 1015, Switzerland; nicolas.ackermann@gamaya.com
* Correspondence: m.stelmas@uni-jena.de; Tel.: +49-364-1948-978; Fax: +49-364-1948-882

Academic Editors: Francisco Rovira-Más and Gonzalo Pajares Martinsanz
Received: 30 October 2015; Accepted: 15 December 2015; Published: 25 December 2015

Abstract: The main objective of this paper is to investigate the effectiveness of two recently popular non-parametric models for aboveground biomass (AGB) retrieval from Synthetic Aperture Radar (SAR) L-band backscatter intensity and coherence images. An area in Siberian boreal forests was selected for this study. The results demonstrated that relatively high estimation accuracy can be obtained at a spatial resolution of 50 m using the MaxEnt and the Random Forests machine learning algorithms. Overall, the AGB estimation errors were similar for both tested models (approximately 35 t·ha^{-1}). The retrieval accuracy slightly increased, by approximately 1%, when the filtered backscatter intensity was used. Random Forests underestimated the AGB values, whereas MaxEnt overestimated the AGB values.

Keywords: SAR; MaxEnt; random forests; estimation error; forest; biomass; carbon

1. Introduction

Aboveground biomass (AGB) is an important variable in carbon accounting and climate science. In particular, forest AGB is relevant because forests constitute approximately 70%–90% of the Earth's aboveground biomass [1].

The AGB is defined as the mass of living organic matter growing above ground level per unit area at a particular time. The difference in AGB over time allows for measurement of carbon sequestration (excluding root growth) and carbon emission from deforestation, forest degradation, and forest fires. The estimation of AGB in the boreal forest is of special concern as it constitutes the largest biome in the world and has substantial carbon accumulation capability. Russia, as the country with the largest forested area in the world (809 million ha [2]), provided more than 90% of the carbon sink of the world's boreal forests between the years 2000 and 2007 [3]. Despite this importance, Russia's boreal forest has the highest uncertainty in carbon stock calculations [4,5]. This is mostly due to poor measurements of biomass stocks, forest degradation, deforestation, and forest growth. Additionally, due to the lack of financial support, some forested regions in Siberia have not been inventoried for more than 20 years [6]. Therefore, there is a strong need for earth observation-based methods to reduce costs and improve biomass estimations.

The most common method of measuring AGB is estimation from field measurements, such as stem diameter and tree height, using allometric models. However, due to the sampling nature of the field measurements and their high acquisition costs, they can only be collected over small areas. Satellite technology together with reliable *in situ* measurements allows for accurate and relatively cost-efficient wall-to-wall AGB estimates.

There are different remote sensing techniques for AGB retrieval. Several publications provide a comprehensive review of use of remote sensing techniques for biomass estimation including sourcebooks of recommended methods and data sources [7–16]. The estimates using optical sensors are feasible at low biomass levels using vegetation indices, bidirectional reflectance distribution function (BRDF), and texture [17–22]. The latest results of biomass estimation using Landsat data showed that an accuracy of ±36% can be measured in the boreal zones [23]. The most accurate estimates are gathered from airborne light detection and ranging (LiDAR) systems. The only archive data from the satellite profiling LiDAR for measuring and monitoring vegetation are from the Ice, Cloud, and land Elevation (ICESat) Geoscience Laser Altimeter System (GLAS). However, the data have some limitations related to the large footprint, sparse coverage, and sensitivity to terrain variability [24–28]. Another technique that has a potential for forest AGB estimation is synthetic aperture radar (SAR). Similar to LiDAR sensors, radar systems are sensitive to the geometrical properties of observed objects; on the contrary, SAR platforms are imaging sensors. SAR is an active system that transmits microwave energy at wavelengths ranging from 3.1 cm (the X band) to 23.6 cm (the L band). Longer wavelengths, *i.e.*, L-band, are preferable due to its deeper penetration into the forest canopy and due to saturation of radar signal at higher biomass levels [29–31]. Especially, SAR data with long wavelength, horizontal and vertical polarizations, interferometric capabilities, and global acquisition strategy are of great value for biomass retrieval. Data from those sensors are already available as archive data from L-band Japanese Earth Resources Satellite 1 (JERS-1) and Advanced Land Observing Satellite (ALOS) Phased Array L-band Synthetic Aperture Radar (PALSAR), or as new acquisitions from ALOS-2 PALSAR-2. Moreover, new SAR missions are planned that will ensure data continuity, e.g., the European Space Agency's P-band (68.9 cm) BIOMASS mission [32] and the UK NovaSAR S-band mission [33].

SAR data have been investigated for more than 30 years and were proved to be physically related to forest parameters until the saturation point showing correlation coefficients of up to approximately 0.9 [29,34–36]. Many SAR properties have been successfully exploited to quantify biomass: backscattering intensity [37,38], interferometric phase [39,40], interferometric correlation [41–43], polarimetric signature [44,45], SAR tomography, and radargrammetry [46–48]. In particular, satellite SAR data at the L-band have provided low AGB or growing stock volume (GSV) estimation error down to approximately 20% when forest stands greater than 10 ha were used and/or a multi-temporal approach was implemented [31,49–52]. Until now, only few retrieval results using interferometric coherence synergistically with backscatter over Russian boreal forests have been published [50,53,54].

Table 1 [50–52,55–60] presents a summary of the SAR retrieval statistics reported in the literature over Russian boreal forests, the same type of forest that was used in this study. The estimation error is represented as the root mean square error (RMSE) and the relative RMSE, *i.e.*, the RMSE divided by the mean GSV or AGB.

Table 1. Summary of previous studies of growing stock volume (GSV) or aboveground biomass (AGB) estimation over Siberian boreal forests using SAR data. The estimation error is given as the root mean square error (RMSE) and the relative RMSE, *i.e.*, the RMSE divided by the mean GSV or AGB.

Reference	SAR Sensor Wavelength	Estimated Variable	Model	Predictors	Spatial Resolution	Estimation Error RMSE (Units As Estimated Variable)/ Relative RMSE (%)
[55]	JERS-1 L-band	GSV ($m^3 \cdot ha^{-1}$)	Semi-empirical Water-Cloud type model	Backscatter in σ°HH, 9 images	Stand-wise > 8 ha	57–87/33–51
[50]	ERS-1/2 C-band	GSV ($m^3 \cdot ha^{-1}$)	Semi-empirical model Interforemetric Water Cloud Model	Interferometric coherence adjusted by environmental conditions (relative stocking, stand size, topography)	Stand-wise > 3 ha	45–75/25–41
[56]	Envisat ASAR C-band	GSV ($m^3 \cdot ha^{-1}$)	BIOMASAR—semi-empirical Water-Cloud type model	Backscatter in γ°HH and γ°VV polarization, tens of images	1 km	-/34.2
[57]	Envisat ASAR C-band	GSV ($m^3 \cdot ha^{-1}$)	BIOMASAR—semi-empirical Water-Cloud type model	Backscatter in γ°HH and γ°VV, mean 93 observations	0.5°	-/15
[51]	ALOS PALSAR L-band	AGB ($t \cdot ha^{-1}$)	Single and multivariate regression, semi-empirical Water-Cloud type model	Backscatter in σ°HH and σ°VV, 1-5 images	Stand-wise > 10 ha	46–55/25–32
[58,59]	ALOS PALSAR L-band	GSV ($m^3 \cdot ha^{-1}$)	Machine learning approach—Random Forest	4 ALOS PALSAR mosaics, backscatter in γ°HH and γ°HV	25 m	54.4/39.4
[52]	ALOS PALSAR L-band	GSV ($m^3 \cdot ha^{-1}$)	Empirical exponential model	Polarimetric coherence, HHVV-coherence	Stand-wise > 2 ha	33–51/-
[60]	ALOS PALSAR L-band	AGB ($t \cdot ha^{-1}$)	Machine learning approach—MaxEnt	ALOS PALSAR mosaic, backscatter in γ°HH and γ°HV, Landsat bands, and categorical data	50 m	36.4/-

The study presented in this paper is based on multi-temporal ALOS PALSAR L-band backscatter intensity and coherence data. Both types of data were used as explanatory variables for AGB retrieval at a local scale of 0.25 ha. The AGB was estimated using two non-parametric machine learning algorithms: maximum entropy (MaxEnt) and Random Forests. Both models are popular in applied research. The MaxEnt approach [61] is particularly popular in species distribution modeling [62,63]. However, the method was also successfully implemented for AGB estimation at regional and global scale [60,64]. The Random Forests [65] are widely used for classification in ecology [66–68] as well as for AGB estimation [27,58,69–74]. The Random Forests were found to be superior to other methods such as support vector machine (SVM), k-nearest neighbour (KNN), Gaussian processes (GP), and stepwise linear models [75]. So far Random Forests have not been compared with the MaxEnt approach.

In summary, the aim of this paper is to:

1. use SAR L-band backscatter and coherence data synergistically to improve AGB estimation at a local scale;
2. compare AGB retrieval results using two recently popular machine learning algorithms.

2. Test Site and Available Data

2.1. Study Site

The study site is located in Krasnoyarskiy Kray in the Southern part of Central Siberia, Russia, approximately 120 km northeast of the city Krasnoyarsk—part of the Bolshe Murtinsky forest enterprise (center coordinates: 57°12′N and 93°49′E, Figure 1). The area is characterized by a continental climate with long, severe winters and short, warm, and wet summers. From mid-October until the beginning of April, the mean temperature is approximately −15 °C; in summer the mean temperature is approximately +15 °C. The annual precipitation is below 450 millimeters.

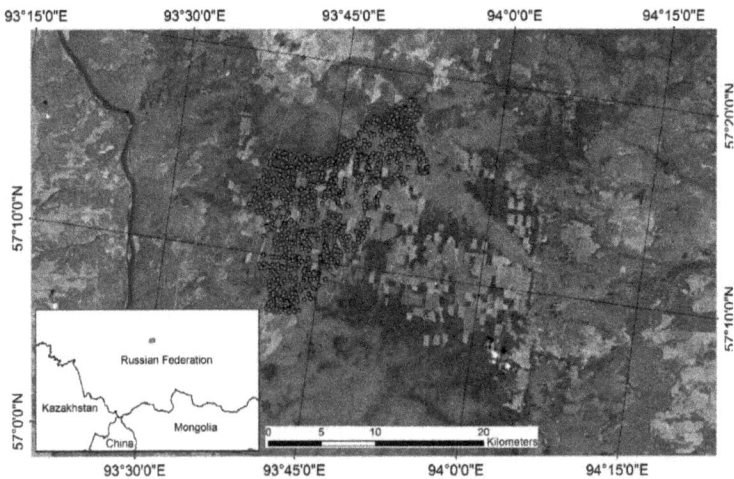

Figure 1. Extent of the study area and the spatial distribution of the reference points. Background image acquired by the Landsat 5 TM satellite (data available from the U.S. Geological Survey Earth Explorer).

The research area covers almost 2000 km². The area is mainly characterized by needleleaf, coniferous forests. The dominant trees are pine, spruce, fir, and larch. The main disturbances are logging activities and fire events. The study site is characterized by gentle topography with heights from 90 m to 572 m above sea level (a.s.l.) with an average height of 243 m a.s.l. The slopes range from 0° to 54° (riverside), with a mean value of 6°.

2.2. Available Data

The ALOS PALSAR L-band data were used for the explanatory variables. The data were delivered in Single Look Complex (SLC) Level 1.1 format. The data were provided by the Japan Aerospace Exploration Agency (JAXA) within the third phase of the Kyoto and Carbon Initiative [76,77]. In total, 19 scenes were available for this study area (Table 2). The weather data were downloaded from http://www.sibessc.uni-jena.de/ [78].

Ten scenes were acquired between 2006 and 2011 in fine beam single (FBS) mode and nine scenes were obtained in fine beam dual (FBD) mode. In the case of the FBS mode, the data were collected in horizontally transmitted and horizontally received polarization (HH), and in the case of the FBD mode, the data were given in HH and horizontally transmitted and vertically received (HV) polarizations.

The spatial resolution of the single-look image is 9.37 m in range and 3.14 m in azimuth for data acquired in the FBD mode and 4.68 m in range and 3.14 m in azimuth for data acquired in the FBS mode. The acquisition angle is $34.3°$.

The forest inventory data were used as the dependent variable. The field data were provided by the Russian State Forest Inventory within the SIBERIA project [79]. The inventory dates back to 1998. Because of the time difference between inventory data and the SAR data acquisitions (>10 years) the field data were updated using semi-empirical phytomass models [80] and growth (yield) tables [81]. The latter were developed by the International Institute for Applied Systems Analyses (IIASA) in collaboration with the V.N. Sukachev Institute of Forest, Siberian Branch, Russian Academy of Sciences, and Moscow State Forest University. Those models and tables are recommended for use in forestry and forest management in Russia (Protocol of the Council of Federal Agency of Forest Management No. 2, dated 8 June 2006) [81].

The original data were gathered in the framework of the Russian forest inventory and planning (FIP) and are available in GIS vector format. Attributes such as GSV, age, tree height, diameter at breast height, and species composition were provided. All information is given for a forest stand, *i.e.* a group of trees occupying a specific area uniform in species composition, size, age, and management strategy. In total, information about 1604 stands was available.

Table 2. Summary of SAR data available for the test site. The data acquisition time was approximately 16 GMT—10 PM local time in summer or 11 PM local time in winter. The weather parameters are given as a mean from eight daily measurements [78]. The weather station is located approximately 50 km southwest from the center of the test site.

Image Name	Acquisition Date	Acquisition Mode	Weather Conditions: Mean Temperature (°C)/Wind Speed (m/s)/Precipitation (mm)/Snow Depth (mm)
ALPSRP049391140	28 December 2006	FBS	dry frozen conditions, −17.8/0.5/0/238.8
ALPSRP056101140	12 February 2007	FBS	dry frozen conditions, −18.9/1.7/0/340.4
ALPSRP082941140	15 August 2007	FBD	wet unfrozen conditions, 13.6/1.2/0/0 (2 days before heavy rain)
ALPSRP089651140	30 September 2007	FBD	wet unfrozen conditions, 13.1/3.9/-/0 (3 days before heavy rain)
ALPSRP103071140	31 December 2007	FBS	dry frozen conditions, −5.9/4.4/0/279.4
ALPSRP109781140	15 February 2008	FBD	dry frozen conditions, −17.2/0.4/0/381
ALPSRP129911140	2 July 2008	FBD	wet unfrozen conditions, 19.7/1.7/0.3/0
ALPSRP136621140	17 August 2008	FBD	wet unfrozen conditions, 16.8/1.6/0.8/0
ALPSRP156751140	2 January 2009	FBS	dry frozen conditions, −12.7/1.1/0/299.7
ALPSRP163461140	17 February 2009	FBS	dry frozen conditions, -31.3/0.6/0/459.7
ALPSRP190301140	20 August 2009	FBD	wet unfrozen conditions,14.7/0.8/1/0 (6 days before heavy rain)
ALPSRP197011140	5 October 2009	FBD	dry unfrozen conditions,11.5/1.6/0/0
ALPSRP210431140	5 January 2010	FBS	dry frozen conditions, −33.7/1.5/0/589.3
ALPSRP217141140	20 February 2010	FBS	dry frozen conditions, −22.9/1.2/0/599.4
ALPSRP243981140	23 August 2010	FBD	dry unfrozen conditions, 22.6/2.3/0/0
ALPSRP250691140	8 October 2010	FBD	wet unfrozen conditions, 1.4/1.7/0.5/10.2
ALPSRP257401140	23 November 2010	FBD	dry frozen conditions, −12.9/1.7/0/119.4
ALPSRP264111140	8 January 2011	FBS	dry frozen conditions, −23.3/1.5/0/360.7
ALPSRP270821140	23 February 2011	FBS	dry frozen conditions, −27.6/0.5/0/429.3

3. Processing Methods

3.1. Forest Inventory Data

The available forest inventory data are for 1998. Therefore, the first step of AGB retrieval was an update of the reference data. The data were improved to the year 2010. The data update consisted of four stages. First, the forest stands that changed from forest to non-forest were excluded by visual interpretation using very high and high resolution optical data. Cloud-free images were selected from KOMPSAT-2 and RapidEye. The data were acquired from spring to autumn in 2010, 2011, and 2012. Then, from the resulting stands only those were selected in which at least 60% of the trees belong to a single species. The reason for the selection is that the growth (yield) tables were done for dominant tree species. In the second step, the stands were used in the semi-empirical phytomass models. The basis for the improvement of the old inventory data is a site index (SI). SI is defined as the edaphic and climatic characteristics of a site that have an impact on the growth and yield of a given tree species [82]. Usually SI classes are determined by the relationship between the mean tree height and the mean age of a stand. As the SI was not available in the original forest inventory data it was calculated from the following equation [83]:

$$H_t = 22.47 \left(1 - e^{-0.0234A(1-e^{-0.0057A})}\right)^{0.548} + (3 - SI)\,\Delta \tag{1}$$

$$\Delta = 4.35 \left(1 - e^{-0.0205A}\right)^{0.957} \tag{2}$$

where H_t represents tree height, A denotes forest stand age, and Δ is the interval between site indexes.

In the case of Russian forests, the SI from Ib, Ia, I–V, Va, and Vb are denoted as where the class with the lowest SI indicates the best site conditions for forest trees to grow. The site indexes II and III are dominant in the study area with fir, birch, and aspen as the dominant species (Figure 2).

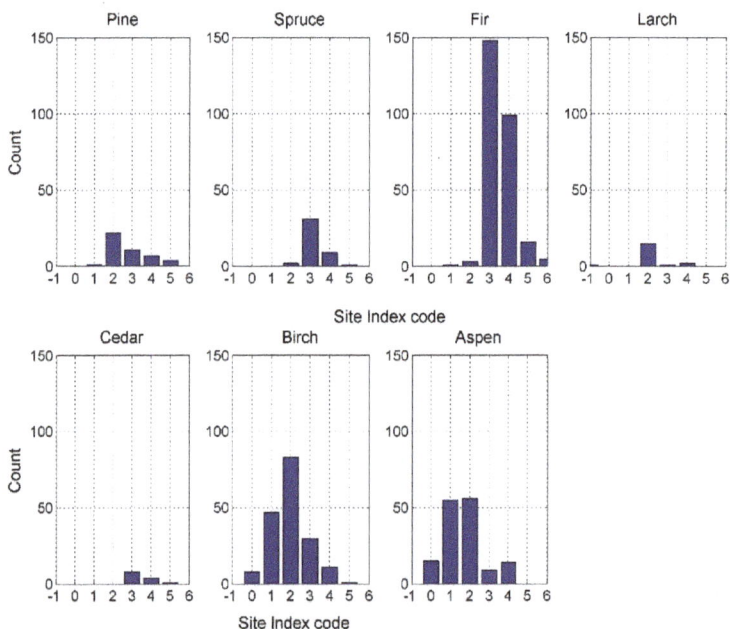

Figure 2. The number of stands with more than 60% of tree species per site index; site index code: −1-Ib, 0–Ia, 1–I, 2-II, 3-III, 4-IV, 5-V, 6-Va.

After calculating the SI, the growth rate was derived by implementing the Richards–Chapman growth function in polynomial quadratic form [81]:

$$GSV_{growth\ rate} = c_1 c_2 c_3 \left(1 - e^{-c_2 A}\right)^{c_3 - 1} e^{-c_2 A}, \tag{3}$$

where A denotes age and c_i are the parameters that have an ecological interpretation and depend on site index (SI) [81]:

$$c_i = c_{i3} SI^2 + c_{i2} SI + c_{i1}. \tag{4}$$

After obtaining the GSV growth rate, the correction coefficient GSV_{cc} was calculated:

$$GSV_{cc} = \frac{GSV_{in-situ}}{GSV_{model}} \tag{5}$$

where $GSV_{in-situ}$ represents the GSV measured in the field, whereas GSV_{model} is the GSV calculated according to the models [81] for a particular site index and forest stand age. A new GSV was then derived:

$$GSV_{new} = GSV_{in-situ} + A_{diff} GSV_{growth\ rate} GSV_{cc} \tag{6}$$

where A_{diff} is the age difference between the old inventory and reference year 2010 and equals 12.

Thirdly, GSV were converted to AGB. Based on freely available *in situ* measurements of forest live biomass (phytomass) [83], a regional allometric model relating GSV to AGB was developed (Figure 3).

The final step of the reference data processing was to rasterize the inventory data to 50 m spatial resolution, and then to erode the stands. An erosion of two pixels was used (100 m) to avoid border effects in the SAR products. Then, the stands were converted into points using the center of gravity (centroid) of the stand. The final updated AGB values ranged from 0 to 224 t·ha^{-1}, with a mean value of 98 t·ha^{-1}. The stand size with erosion varied from 0.5 to 130 ha, with a mean value of 16 ha that corresponds to 64 pixels. The stand age varied from 20 to 290 years. Six hundred forty-one stands remained for the investigations.

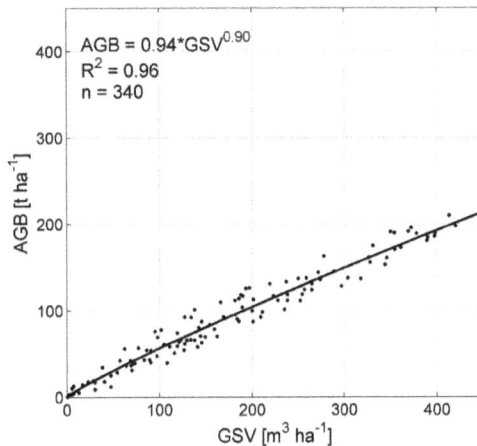

Figure 3. Allometric model relating AGB to GSV using *in situ* measurements of forest phytomass.

3.2. SAR Data Processing and Data Selection

The first pre-processing steps of SAR data included data calibration and multi-looking. In order to obtain data with squared pixels of approximately 50 m spatial resolution, the following multi-looking factors (range x azimuth) were used: FBS mode 6 × 15, FBD mode 3 × 15. Thereafter, the SAR backscattering coefficient was calculated as $\gamma°$, which includes a correction of the backscatter for the local incidence angle θ_i [84]:

$$\gamma^0 = \sigma^0 \frac{A_{flat}}{A_{slope}} \left(\frac{\cos(\theta)}{\cos(\theta_i)} \right)^n$$ (7)

where σ^0 is the backscattering coefficient and θ the incidence angle measured at mid-swathe (34.3°). A_{flat} and A_{slope} represent the local and the true pixel area, respectively. The cosine of the local incidence angle θ_i corrects the radiometry of backscatter for local slopes and converts the data from σ^0 to γ^0. This correction is known as a topographic normalization based on the local incidence angle and pixel area. In total eight backscatter images acquired in FBD mode were generated. The data selection was based on the recommendations from previous studies, in which it was reported that the backscatter data acquired during summer and autumn were superior to the winter acquisitions with regards to the AGB/GSV retrieval [31,85,86].

To reduce the speckle effect in the backscatter data, a filtering approach was implemented. The method was based on a multi-temporal speckle filtering calculated according to [87]:

$$J_k(x,y) = \frac{\langle I_k \rangle}{N} \sum_{i=1}^{N} \frac{I_i(x,y)}{I_i}$$ (8)

where I is the local mean value of pixels in a window with a center at (x, y) in image I, $k = 1,..,N$ and N represents the number of multi-temporal images, with intensity at position (x, y) in the k image—$I_i(x,y)$. J_k filters uncorrelated speckle between the images. In this study, N was equal to 8 for filtering of intensity images and window size was 5×5. No larger filtering window size was used because already large multi-looking factors were implemented.

The coherence images were calculated according to [88]:

$$\gamma = |\gamma| e^{j\phi} = \frac{E\{g_1 g^*_2\}}{\sqrt{E\{|g_1|^2\}E\{|g_2|^2\}}}$$ (9)

where ϕ is the phase and E{} represents expected value.

Coherence measures the degree of correlation between two SAR images and takes values between 0 for total decorrelation and 1 for perfect correlation. Data processing consisted of co-registration at sub-pixel level (less than 0.05 pixel), common range and azimuth band filtering, and interferogram calculation and flattening. Coherence was estimated by spatial averaging within a two-dimensional window. In this study an adaptive estimation window was used with a window size between 3×3 and 5×5. A larger window size was used in the areas of low coherence in order to reduce the coherence bias. The resulting coherence was computed using the same number of looks as in the case of backscatter data (FBS mode 6×15, FBD mode 3×15 in range, and azimuth). This number of looks significantly reduced the coherence overestimation for low coherence values, as described in [89]. Therefore, the coherence bias was calculated to be close to zero (for coherence equal to 0 the values were ~0.002). The coherence was generated for 19 SLC pairs (Table 3). As a master image the image acquired in 2010 was used, additionally coherence for data acquired in 2011 was calculated. The perpendicular baselines B_n were between 224 and 3829 m and thus shorter than the critical baselines: 14.7 and 7.3 km for the FBS and FBD modes, respectively. Five coherence images were considered for the AGB retrieval. The latter were selected according to the stable weather and environmental conditions during the data acquisition [50] and a simple visual interpretation. To avoid coherence variability due to topography, slopes greater than 5° were masked out (mainly riverside areas). The mask was implemented to all SAR products used for AGB estimation. The SAR images were geocoded and normalized using the Shuttle Radar Topography Mission (SRTM) 90 m digital elevation model version 4.1 [90,91]. The SAR pre-processing was performed with the GAMMA Interferometric SAR Processor [92]. The backscatter data were geocoded using the bicubic-log spline resampling method while the coherence data were processed with the bicubic spline interpolation approach. The final spatial resolution of the SAR products was 50 m.

Table 3. Summary of coherence images generated for the test site.

PALSAR Data	Perpendicular Baseline $|B_n|$ (m)	Temporal Baseline B_t (days)
5 January 2010 & 20 February 2010	789	46
23 August 2010 & 8 October 2010	461	46
8 October 2010 & 23 November 2010	224	46
28 December 2006 & 5 January 2010	2390	1104
12 February 2007 & 5 January 2010	1111	1058
12 February 2007 & 20 February 2010	1899	1104
31 December 2007 & 5 January 2010	1086	736
31 December 2007 & 20 February 2010	297	782
15 February 2008 & 5 January 2010	2190	690
15 February 2008 & 20 February 2010	1401	736
2 January 2009 & 5 January 2010	3041	368
2 January 2009 & 20 February 2010	3829	414
17 February 2009 & 5 January 2010	2339	322
17 February 2009 & 20 February 2010	3126	368
5 January 2010 & 8 January 2011	2978	368
20 February 2010 & 8 January 2011	2190	322
5 January 2010 & 23 February 2011	3780	414
20 February 2010 & 23 February 2011	2992	368
8 January 2011 & 23 February 2011	803	46

In addition to the backscatter and coherence data, the normalized ratio of the backscatter in linear scale and coherence, R_n, was calculated:

$$R_n = \frac{\gamma^0/\gamma - min\,(\gamma^0/\gamma)}{max\,(\gamma^0/\gamma) - min\,(\gamma^0/\gamma)}. \tag{10}$$

The rationale behind the ratio calculation is purely statistical. The ratio was introduced to enhance the backscatter relation to AGB and to reduce the number of potential outliers influencing the AGB retrieval error. In this approach, coherence γ values are considered as weighting factors for the backscatter γ^0 (linear scale) and strengthen the response over forested areas. The ratio calculation was performed using a reference coherence image with a temporal baseline of 46 days. The SLC data for the coherence calculation were acquired in winter 2010, namely on 5 January 2010 and 20 February 2010. This coherence was selected due to the highest dynamic range of coherence values resulted from optimal and similar environmental conditions during the data acquisition, as well as the short perpendicular baseline B_n. The resulting ratio values over forested areas are relatively high compared to the values over non-forested and sparsely forested areas. For example, a value of 0.16 of backscatter on a linear scale with a corresponding coherence value of 0.2 (dense forest) is approximately 0.8, whereas a backscatter value of 0.07 with a coherence value of 0.6 (non-forest) is approximately 0.1. In order to adjust the ratio values to a single scale, the values were normalized and ranged from 0 to 1. A normalized ratio between available backscatter data (in linear scale) acquired on 23 August 2010 and 8 October 2010 in both polarizations and the coherence between acquisitions taken in 2010 (5 January and 20 February) were calculated. The resulting range of calculated values is illustrated by box plots (Figure 4, plot A).

On each box the central mark is the median, the edges of the box are the 25th and 75th percentiles, and the whiskers extend to the extreme values. Most of the values varied from 0 to approximately 0.2 show an almost linear relationship up to approximately 60 t·ha^{-1} and a non-linear relationship for higher AGB values (Figure 4B,C). The values close to the upper boundary represented small heterogeneous stands in the forest inventory data (<8 ha).

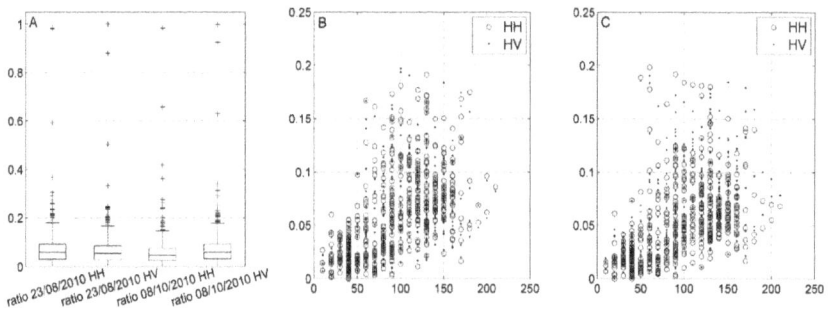

Figure 4. (**A**) presents box plots for four calculated ratios between backscatter images (acquired on 23 August 2010 and 8 October 2010) and coherence acquired in winter 2010 (5 January and 20 February); (**B**) and (**C**) present the ratio as a function of aboveground biomass (AGB).

In total, 13 SAR products were selected for AGB retrieval. The examples of the generated SAR products are presented in Figure 5.

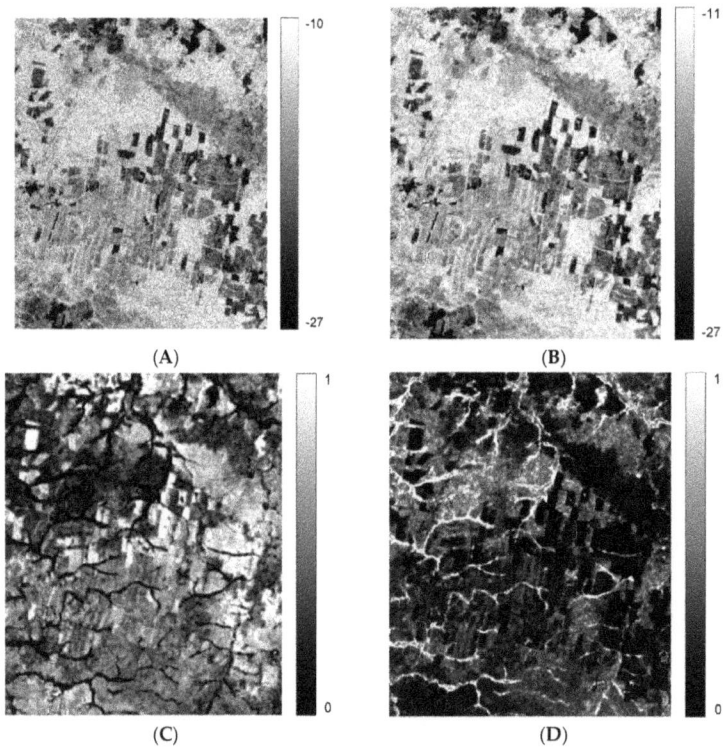

Figure 5. Subset of SAR products used for AGB retrieval. (**A**): backscatter data acquired on 23 August 2010 in HV polarization in dB; (**B**): filtered backscatter image in dB; (**C**): coherence between 5 January 2010 and 20 February 2010; (**D**): normalized ratio between acquired on 23 August 2010 in HV polarization and coherence (5 January & 20 February 2010).

3.3. AGB Retrieval Models

In this study, two non-parametric data fusion machine learning algorithms were considered: maximum entropy (MaxEnt) and Random Forests. In both cases the updated forest inventory was used as the response data. The model's training was done on 90% of the sample size (577 samples), whereas 10% was used for independent validation (64 samples). For the selection of training and validation data, a stratified sampling was implemented. Only 10% of the response data was used for the independent validation as the mentioned algorithms calculate unbiased model error using 25% in the case of MaxEnt and approximately 1/3 in the case of Random Forests of the data randomly excluded from the training data.

The first approach was a model based on the MaxEnt algorithm. The model was run using the MaxEnt program for maximum entropy modeling version 3.3.3 k (under Java Runtime Environment). The MaxEnt is an exponential model that can be compared with the generalized linear (GLM) and generalized additive (GAM) models. The concept of the method is to estimate the probability distribution of maximum entropy constrained by a set of remote sensing variables. Through numerous iterations, the weights of these variables are adjusted to maximize the average sample likelihood (training gain). The weights are then used to estimate the distribution over the whole space for each of the AGB classes. In this study, AGB reference data were divided into seven AGB classes, in $t \cdot ha^{-1}$: 0–40, 40–60, 60–80, 80–100, 100–120, 120–140, and >140. The values were grouped such that each class could be represented equally in terms of occurrences. The classes' representation is illustrated as a histogram (Figure 6).

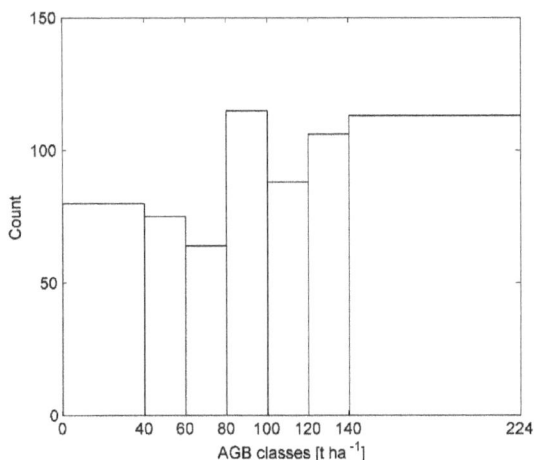

Figure 6. Number of measurements in selected AGB class.

Then, the AGB value was calculated for each pixel using the maximum probability weighted average [64]:

$$\widehat{AGB} = \frac{\sum_{i=1}^{N} P_i^n AGB_i}{\sum_{i=1}^{N} P_i^n} \tag{11}$$

where i refers to the class number, P_i is the MaxEnt probability, AGB_i is the estimated average biomass range, and $\widehat{AGB}$ is the predicted value of AGB for each pixel.

As an input for the MaxEnt algorithm, all available variables were used and two groups of data were distinguished. The first one contained unfiltered data, whereas the second one was with backscatter data filtered according to Equation (8).

In the MaxEnt program the resampling of the data for each replication is done by bootstrapping, whereas the jackknife tests are used for calculation of variable contribution. The jackknife tests are generated using regularized gain and AUC. The jackknife test is used in two cases: withhold one predictor and refit model and withhold all predictors but one and refit the model. To determine the variable percent contribution, in each iteration of the training algorithm the increase in regularized gain was added to the contribution of the corresponding variable, or subtracted from it if the resulted value was negative. At the end the values were converted to percentages.

The second approach was based on a supervised Random Forests® regression approach, available as the randomForest package in the R software [93]. It is an ensemble learning method that operates by constructing a large number of trees by randomly selecting the predictors and then calculating a mean prediction from all individual regression trees. Each tree is constructed using a different randomly permuted sample from the input dataset. One-third of the data are left out of the bootstrap sample and not used in the construction of the tree. This sample is called out-off-bag (OOB) and is used to obtain an unbiased estimate for the retrieval error, OOB error.

The randomForest package also provides measures for evaluating the importance of the different predictors in the model development. In this study, we used the measure that is computed from permuting OOB data and permuting each predictor variable for each tree. The difference between the OOB MSE and predictor MSE are then averaged over all trees, and normalized by the standard deviation of the differences.

In order to evaluate the accuracy and performance of the implemented models, the following quantitative measures were considered for the independent validation:

a. root mean square error (RMSE) is defined as:

$$RMSE = \sqrt{\frac{\sum_{i=1}^{n}\left(AGB_{ref(i)} - \widehat{AGB}_i\right)^2}{n}} \tag{12}$$

where $AGB_{ref(i)}$ represents AGB reference value for stand i, $\widehat{AGB}_i$ predicted AGB and n a number of AGB observations.

b. corrected root-mean-square error ($rRMSE_{cor}$) is defined as:

$$rRMSE_{cor} = \sqrt{RMSE_{Sat}^2 - RMSE_{Ref}^2} \tag{13}$$

where $RMSE_{Sat}$ represents the root mean square error in a satellite-derived estimation of AGB and $RMSE_{Ref}$ is the root mean square error in the forest inventory data. According to the manual on forest inventory and planning in Russian forests, the maximum error of GSV is expected to be 15% [94]. This value was also considered in the AGB estimation because GSV is the main component of AGB. According to [95], GSV constitutes in our test site 73% of AGB.

Additionally, relative RMSEs were calculated by dividing the RMSEs by the mean AGB and multiplying by 100%.

c. bias is defined as the mean of estimation error:

$$bias = \frac{\sum_{i=1}^{n}\left(\widehat{AGB}_i - AGB_{ref(i)}\right)}{n} \tag{14}$$

where positive values of bias expresses overestimation, and negative values underestimation.

The models' predictive performance was evaluated using pseudo R-squared (1-MSE/variance) [93] in the case of the Random Forests and an area under receiver operating characteristic curves (AUC) [96] in the case of the MaxEnt. The parameters were generated using the models' testing samples.

4. Retrieval Results

4.1. MaxEnt Performance

As an output of MaxEnt seven continuous probability distribution maps with pixel values from 0 to 1 were obtained, where 1 is a high predicted probability of being classified to a specified AGB class. The probability values were then used to calculate the final AGB map according to Equation (11).

To assess the MaxEnt algorithm performance, AUC was calculated by bootstrapping 25% of the training data. The AUC was computed for each AGB class (Figure 7).

The AUC values are higher than 0.7, which shows that the model performed well. Only in the case of filtered data or the 60–80 t·ha^{-1} AGB class was the AUC was lower than 0.6. The highest predictive power of the model was observed for the high and the low AGB ranges. The mean AUC for all AGB classes was calculated to be 0.76 and 0.77 for unfiltered and filtered datasets, respectively.

To assess the importance of the predictors, the variable percent of contribution was calculated. A high value indicates that the model depends more on that variable (Figure 8A,B).

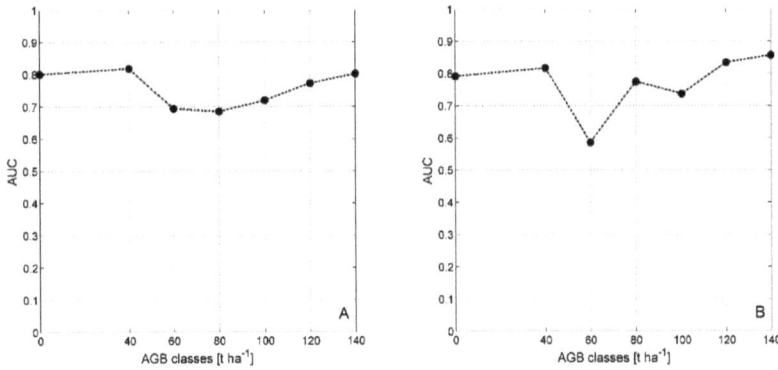

Figure 7. Evaluation of MaxEnt model performance using Area Under the Receiver Operator Curve (AUC) for testing dataset (**A**) with unfiltered and (**B**) with filtered backscatter data; aboveground biomass (AGB) classes 0: 0–40, 40: 40–60, 60: 60–80, 80: 80–100, 100: 100–120, 120: 120–140, 140: >140 t·ha^{-1}.

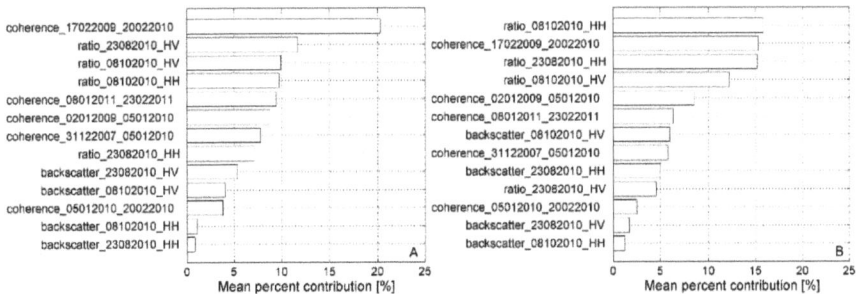

Figure 8. Predictor importance presented as percent contribution (**A**) dataset with unfiltered and (**B**) with filtered backscatter data.

In the case of the dataset with the unfiltered backscatter, the variable that mostly decreased model performance when it was omitted was coherence between data acquired on 17 February 2009 and 20 February 2010. For the filtered dataset, it was the ratio of backscatter in HH polarization acquired on 8

October 2010. Therefore, the coherence and normalized ratio appear to contain the most information that is not present in the other variables. The backscatter acquired on 23 August 2010 in HH and on 8 October 2010 in HH polarization seem to be of less importance for AGB prediction.

The variable importance was also analyzed per AGB class. To better illustrate how the variable contribution changes among the seven AGB classes, the data were distinguished between coherence, backscatter, and normalized ratio (Figure 9).

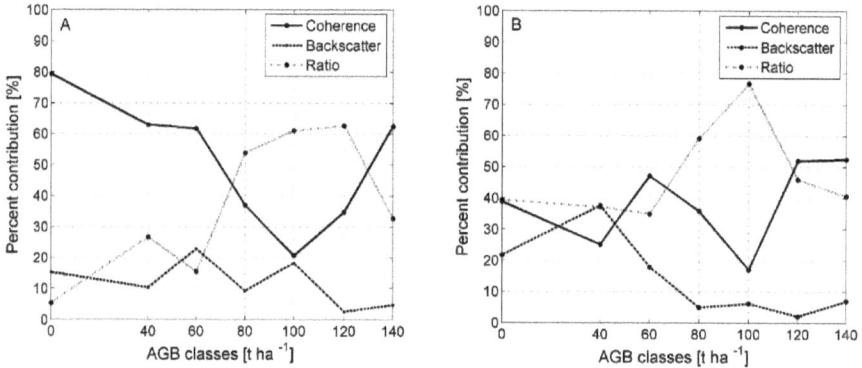

Figure 9. Coherence, backscatter, and normalized ratio contributions *versus* AGB classes. (**A**) Dataset with unfiltered or (**B**) filtered backscatter data. Class 0 represents AGB values from 0 to 40, 40: 40–60, 60: 60–80, 80: 80–100, 100: 100–120, 120: 120–140, 140: >140 t· ha^{-1}.

Figure 9 confirms with a mean percent contribution value of 51.2% that in the case of unfiltered data the most important variables were coherence data. The data were the most important in four AGB classes: 0–40, 40–60, 60–80, and >140 t· ha^{-1}. The mean percent contribution for the ratio products was 36.8% and for the backscatter products was 12%. Therefore, the backscatter seems to provide the least information for MaxEnt. When the filtered data were used, the most important group of data was the ratio layers, with a mean percent contribution of 47.7%. The data were the most important in two classes, *i.e.*, AGB classes 80–100 and 100–120 t· ha^{-1}. The second most important type of data was the coherence products, with a mean percent contribution of 38.3%. The data contributed most to the AGB retrieval for three AGB classes 60–80, 120–140, and >140 t· ha^{-1}. The backscatter data provided the least information for the retrieval. The mean percent contribution was 13.9%. In both plots the rise of the ratio importance is related to the decrease of the influence of the backscatter and coherence.

4.2. Random Forests Performance

The higher percent of variance explained was calculated for the dataset with filtered backscatter, the value of the pseudo R-squared was 38.4%. In the case of the unfiltered dataset the pseudo R-squared was 36.4%. The values were calculated based on the testing sample (OOB sample). Overall, the Random Forests predictor importance ranking (Figure 10) revealed only small differences between the datasets with unfiltered and filtered backscatter data. The ranking showed that of the 13 predictors the normalized ratio between the backscatter in HV polarization acquired on 8 October 2010 and coherence (5 January & 20 February 2010) was the most important for the retrieval of AGB with values of 26% and 28.7% for the unfiltered and filtered datasets, respectively. The mean value of the increase in MSE in the case of all ratio products was 22.2% and 24.9%.

The second most important data group in the case of the both datasets was backscatter products. The mean value of the increase in MSE was 17.4% and 17.9%. Random Forests suggested that coherence products had the smallest influence on AGB retrieval with a mean value of 16.4% and 15.4% for unfiltered and filtered datasets, respectively.

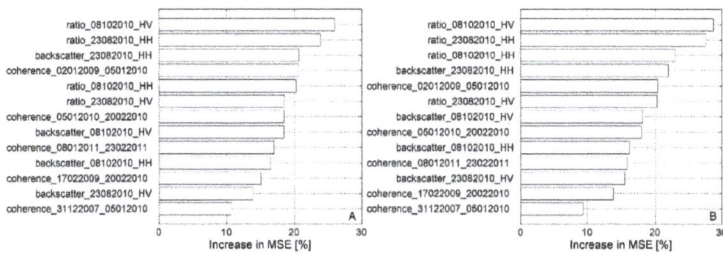

Figure 10. Predictor importance presented as increase in MSE for (**A**) unfiltered dataset and (**B**) dataset with filtered backscatter data.

4.3. AGB Mapping Results

Figures 11–14 show SAR-derived AGB maps with a spatial resolution of 50 m. Each map presents AGB retrieval results expressed in t· ha^{-1}. The first two maps show AGB values derived by the MaxEnt algorithm; the other two are by the Random Forests. Examining the results with a simple visual interpretation, it can be noticed that the maps differ taking into account the spatial variability of AGB values. In the case of the AGB maps generated by MaxEnt, the values seem to be more heterogonous both in high and low biomass ranges.

The range and the mean value of the retrieved AGB for each map are given in Table 4.

Table 4. Range of derived AGB.

Method	MaxEnt Unfiltered Dataset	Random Forests Unfiltered Dataset	Maxent Filtered Dataset	Random Forests Filtered Dataset
Range t· ha^{-1}	0–140	0–160	0–150	0–170
Mean AGB t ha^{-1}	87	95	89	97

In the case of the MaxEnt algorithm, the AGB ranged from 0 to 140 (150 t· ha^{-1}), whereas the values computed by the Random Forests were higher by 10 t· ha^{-1}. The mean values derived by MaxEnt were lower than 90 t· ha^{-1} and in the case of Random Forests greater than 95 t· ha^{-1}.

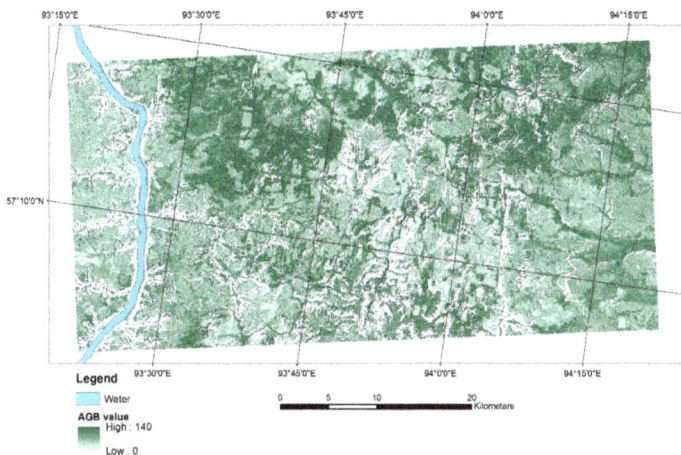

Figure 11. AGB map generated using the MaxEnt algorithm and a dataset with unfiltered backscatter data.

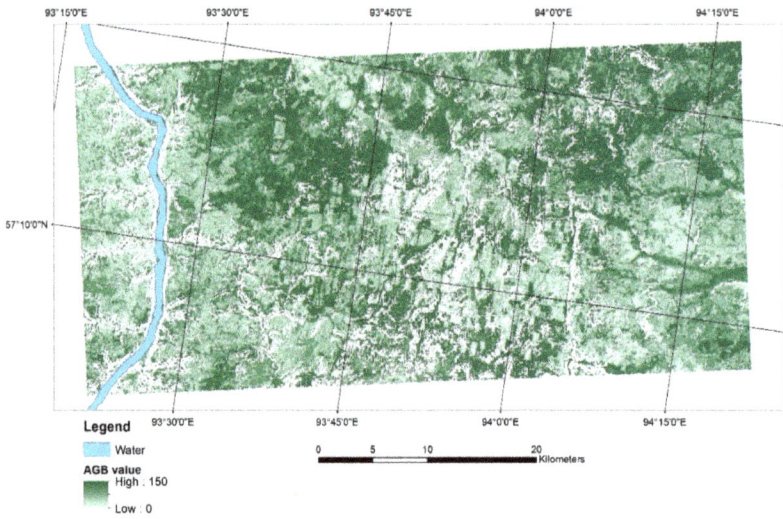

Figure 12. AGB map generated using the MaxEnt algorithm and a dataset with filtered backscatter data.

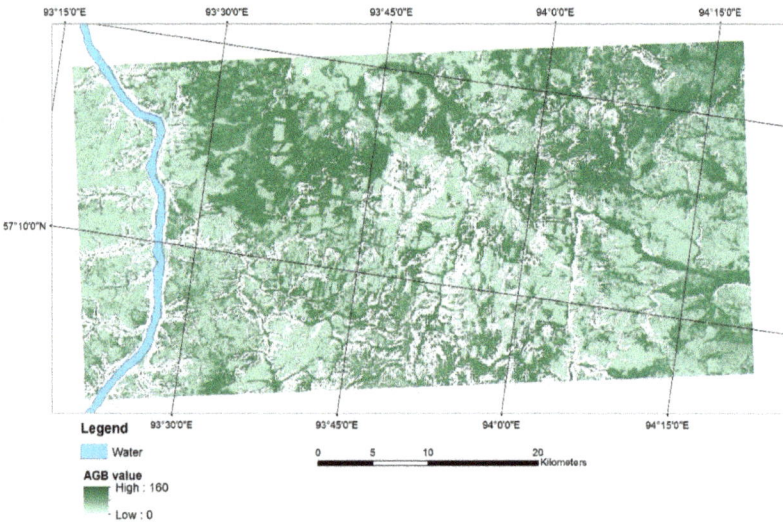

Figure 13. AGB map generated using the Random Forests algorithm and a dataset with unfiltered backscatter data.

To better observe the differences in spatial distribution, the difference maps between updated forest inventory (*in situ* data) and SAR-derived AGB were calculated (Figure 15). Green represents overestimation, whereas red is underestimation. In yellow are the AGB values estimated correctly.

In general, there are almost no differences between maps generated using unfiltered and filtered datasets. In the case of MaxEnt, the retrieved AGB values are displayed in green, which means overestimation. The Random Forests tends to underestimate. The AGB values are displayed in red and orange. In both cases, the overestimation can be seen on the borders of the stands. The underestimation is observed for stands with high AGB values.

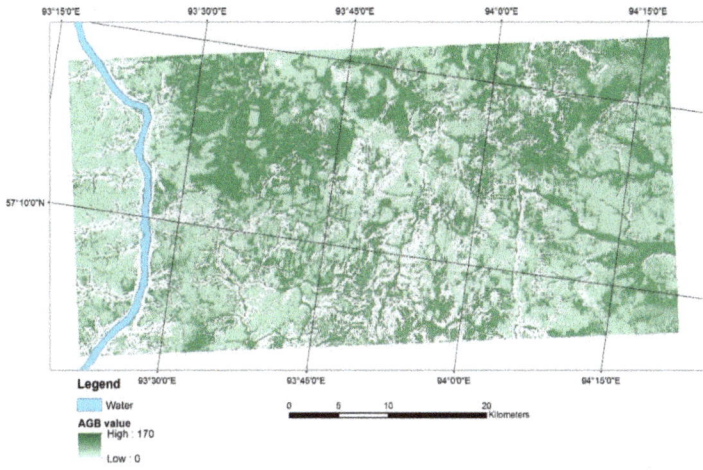

Figure 14. AGB map generated using the Random Forests algorithm and a dataset with filtered backscatter data.

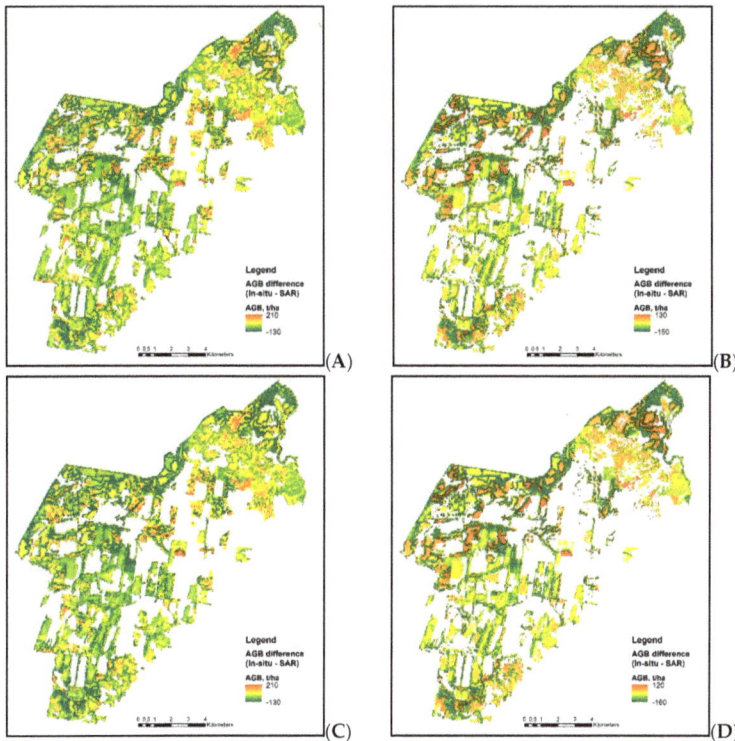

Figure 15. AGB difference maps between updated forest inventory (*in situ*) and SAR-derived AGB. (**A**) MaxEnt model—unfiltered data; (**B**) Random Forests—unfiltered data; (**C**) MaxEnt model with filtered backscatter; (**D**) Random Forests with filtered backscatter.

4.4. Validation

Table 5 summarizes the accuracies of the MaxEnt and the Random Forests AGB retrieval when using backscatter data (filtered and unfiltered), coherence images, and normalized ratio products. The corrected RMSE and relative corrected RMSE were calculated for a training sample and an independent sample.

Table 5. Validation of AGB retrieval results. RMSE values are given for training and validation samples (training/validation).

Model	Dataset	$RMSE_{cor}$ (t· ha^{-1})	$rRMSE_{cor}$ (%)	Bias (t· ha^{-1})
MaxEnt	Unfiltered	33.3/36.4	34.3/39.5	12.0/5.2
	Filtered	28.7/35.8	29.6/38.8	6.9/4.3
Random Forests	Unfiltered	21.6/35.4	22.3/38.4	3.2/−4.4
	Filtered	21.3/35.0	22.0/36.9	0.9/−4.5

In the case of MaxEnt, the corrected RMSE was 36.4 t· ha^{-1} (33.3 t· ha^{-1} for the training sample) and 28.7 t· ha^{-1} (35.8 t· ha^{-1} for the training sample) for unfiltered and filtered datasets, respectively. Overall, the estimation error of 39.5% (34.3% for training sample) was calculated for an unfiltered dataset and 38.8% (29.6% for the training sample) for a dataset with filtered backscatter data. The bias of 5.2 t· ha^{-1} (12 t· ha^{-1} for the training sample) was calculated in the case of unfiltered data and of 4.3 t· ha^{-1} (6.9 t· ha^{-1} for the training sample) in the case of the dataset with filtered backscatter data.

The Random Forests results show similar estimation error. The corrected RMSE was 35.4 t· ha^{-1} (21.6 t· ha^{-1} for training sample) and 35.0 t· ha^{-1} (21.3 t· ha^{-1} for training sample) for unfiltered and filtered datasets, respectively. The relative corrected RMSEs of 38.4% (22.3% for the training sample) in the case of unfiltered datasets and of 36.9% (22.0% for the training sample) were calculated. The bias of −4.4 t· ha^{-1} (3.2 t· ha^{-1} for the training sample) and of −4.5 t· ha^{-1} (0.9 t· ha^{-1} for the training sample) was calculated in the case of unfiltered and filtered data, respectively.

It should be noted that a difference of more than 10 t· ha^{-1} in the case of the Random Forests between corrected RMSE calculated using training and validation datasets could result from the small size of validation sample.

5. Discussion

In general, the MaxEnt machine learning algorithm as well as the Random Forests regression approach provided good AGB retrieval results compared to previous studies [51,52,55,58,60]. The reported analyses also used ALOS PALSAR L-band data for AGB/GSV retrieval over Siberian boreal forests. The researchers reported RMSEs between 33 and 87 m^3· ha^{-1} and 55 t· ha^{-1} at the coarser scales, and 36.4 t· ha^{-1} at 0.25 ha scale. Those results are similar to or worse than those derived in this study. The results presented in this paper showed that a relatively high estimation accuracy (down to 30%) can be obtained at a local scale. The AGB estimation showed only slightly better results when a dataset with filtered backscatter intensity was used.

When the model performance is taken into account, the MaxEnt performed better than the Random Forests. The area under the receiver operator curve (AUC) was higher than 0.7, except the AGB range from 60 to 80 t· ha^{-1}, whereas Random Forests reached an R^2 of 38.4%. MaxEnt generated AGB maps with dominantly overestimated AGB values, whereas Random Forests provided slightly underestimated values. The range of derived AGB values was underestimated but similar and differs by 10 t· ha^{-1} between the applied models. The mean AGB provided by Random Forests was comparable to the reference AGB mean. MaxEnt showed an underestimation of approximately 10 t· ha^{-1}. The estimation bias was lower in the case of Random Forests.

MaxEnt and Random Forests provided measures for evaluating the importance of the predictors used in the model construction. The algorithms showed no agreement between derived variable rankings. In the case of MaxEnt, the coherence together with the ratio products were the most important for model construction, whereas ratio and backscatter products provided the most information in the case of modeling with Random Forests. In both cases, the ratio products seem to provide important information for AGB retrieval. Within the data groups, in the case of the MaxEnt the most important variables were the coherence between images acquired on 17 February 2009 and 20 February 2010, the normalized ratio between the backscatter from 23 August 2010 in HV and coherence, the backscatter from 23 August 2010 in HV polarization for unfiltered dataset. For filtered dataset, the most important variables within the data groups were the normalized ratio between the backscatter from 8 October 2010 in HH polarization and coherence, the coherence between images acquired on 17 February 2009 and 20 February 2010, and the backscatter image acquired on 23 August 2010 in HV polarization. In the case of Random Forests, the most important variables were the normalized ratio between the backscatter acquired on 8 October 2010 in HV polarization and coherence, the backscatter acquired on 23 August 2010 in HH polarization, and the coherence between images from 2 January 2009 and 5 January 2010 for both unfiltered and filtered datasets. The coherence images generated with the temporal baseline (B_t) greater than 46 days were superior to those derived with shorter B_t.

6. Conclusions

In this study we have demonstrated the feasibility of synergistic usage of backscatter and coherence for aboveground biomass (AGB) retrieval for boreal forests in Siberia at a local scale of 0.25 ha. This research was focused on the further exploitation of SAR data. The ALOS PALSAR L-band backscatter was combined with coherence, introducing the backscatter–coherence normalized ratio. The latter was developed based on the statistical data analysis. In total 13 variables were used for the AGB estimation. For the AGB retrieval two popular machine learning algorithms were implemented. The MaxEnt and the Random Forests performed well, showing promising AGB estimations and demonstrating the model's robustness. The corrected RMSEs were between 35.8 and 36.4 t·ha^{-1} and between 35.8 and 35.0 t·ha^{-1} for MaxEnt and Random Forests, respectively. The estimation error slightly decreased, by approximately 1%, when the filtered backscatter intensity was used. In this study, the retrieval of AGB using the SAR products was demonstrated only for Siberian unmanaged forests. It is expected that the estimation error over well-managed forests could be further reduced. Estimation improvement is also foreseen at the stand level due to the reduction of the spatial variability in the SAR data. Another issue that could have an influence on the retrieval accuracy is the reference data. Using optical remote sensing data and recommended yield tables and semi-empirical phytomass models in Russian forestry and forest management, the authors updated the old inventory data. Unfortunately, it was not possible to fully validate the obtained values in the field, hence the error in the reference data was only partially known.

The models provided different variable importance rankings. However, in both cases the normalized ratio products seem to contain important information for the model development. The coherence data were the most important in the low and high AGB ranges, whereas the ratio was most important for middle to high AGB ranges. Thus, a strategy of using different datasets for estimation of low, medium, and high AGB values could further increase biomass retrieval accuracy. It was observed that the backscatter data increased their contribution in the model construction after filtering. In terms of the retrieved AGB values, the Random Forests algorithm provided AGB mean estimation almost the same as the reference value. MaxEnt provided slightly overestimated AGB values, whereas Random Forests tended to underestimate the AGB values.

The MaxEnt and Random Forests machine learning algorithms demonstrated their potential use for forestry applications, especially for estimations in remote areas. Often no information about AGB is available in those regions. The models could be used to provide AGB estimations with relatively low estimation errors. Moreover, the results generated for different time spans could easily be applied for AGB change monitoring, which is very important from the carbon account calculations perspective.

Acknowledgments: Acknowledgments: This work was undertaken within the framework of the GMES Initial Operations—Network for Earth Observation Research Training GIONET project, grant agreement PITN-GA-2010-264509. This work was partly undertaken within the framework of the JAXA Kyoto & Carbon Initiative. H. Balzter was supported by the Royal Society Wolfson Research Merit Award, 2011/R3 and the NERC National Centre for Earth Observation. Pedro Rodriguez-Veiga was also supported by the NERC National Centre for Earth Observation. KOMPSAT-2 data were provided by the European Space Agency (project id: 13300).

The authors would like to acknowledge the support of Dmitry Schepaschenko and Anatoly Shvidenko from the International Institute for Applied Systems Analysis in updating the reference data. The authors would like to acknowledge the comments and suggestions of the reviewers.

Author Contributions: Author Contributions: Martyna A. Stelmaszczuk-Górska processed and analyzed in-situ and SAR data, modeled with the Random Forests approach, wrote the manuscript, and coordinated manuscript revisions. Pedro Rodriguez-Veiga assisted with the allometry and modelled using the MaxEnt algorithm. Nicolas Ackermann provided input for the backscatter–coherence ratio. Christian Thiel, Heiko Balzter, and Christiane Schmullius provided feedback for the study design and contributed to the manuscript.

Conflicts of Interest: Conflicts of Interest: The authors declare no conflict of interest.

References

1. Houghton, R.A.; Hall, F.; Goetz, S.J. Importance of biomass in the global carbon cycle. *J. Geophys. Res.* **2009**, *114*, 1–13. [CrossRef]
2. FAO. *Global Forest Resources. Assesment 2010. Main report*; FAO: Rome, Italy, 2010.
3. FAO. *State of the World's Forests*; FAO: Rome, Italy, 2012.
4. Pan, Y.; Birdsey, R.A.; Fang, J.; Houghton, R.; Kauppi, P.E.; Kurz, W.A.; Phillips, O.L.; Shvidenko, A.; Lewis, S.L.; Canadell, J.G.; *et al.* A large and persistent carbon sink in the world's forests. *Science* **2011**, *333*, 988–993. [CrossRef] [PubMed]
5. Nilsson, S.; Shvidenko, A.; Jonas, M.; McCallum, I.; Thomson, A.; Balzter, H. Uncertainties of a regional terrestrial biota full carbon account: A systems analysis. *Water Air Soil Pollut. Focus* **2007**, *7*, 425–441. [CrossRef]
6. Shvidenko, A.Z.; Schepaschenko, D.G.; Vaganov, E.A.; Sukhinin, A.I.; Maksyutov, S.S.; McCallum, I.; Lakyda, I.P. Impact of wildfire in Russia between 1998–2010 on ecosystems and the global carbon budget. *Dokl. Earth Sci.* **2011**, *441*, 1678–1682. [CrossRef]
7. Kasischke, E.S.; Melack, J.M.; Dobson, M.C. The use of imaging radars for ecological applications—A review. *Remote Sens. Environ.* **1997**, *59*, 141–156. [CrossRef]
8. Rosenqvist, Å.; Milne, A.; Lucas, R.; Imhoff, M.; Dobson, C. A review of remote sensing technology in support of the Kyoto Protocol. *Environ. Sci. Policy* **2003**, *6*, 441–455. [CrossRef]
9. Patenaude, G.; Milne, R.; Dawson, T.P. Synthesis of remote sensing approaches for forest carbon estimation: Reporting to the Kyoto Protocol. *Environ. Sci. Policy* **2005**, *8*, 161–178. [CrossRef]
10. McCallum, I.; Wagner, W.; Schmullius, C.; Shvidenko, A.; Obersteiner, M.; Fritz, S.; Nilsson, S. Satellite-based terrestrial production efficiency modeling. *Carbon Balance Manag.* **2009**, *4*. [CrossRef] [PubMed]
11. Goetz, S.; Baccini, A.; Laporte, N.; Johns, T.; Walker, W.; Kellndorfer, J.; Houghton, R.; Sun, M. Mapping and monitoring carbon stocks with satellite observations: A comparison of methods. *Carbon Balance Manag.* **2009**, *4*, 1–7. [CrossRef] [PubMed]
12. Frolking, S.; Palace, M.W.; Clark, D.B.; Chambers, J.Q.; Shugart, H.H.; Hurtt, G.C. Forest disturbance and recovery: A general review in the context of spaceborne remote sensing of impacts on aboveground biomass and canopy structure. *J. Geophys. Res.* **2009**, *114*. [CrossRef]
13. Thiel, C.; Santoro, M.; Cartus, O.; Thiel, C.; Riedel, T.; Schmullius, C. Perspectives of SAR based forest cover, forest cover change and biomass mapping. In *The Kyoto Protocol: Economic Assessments, Implementation Mechanisms, and Policy implications*; Vasser, C.P., Ed.; Nova Science Publishers, Inc.: New York, 2009; pp. 13–56.
14. Goetz, S.; Dubayah, R. Advances in remote sensing technology and implications for measuring and monitoring forest carbon stocks and change. *Carbon Manag.* **2011**, *2*, 231–244. [CrossRef]
15. Zolkos, S.G.; Goetz, S.J.; Dubayah, R. A meta-analysis of terrestrial aboveground biomass estimation using lidar remote sensing. *Remote Sens. Environ.* **2013**, *128*, 289–298. [CrossRef]
16. GOFC-GOLD. *A Sourcebook of Methods and Procedures for Monitoring and Reporting Anthropogenic Greenhouse Gas Emissions and Removals Associated with Deforestation, Gains and Losses of Carbon Stocks in Forests Remaining Forests, and Forestation. GOFC-GOLD Report Versio*; GOFC-GOLD: Wageningen, The Netherlands, 2014.

17. Sader, S.A.; Waide, R.B.; Lawrence, W.T.; Joyce, A.T. Tropical Forest Biomass and Successional Age Class Relationships to a Vegetation Index Derived from Landsat TM Data. *Remote Sens. Environ.* **1989**, *28*, 143–156. [CrossRef]

18. Wasseige, C.; de Defourny, P. Retrieval of tropical forest structure characteristics from bi-directional reflectance of SPOT images. *Remote Sens. Environ.* **2002**, *83*, 362–375. [CrossRef]

19. Dong, J.; Kaufmann, R.K.; Myneni, R.B.; Tucker, C.J.; Kauppi, P.E.; Liski, J.; Buermann, W.; Alexeyev, V.; Hughes, M.K. Remote sensing estimates of boreal and temperate forest woody biomass: Carbon pools, sources, and sinks. *Remote Sens. Environ.* **2003**, *84*, 393–410. [CrossRef]

20. Joshi, C.; Leeuw, J.; de Skidmore, A.K.; Duren, I.C.; van Oosten, H. Remotely sensed estimation of forest canopy density: A comparison of the performance of four methods. *Int. J. Appl. Earth Obs. Geoinf.* **2006**, *8*, 84–95. [CrossRef]

21. Avitabile, V.; Herold, M.; Henry, M.; Schmullius, C. Mapping biomass with remote sensing: a comparison of methods for the case study of Uganda. *Carbon Balance Manag.* **2011**, *6*. [CrossRef] [PubMed]

22. Houghton, R.A. Balancing the Global Carbon Budget. *Annu. Rev. Earth Planet. Sci.* **2007**, *35*, 313–347. [CrossRef]

23. Ji, L.; Wylie, B.K.; Nossov, D.R.; Peterson, B.; Waldrop, M.P.; McFarland, J.W.; Rover, J.; Hollingsworth, T.N. Estimating aboveground biomass in interior Alaska with Landsat data and field measurements. *Int. J. Appl. Earth Obs. Geoinf.* **2012**, *18*, 451–461. [CrossRef]

24. Lefsky, M.A.; Harding, D.J.; Keller, M.; Cohen, W.B.; Carabajal, C.C.; Del Bom Espirito-Santo, F.; Hunter, M.O.; de Oliveira, R. Estimates of forest canopy height and aboveground biomass using ICESat. *Geophys. Res. Lett.* **2005**, *32*, 22–25. [CrossRef]

25. Nelson, R.; Ranson, K.J.; Sun, G.; Kimes, D.S.; Kharuk, V.; Montesano, P. Estimating Siberian timber volume using MODIS and ICESat/GLAS. *Remote Sens. Environ.* **2009**, *113*, 691–701. [CrossRef]

26. Duncanson, L.I.; Niemann, K.O.; Wulder, M.A. Estimating forest canopy height and terrain relief from GLAS waveform metrics. *Remote Sens. Environ.* **2010**, *114*, 138–154. [CrossRef]

27. Simard, M.; Pinto, N.; Fisher, J.B.; Baccini, A. Mapping forest canopy height globally with spaceborne lidar. *J. Geophys. Res.* **2011**, *116*. [CrossRef]

28. Khalefa, E.; Smit, I.P.J.; Nickless, A.; Archibald, S.; Comber, A.; Balzter, H. Retrieval of savanna vegetation canopy height from ICESat-GLAS spaceborne LiDAR with terrain correction. *IEEE Geosci. Remote Sens. Lett.* **2013**, *10*, 1439–1443. [CrossRef]

29. Le Toan, T.; Beaudoin, A.; Riom, J.; Guyon, D. Relating forest biomass to SAR data. *IEEE Trans. Geosci. Remote Sens.* **1992**, *30*, 403–411. [CrossRef]

30. Kurvonen, L.; Pulliainen, J.; Hallikainen, M. Retrieval of biomass in boreal forests from multitemporal ERS-1 and JERS-1 SAR images. *IEEE Trans. Geosci. Remote Sens.* **1999**, *37*, 198–205. [CrossRef]

31. Rignot, E.; Way, J.; Williams, C.; Viereck, L. Radar estimates of aboveground biomass in boreal forests of interior Alaska. *IEEE Trans. Geosci. Remote Sens.* **1994**, *32*, 1117–1124. [CrossRef]

32. Le Toan, T.; Quegan, S.; Davidson, M.W.J.; Balzter, H.; Paillou, P.; Papathanassiou, K.; Plummer, S.; Rocca, F.; Saatchi, S.; Shugart, H.; *et al.* The BIOMASS mission: Mapping global forest biomass to better understand the terrestrial carbon cycle. *Remote Sens. Environ.* **2011**, *115*, 2850–2860. [CrossRef]

33. Natale, A.; Guida, R.; Bird, R.; Whittaker, P.; Cohen, M.; Hall, D. Demonstration and analysis of the applications of S-band SAR. In Proceedings of the APSAR (The Asia-Pacific Conference on Synthetic Aperture Radar), Seoul, Korea, 26–30 September 2011.

34. Beaudoin, A.; Le Toan, T.; Goze, S.; Nezry, E.; Lopes, A.; Mougin, E.; Hsu, C.C.; Han, H.C.; Kong, J.A.; Shin, R.T. Retrieval of forest biomass from SAR data. *Int. J. Remote Sens.* **1994**, *15*, 2777–2796. [CrossRef]

35. Dobson, M.C.; Ulaby, F.T.; Pierce, L.E.; Sharik, T.L.; Bergen, K.M.; Kellndorfer, J.; Kendra, J.R.; Li, E.; Lin, Y.C.; Nashashibi, A.; *et al.* Estimation of forest biophysical characteristics in Northern Michigan with SIR-C/X-SAR. *IEEE Trans. Geosci. Remote Sens.* **1995**, *33*, 877–895. [CrossRef]

36. Ranson, K.J.; Sun, G. Mapping biomass of a northern forest using multifrequency SAR data. *IEEE Trans. Geosci. Remote Sens.* **1994**, *32*, 388–396. [CrossRef]

37. Pulliainen, J.T.; Heiska, K.; Hyyppa, J.; Hallikainen, M.T. Backscattering properties of boreal forests at the C- and X-bands. *IEEE Trans. Geosci. Remote Sens.* **1994**, *32*, 1041–1050. [CrossRef]

38. Fransson, J.E.S.; Israelsson, H. Estimation of stem volume in boreal forests using ERS-1 C- and JERS-1 L-band SAR data. *Int. J. Remote Sens.* **1999**, *20*, 123–137. [CrossRef]

39. Solberg, S.; Astrup, R.; Gobakken, T.; Næsset, E.; Weydahl, D.J. Estimating spruce and pine biomass with interferometric X-band SAR. *Remote Sens. Environ.* **2010**, *114*, 2353–2360. [CrossRef]

40. Askne, J.; Fransson, J.; Santoro, M.; Soja, M.; Ulander, L. Model-based biomass estimation of a hemi-boreal forest from multitemporal TanDEM-X acquisitions. *Remote Sens.* **2013**, *5*, 5574–5597. [CrossRef]

41. Koskinen, J.T.; Pulliainen, J.T.; Hyyppä, J.M.; Engdahl, M.E.; Hallikainen, M.T. The seasonal behavior of interferometric coherence in boreal forest. *IEEE Trans. Geosci. Remote Sens.* **2001**, *39*, 820–829. [CrossRef]

42. Santoro, M.; Askne, J.; Smith, G.; Fransson, J.E.S. Stem volume retrieval in boreal forests from ERS-1/2 interferometry. *Remote Sens. Environ.* **2002**, *81*, 19–35. [CrossRef]

43. Næsset, E.; Bollandsås, O.M.; Gobakken, T.; Solberg, S.; McRoberts, R.E. The effects of field plot size on model-assisted estimation of aboveground biomass change using multitemporal interferometric SAR and airborne laser scanning data. *Remote Sens. Environ.* **2015**, *168*, 252–264. [CrossRef]

44. Papathanassiou, K.P.; Cloude, S.R. Single-baseline polarimetric SAR interferometry. *IEEE Trans. Geosci. Remote Sens.* **2001**, *39*, 2352–2363. [CrossRef]

45. Neumann, M.; Saatchi, S.S.; Ulander, L.M.H.; Fransson, J.E.S. Assessing performance of L- and P-Band polarimetric interferometric SAR data in estimating boreal forest above-ground biomass. *IEEE Trans. Geosci. Remote Sens.* **2012**, *50*, 714–726. [CrossRef]

46. Tebaldini, S.; Rocca, F. Multibaseline polarimetric SAR tomography of a boreal forest at P- and L-bands. *IEEE Trans. Geosci. Remote Sens.* **2012**, *50*, 232–246. [CrossRef]

47. Persson, H.; Fransson, J. Forest variable estimation using radargrammetric processing of TerraSAR-X images in boreal forests. *Remote Sens.* **2014**, *6*, 2084–2107. [CrossRef]

48. Vastaranta, M.; Niemi, M.; Karjalainen, M.; Peuhkurinen, J.; Kankare, V.; Hyyppä, J.; Holopainen, M. Prediction of forest stand attributes using TerraSAR-X stereo imagery. *Remote Sens.* **2014**, *6*, 3227–3246. [CrossRef]

49. Askne, J.; Santoro, M.; Smith, G.; Fransson, J.E. S. Multitemporal repeat-rass SAR interferometry of boreal forests. *IEEE Trans. Geosci. Remote Sens.* **2003**, *41*, 1540–1550. [CrossRef]

50. Santoro, M.; Shvidenko, A.; Mccallum, I.; Askne, J.; Schmullius, C. Properties of ERS-1/2 coherence in the Siberian boreal forest and implications for stem volume retrieval. *Remote Sens. Environ.* **2007**, *106*, 154–172. [CrossRef]

51. Peregon, A.; Yamagata, Y. The use of ALOS/PALSAR backscatter to estimate above-ground forest biomass: A case study in Western Siberia. *Remote Sens. Environ.* **2013**, *137*, 139–146. [CrossRef]

52. Chowdhury, T.A.; Thiel, C.; Schmullius, C. Growing stock volume estimation from L-band ALOS PALSAR polarimetric coherence in Siberian forest. *Remote Sens. Environ.* **2014**, *155*, 129–144. [CrossRef]

53. Balzter, H.; Talmon, E.; Wagner, W.; Gaveau, D.; Plummer, S.; Yu, J.J.; Quegan, S.; Davidson, M.; Le Toan, T.; Gluck, M.; Shvidenko, A.; Nilsson, S.; Tansey, K.; Luckman, A.; Schmullius, C. Accuracy assessment of a large-scale forest cover map of central Siberia from synthetic aperture radar. *Can. J. Remote Sens.* **2002**, *28*, 719–737. [CrossRef]

54. Wagner, W.; Luckman, A.; Vietmeier, J.; Tansey, K.; Balzter, H.; Schmullius, C.; Davidson, M.; Gaveau, D.; Gluck, M.; Le, T.; *et al.* Large-scale mapping of boreal forest in SIBERIA using ERS tandem coherence and JERS backscatter data. *Remote Sens.* **2003**, *85*, 125–144. [CrossRef]

55. Santoro, M.; Eriksson, L.; Askne, J.; Schmullius, C. Assessment of stand-wise stem volume retrieval in boreal forest from JERS-1 L-band SAR backscatter. *Int. J. Remote Sens.* **2006**, *27*, 3425–3454. [CrossRef]

56. Santoro, M.; Beer, C.; Cartus, O.; Schmullius, C.; Shvidenko, A.; Mccallum, I.; Wegmüller, U.; Wiesmann, A. Retrieval of growing stock volume in boreal forest using hyper-temporal series of Envisat ASAR ScanSAR backscatter measurements. *Remote Sens. Environ.* **2011**, *115*, 490–507. [CrossRef]

57. Santoro, M.; Cartus, O.; Fransson, J.; Shvidenko, A.; McCallum, I.; Hall, R.; Beaudoin, A.; Beer, C.; Schmullius, C. Estimates of Forest Growing Stock Volume for Sweden, Central Siberia, and Québec Using Envisat Advanced Synthetic Aperture Radar Backscatter Data. *Remote Sens.* **2013**, *5*, 4503–4532. [CrossRef]

58. Wilhelm, S.; Hüttich, C.; Korets, M.; Schmullius, C. Large area mapping of boreal Growing Stock Volume on an annual and multi-temporal level using PALSAR L-band backscatter mosaics. *Forests* **2014**, *5*, 1999–2015. [CrossRef]

59. Hüttich, C.; Korets, M.; Bartalev, S.; Zharko, V.; Schepaschenko, D.; Shvidenko, A.; Schmullius, C. Exploiting Growing Stock Volume Maps for Large Scale Forest Resource Assessment: Cross-Comparisons of ASAR- and PALSAR-Based GSV Estimates with Forest Inventory in Central Siberia. *Forests* **2014**, *5*, 1753–1776. [CrossRef]

60. Rodriguez-Veiga, P.; Stelmaszczuk-Górska, M.; Hüttich, C.; Schmullius, C.; Tansey, K.; Balzter, H. Aboveground Biomass Mapping in Krasnoyarsk Kray (Central Siberia) using Allometry, Landsat, and ALOS PALSAR. In Proceedings of the RSPSoc Annual Conference; Remote Sensing and Photogrammetry Society, Aberystwyth, Wales, 2–5 September 2014.

61. Phillips, S.J.; Anderson, R.P.; Schapire, R.E. Maximum entropy modeling of species geographic distributions. *Ecol. Model.* **2006**, *190*, 231–259. [CrossRef]

62. Elith, J.; Kearney, M.; Phillips, S. The art of modelling range-shifting species. *Methods Ecol. Evol.* **2010**, *1*, 330–342. [CrossRef]

63. Elith, J.; Phillips, S.J.; Hastie, T.; Dudík, M.; Chee, Y.E.; Yates, C.J. A statistical explanation of MaxEnt for ecologists. *Divers. Distrib.* **2011**, *17*, 43–57. [CrossRef]

64. Saatchi, S.S.; Harris, N.L.; Brown, S.; Lefsky, M.; Mitchard, E.T.A; Salas, W.; Zutta, B.R.; Buermann, W.; Lewis, S.L.; Hagen, S.; *et al.* Benchmark map of forest carbon stocks in tropical regions across three continents. *Proc. Natl. Acad. Sci. USA* **2011**, *108*, 9899–9904. [CrossRef] [PubMed]

65. Breiman, L. Random Forests. *Mach. Learn.* **2001**, *45*, 5–32. [CrossRef]

66. Cutler, D.R.; Edwards, T.C.; Beard, K.H.; Cutler, A.; Hess, K.T.; Gibson, J.; Lawler, J.J. Random forests for classification in ecology. *Ecology* **2007**, *88*, 2783–2792. [CrossRef] [PubMed]

67. Hüttich, C.; Herold, M.; Strohbach, B.J.; Dech, S. Integrating in-situ, Landsat, and MODIS data for mapping in Southern African savannas: Experiences of LCCS-based land-cover mapping in the Kalahari in Namibia. *Environ. Monit. Assess.* **2011**, *176*, 531–547. [CrossRef] [PubMed]

68. Prasad, A.M.; Iverson, L.R.; Liaw, A. Newer classification and regression tree techniques: Bagging and random forests for ecological prediction. *Ecosystems* **2006**, *9*, 181–199. [CrossRef]

69. Walker, W.S.; Kellndorfer, J.M.; Lapoint, E.; Hoppus, M.; Westfall, J. An empirical InSAR-optical fusion approach to mapping vegetation canopy height. *Remote Sens. Environ.* **2007**, *109*, 482–466. [CrossRef]

70. Houghton, R.A.; Butman, D.; Bunn, A.G.; Krankina, O.N.; Schlesinger, P.; Stone, T.A. Mapping Russian forest biomass with data from satellites and forest inventories. *Environ. Res. Lett.* **2007**, *2*. [CrossRef]

71. Avitabile, V.; Baccini, A.; Friedl, M.A.; Schmullius, C. Capabilities and limitations of Landsat and land cover data for aboveground woody biomass estimation of Uganda. *Remote Sens. Environ.* **2012**, *117*, 366–380. [CrossRef]

72. Cartus, O.; Kellndorfer, J.; Rombach, M.; Walker, W. Mapping Canopy Height and Growing Stock Volume Using Airborne Lidar, ALOS PALSAR and Landsat ETM+. *Remote Sens.* **2012**, *4*, 3320–3345. [CrossRef]

73. Cartus, O.; Kellndorfer, J.; Walker, W.; Franco, C.; Bishop, J.; Santos, L.; Fuentes, J. A National, Detailed Map of Forest Aboveground Carbon Stocks in Mexico. *Remote Sens.* **2014**, *6*, 5559–5588. [CrossRef]

74. Baccini, A.; Laporte, N.; Goetz, S.J.; Sun, M.; Dong, H. A first map of tropical Africa's above-ground biomass derived from satellite imagery. *Environ. Res. Lett.* **2008**, *3*. [CrossRef]

75. Fassnacht, F.E.; Hartig, F.; Latifi, H.; Berger, C.; Hernández, J.; Corvalán, P.; Koch, B. Importance of sample size, data type and prediction method for remote sensing-based estimations of aboveground forest biomass. *Remote Sens. Environ.* **2014**, *154*, 102–114. [CrossRef]

76. JAXA Earth Observation Research Center (EORC) ALOS Kyoto & Carbon Initiative. Available online: http://www.eorc.jaxa.jp/ALOS/en/kyoto/kyoto_index.htm (accessed on 13 April 2015).

77. Schmullius, P.C.; (Friedrich-Schiller-University Jena, Germany); Thiel, C.; (Friedrich-Schiller-University Jena, Germany). *Proposal to JAXA for K & C Phase 3 PALSAR Intensities and Coherence for Forest Cover and Forest Cover Change Mapping and Biomass Retrieval*; Unpublished, 2011.

78. Friedrich-Schiller-University Siberian Earth System Science Cluster. Available online: http://www.sibessc.uni-jena.de/ (accessed on 20 June 2012).

79. Schmullius, C.; Baker, J.; Balzter, H.; Davidson, M.; Eriksson, L.; Gaveau, D.; Gluck, M.; Holz, A.; Le Toan, T.; Luckman, A.; *et al.* *SAR Imaging for Boreal Ecology and Radar Interferometry Applications SIBERIA project (Contract No. ENV4-CT97-0743-SIBERIA) – Final Report*; Friedrich-Schiller-University: Jena, Germany, 2001. Available online: http://www.siberia1.uni-jena.de/pdf_files/final_report.pdf (accessed on 11 January 2012).

80. Shvidenko, A.; Schepaschenko, D.; Nilsson, S.; Bouloui, Y. Semi-empirical models for assessing biological productivity of Northern Eurasian forests. *Ecol. Modell.* **2007**, *204*, 163–179. [CrossRef]
81. Shvidenko, A.; Schepaschenko, D.; Nilsson, S.; Boului, Y. *Tables and Models of Growth and Productivity of Forests of Major Forming Species of Northern Eurasia (standard and reference materials)*; Federal Agency of Forest Management: Moscow, Russia, 2008.
82. Van Laar, A.; Akca, A. *Forest Mensuration*; von Gadow, K., Pukkala, T., Tome, M., Eds.; Springer: Dordrecht, The Netherlands, 2007; Volume 13.
83. IIASA Russian Forests & Forestry. Live Biomass & Net Primary Production—Measurements of Forest Phytomass *in situ*. Available online: http://webarchive.iiasa.ac.at/Research/FOR/forest_cdrom/english/for_prod_en.html (accessed on 20 May 2013).
84. Ulander, L.M.H. Radiometrie slope correction of synthetic-aperture radar images. *IEEE Trans. Geosci. Remote Sens.* **1996**, *34*, 1115–1122. [CrossRef]
85. Rignot, E.J.; Zimmermann, R.; Zyl, J.J. Van Spaceborne applications of P-band imaging radars for measuring forest biomass. *IEEE Trans. Geosci. Remote Sens.* **1995**, *33*, 1162–1169. [CrossRef]
86. Pulliainen, J.T.; Mikkela, P.J.; Hallikainen, M.T.; Ikonen, J. Seasonal dynamics of C-band backscatter of boreal forests with applications to biomass and soil moisture estimation. *IEEE Trans. Geosci. Remote Sens.* **1996**, *34*, 758–770. [CrossRef]
87. Quegan, S.; Le Toan, T.; Yu, J.J.; Ribbes, F.; Floury, N. Multitemporal ERS SAR analysis applied to forest mapping. *IEEE Trans. Geosci. Remote Sens.* **2000**, *38*, 741–753. [CrossRef]
88. Born, M.; Wolf, E. *Principles of Optics*, 7th ed.; Cambridge University Press: Cambridge, UK, 1999.
89. Bamler, R.; Hartl, P. Synthetic aperture radar interferometry. *Inverse Probl.* **1998**, *14*, R1–R54. [CrossRef]
90. Jarvis, A.; Reuter, H.I.; Nelson, A.; Guevara, E. Hole-Filled Seamless SRTM Data V4, International Centre for Tropical Agriculture (CIAT). Available online: http://srtm.csi.cgiar.org (accessed on 4 February 2013).
91. Reuter, H.I.; Nelson, A.; Jarvis, A. An evaluation of void filling interpolation methods for SRTM data. *Int. J. Geogr. Inf. Sci.* **2007**, *21*, 983–1008. [CrossRef]
92. Wegmüller, U.; Werner, C.; Strozzi, T. SAR Interferometric and Differential Interferometric Processing Chain. In Proceedings of the IGARSS'98; IEEE Publications: Seattle, WA, USA, 1998; pp. 1106–1108. [CrossRef]
93. Breiman, L.; Cutler, A. Breiman and Cutler's random forests for classification and regression. Available online: https://cran.r-project.org/web/packages/randomForest/randomForest.pdf (accessed on 18 February 2014).
94. Federal Forestry Agency. *Manual on Forest Inventory and Planning in Russian Forest*; Federal Forestry Agency: Moscow, Russia, 1995.
95. Shvidenko, A.Z.; Gustafson, E.; Mcguire, A.D.; Kharuk, V.I.; Schepaschenko, D.G.; Shugart, H.H.; Tchebakova, N.M.; Vygodskaya, N.N.; Onuchin, A.A.; Hayes, D.J.; *et al.* Terrestrial ecosystems and their change. In *Regional Environmental Changes in Siberia and Their Global Consequences*; Groisman, P.Y., Gutman, G., Eds.; Springer Environmental Science and Engineering: Dordrecht, The Netherlands, 2013; pp. 171–249.
96. Fielding, A.H.; Bell, J.F. A review of methods for the assessment of prediction errors in conservation presence/absence models. *Environ. Conserv.* **1997**, *24*, 38–49. [CrossRef]

Journal of
Imaging

MDPI

Article

Imaging for High-Throughput Phenotyping in Energy Sorghum

Jose Batz *, Mario A. Méndez-Dorado and J. Alex Thomasson

Department of Biological & Agricultural Engineering, Texas A&M University, 2117 TAMU, College Station, TX 77843-2117, USA; mario.mendez@tamu.edu (M.A.M.-D.); thomasson@tamu.edu (J.A.T.)
* Correspondence: batz.jose@tamu.edu; Tel.: +1-713-835-8854

Academic Editors: Gonzalo Pajares Martinsanz and Francisco Rovira-Más
Received: 3 November 2015; Accepted: 15 January 2016; Published: 26 January 2016

Abstract: The increasing energy demand in recent years has resulted in a continuous growing interest in renewable energy sources, such as efficient and high-yielding energy crops. Energy sorghum is a crop that has shown great potential in this area, but needs further improvement. Plant phenotyping—measuring physiological characteristics of plants—is a laborious and time-consuming task, but it is essential for crop breeders as they attempt to improve a crop. The development of high-throughput phenotyping (HTP)—the use of autonomous sensing systems to rapidly measure plant characteristics—offers great potential for vastly expanding the number of types of a given crop plant surveyed. HTP can thus enable much more rapid progress in crop improvement through the inclusion of more genetic variability. For energy sorghum, stalk thickness is a critically important phenotype, as the stalk contains most of the biomass. Imaging is an excellent candidate for certain phenotypic measurements, as it can simulate visual observations. The aim of this study was to evaluate image analysis techniques involving K-means clustering and minimum-distance classification for use on red-green-blue (RGB) images of sorghum plants as a means to measure stalk thickness. Additionally, a depth camera integrated with the RGB camera was tested for the accuracy of distance measurements between camera and plant. Eight plants were imaged on six dates through the growing season, and image segmentation, classification and stalk thickness measurement were performed. While accuracy levels with both image analysis techniques needed improvement, both showed promise as tools for HTP in sorghum. The average error for K-means with supervised stalk measurement was 10.7% after removal of known outliers.

Keywords: image analysis; stalk thickness; K-means; minimum distance

1. Introduction

In this age of increasing concerns about population growth, energy demands and environmental risks, the need to expand energy resources in a sustainable manner is intensifying. Non-renewable energy from fossil fuels will be insufficient at a certain point in the future [1,2]. Additionally, carbon dioxide emissions from fossil fuels have led to concerns about rising levels of greenhouse gases [3], increasing the importance of developing alternative energy sources, like dedicated energy crops that reduce net carbon dioxide emissions. One solution to these issues is to increase energy crop production by maximizing crop output per unit area.

Improving crop productivity is possible through improvements in breeding and genetics. Historically, breeders and geneticists relied on manual labor to evaluate crop physical traits (phenotypes) on a relatively small number of plant varieties. The advent of sensing and robotics has provided the potential to drastically accelerate the advances of breeding and genetics through "high-throughput phenotyping" (HTP). This new research field involves using autonomous technologies to rapidly and accurately measure phenotypes on a large number of plants, enabling the selection of promising varieties and genes from a broader genetic pool to make faster strides in crop improvement. With dedicated energy crops, the most important phenotypes are biomass yield, rate of growth and resistance to typical crop stresses.

Certain crops like sorghum (Sorghum bicolor) have been demonstrated to have excellent potential for dedicated bioenergy production. Sorghum is a C4 plant with a high biomass accumulation rate, and it can be economically produced and harvested in four months or less with suitable environmental conditions [3–5]. While some varieties of sorghum are grown mainly for grain (grain sorghum), others are grown mainly for the sugars they produce (sweet sorghum) or for their high levels of biomass (forage sorghum or "energy sorghum") [6].

Plant breeders can potentially accelerate biomass production improvements in energy sorghum by using an HTP approach. Previous studies have pursued automation of phenotype-measurement techniques involving lasers, light curtains, ultrasonic sensors and artificial vision [7–9]. The use of phenotypic data along with genotypes of energy sorghum has been reported in a few recent studies [10–13]. Artificial vision systems are able to extract meaningful information from images, such as qualitative and quantitative morphologic variables [14], like plant height, width and shape, stem diameter and leaf number, area and angle. There is also the potential to use spectral characteristics in images to measure plant health, such as by correlating greenness to chlorophyll content [15–18]. Machine vision is potentially faster, more reliable and more robust than manual phenotyping methods [19]. A needed development in machine vision is the capability of working under difficult imaging conditions, like varying light levels, spectra and angles in crop fields, as well as varying temperature, *etc.* [20].

In a dedicated energy crop like energy sorghum, stalk thickness and plant height are the most significant phenotypes correlated to biomass, because a large majority of the plant dry matter resides in the stalk. Plant height can be measured with ultrasonic sensors, but an automated measuring system for stalk thickness remains a significant challenge because of the possible occlusion of the stalk and the difficulty of differentiating the stalk from connected stems. Numerous imaging techniques for HTP have been developed, and many are available at the Plant Image Analysis website [21]. Two of the available software resources at that website are Root Estimator for Shovelomics Traits (REST; [22]) and Digital Imaging of Root Traits (DIRT; [23]), both of which are dedicated to the analysis of images to measure root morphology, such as width, depth, length, shape and diameter. Neither of these incorporates measurement of distance between camera and plant material. It is clear that currently available software offers length measurement capabilities for plants, particularly root masses, but the authors are not aware of existing software dedicated to plant stalk measurement, particularly when camera distance is included. The first objective of this research was to evaluate common image analysis techniques as potential tools for measuring stalk thickness that could be used in automated HTP. The second objective was to evaluate an active depth camera as a tool for measuring distance between camera and plant, a measurement that would likely be necessary if an image-based stalk thickness measuring system were to be deployed.

2. Experimental Section

2.1. Materials

2.1.1. Energy Sorghum Material

Two genetic varieties of energy sorghum known as R07019 (Variety A) and R07007 (Variety B) were planted in plastic pots. These varieties have been studied as potential dedicated energy crops by Dr. John Mullet of Texas A&M University and were made available for this phenotyping study. Four plants per variety were planted, two plants to a pot. Each pot was supplied with 17 g of Osmocote 14-14-14 slow release fertilizer (Scotts). The plants were grown under well-watered conditions; *i.e.*, the soil surface was kept moist throughout the experiment. All pots were progressively thinned to 1 plant per pot as the seedlings grew. The pots were placed in a greenhouse where environmental conditions were monitored with a TinyTag Data Logger© (Gemini Data Loggers). The average daily minimum temperature and relative humidity were 24.1 °C (±1.1 °C) and 36.0% (±14.0%), respectively. The average daily maximum temperature and relative humidity were 32.3 °C (±1.8 °C) and 69.9% (±11.3%). Plants emerged roughly five days after planting and were without branching nodes for the first few weeks. Thus, the first images were collected about three weeks after planting, when the plants began to exhibit thicker stalks and leaves emerging from the stalk.

2.1.2. Camera

An inexpensive (under $200 USD in 2015) combined camera (Creative Technology Ltd., Milpitas, CA, USA, Model VF0780), composed of a standard RGB camera and a 3D active camera (SoftKinetic Inc., Sunnyvale, CA, USA), was used to capture images of the sorghum plants. The 3D camera employs the time-of-flight (TOF) concept, measuring the time between emitting a pulse of light and receiving the reflected light, to measure depth values in a field-of-view plane. TOF cameras, or "depth cameras", are commonly composed of a near-infrared (NIR) source and a CMOS pixel-array detector. The depth camera used includes a three-chip TOF system designed to be sensitive only in the 850 to 870-nm range, matching the NIR source. This TOF system provides a 320 × 240 resolution image in which pixel values relate to the distance from image objects to the emitter on the camera. Manufacturer specifications list the depth values' precision as in the range of 5 to 300 cm, with the best performance from 15 cm to 100 cm. The field of view (FOV) of the RGB and depth cameras is not identical. The RGB camera resolution can be selected from 320 × 240 to 720 × 1280. Initially, 320 × 240 was maintained in the RGB images for consistency with the depth images, but in later stages of plant growth, the RGB resolution was increased to 720 × 1280 for more precise measurements of stalk thickness.

2.1.3. Imaging Environment

A polyvinyl chloride (PVC) pipe frame was built to hold a black cotton cloth to block sunlight in the greenhouse and to aid in isolating sorghum plants from their surroundings, facilitating plant classification during image analysis. The cloth also helped block possible incoming NIR energy that might interfere with the depth camera.

2.2. Methods

Over a period of ten weeks, images and stalk thickness measurements of the sorghum plants were collected on roughly a weekly basis. According to recommendations by Dr. William Rooney of Texas A&M University [24], stalk thickness was to be measured immediately above the plant's third internode to provide consistent and representative measurements of the entire stalk. In the early stages of growth, sorghum plants did not exhibit internodes, so a measurement height of 7 cm above the soil was used until internodes were defined, a situation that occurred about six weeks after data collection began. Sorghum stalks have a roughly elliptical cross-section, so effective thickness was determined based on an elliptical shape model. Manual thickness measurements were made with a caliper. Both the wider (major-axis) and narrower (minor-axis) dimensions of the stalk were measured for each plant on each data collection date. The wider dimension of the sorghum stalk, which is perpendicular to the direction at which the leaves emerge, was oriented perpendicular to the camera viewing direction. A marker was placed on the pots during the first measurement so that each time measurements were made, the orientation of the plants would be the same. Distance from the camera lens to the background cloth was measured with a ruler to the nearest half cm.

Plants were placed between roughly 25 and 70 cm from the camera, as manually measured with a ruler to the nearest half cm, horizontally from the edge of the camera lens to the center of the sorghum stalk. Distances were varied to provide a range of distances to compare between manual and depth-camera measurements and also, in some cases, to provide adequate distance to image the entire plant. The camera was placed so that the point where caliper measurements were made was just below the image center. In the later stages when plants had developed internodes, the camera was placed so that the third internode was slightly below the center of the image. The camera was horizontally leveled to reduce possible error from image distortion that might occur if the camera were not perpendicular to the plant's frontal plane.

2.2.1. Image-Distance Calibration

Image-distance calibration consisted of producing a conversion factor from pixels to length units (mm). A 47.6-mm diameter PVC tube was placed in front of the camera at distances from 20 to 90 cm in 10-cm increments, and an RGB image was collected at each distance. The number of pixels belonging to the object at each distance was determined, and an equation was developed to convert pixels to distance. The accuracy of the depth camera measurements was considered independently, and this analysis is considered below.

2.2.2. Image Analysis

To measure the stalk thickness of a sorghum plant in an image, pixels associated with the stalk must be segmented from pixels associated with other parts of the plant, as well as from the background. K-means clustering and minimum-distance classification were used in this work.

K-Means Algorithm

The K-means clustering algorithm used in this article was adapted from Peter Corke [25] and performs a color-based segmentation based on a pre-selected C number of clusters. The algorithm maps each RGB value to the color *xy*-chromaticity space and then identifies clusters to segregate the pixels into image classes associated with those clusters. The algorithm initially selects locations in the data space as cluster centers and then calculates the Euclidean distance [26] between each pixel and every cluster center in multiple iterations. At the end of an iteration, each pixel is assigned to the cluster with the closest center, and every cluster center location is recalculated as the average location of the pixels assigned to that cluster. The final result is a classification of the image into C number of classes [25,27–29].

Each image was segmented into ten classes. The user identified the five classes that best represented the plants based on whether or not the class clearly contained almost exclusively plant matter, and cluster center locations were stored to a text file. After segmentation, the noise in the image was reduced with a morphological opening with a 9×9 square kernel. The five selected classes were then combined to produce a binary image that was an initial separation of plant from background. The combined image was then processed by area opening to remove small artifacts. Finally, to reduce the effects of plant lesions and minor stalk deformities, a morphological closing with a 5×5 square kernel was applied to the image. The result was a binary image of the plant segmented from the background; Figure 1a.

(a)

Figure 1. *Cont.*

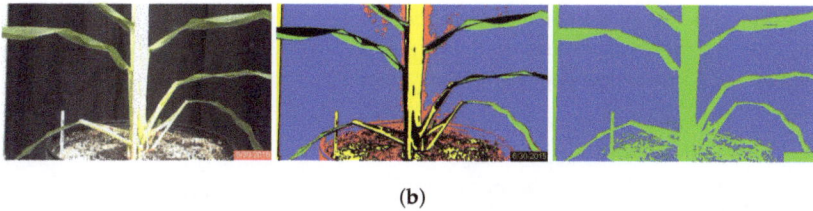

(b)

Figure 1. RGB original, classified and binary images of sorghum plants during the third week of growth of June, Plant 3B. (a) Segmentation using K-means clustering; (b) segmentation using minimum-distance classification.

Minimum Distance Algorithm

Minimum-distance classification of color images considers the three RGB components as orthogonal coordinates in the Euclidean space, and individual pixels are assigned to classes according to their RGB distance from pixel classes that have been defined by training. Four classes were defined by the user according to homogeneous zones in the images, as shown in Figure 2a. In this figure, the four classes are clearly separable [26]: the black cloth background, enclosed by the blue rectangle; the green sorghum plant leaf, enclosed by the green rectangle; the bright sorghum stalk, enclosed in orange; and the brown soil, enclosed in red. The spectral separability is shown in the histograms of Figure 3. The statistical properties shown are distribution, mean and standard deviation, which were used as the references for the minimum-distance classification algorithm. The mean is the center point to which pixel RGB-space distances are compared, while standard deviation is used as a threshold to determine whether or not a pixel is assigned to a defined class. Each image pixel was compared to all class means, and the shortest distance determined whether a pixel would be assigned to one class or another. A threshold of two standard deviations from all four classes was used to label pixels as "unclassified". Once the images were processed, the plant regions were stored as binary images (Figure 1b) and used subsequently to calculate stalk thickness.

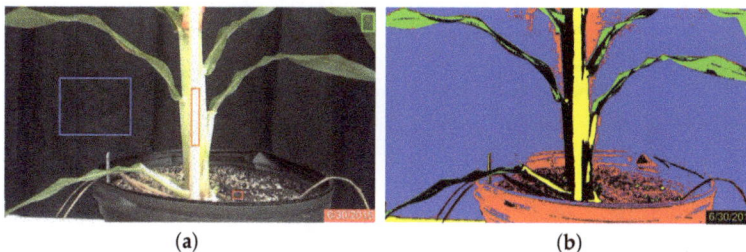

(a)   (b)

Figure 2. Images of sorghum Plant 1A on June 30. The minimum distance classification required training samples to segment the images: (a) original image and the location where the samples were obtained, enclosed by rectangles (blue = background; orange = bright stalk; green = plant leaf; red = soil); (b) image segmented into the five classes, including unclassified pixels.

Stalk Thickness Calculation Algorithm

Stalk-thickness calculation was conducted in "supervised" and "unsupervised" modes. In the supervised mode, the user selected a location in the image to define the stalk center. A 3 × 3 pixel window was then used to automatically scan the image in search of the edges of the stalk. Once the edge pixels were identified by the algorithm, stalk thickness in pixels was calculated by subtracting the edge location on the right from the edge location on the left. In unsupervised mode, the user identified the stalk center and allowed the algorithm to select upper and lower stalk thickness measurement locations. The average thickness based on the upper and lower measurements was taken as the stalk thickness (Figure 4). In both modes, stalk thickness in pixels was multiplied by the pixel-length conversion factor, which was dependent on the distance between plant and camera, to obtain stalk thickness in mm.

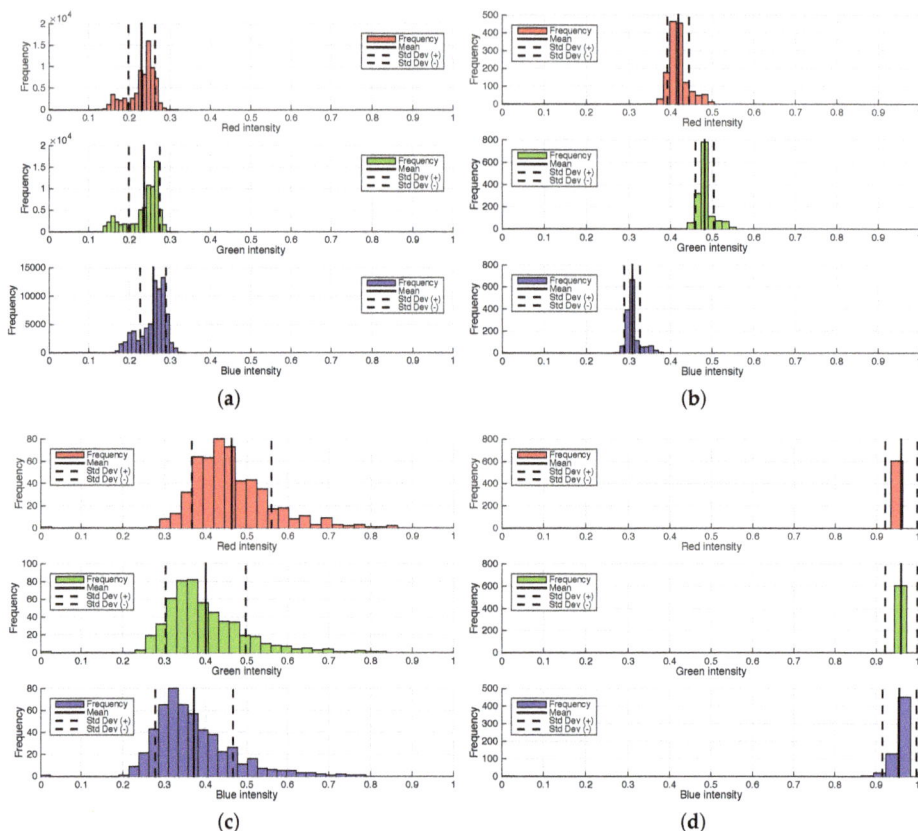

Figure 3. Frequency histograms of RGB values for each training sample. The solid line in the middle of the distribution indicates the location of the mean value, while the dashed lines indicate the location of two times the standard deviation of the each training sample. The information from the histograms was used to define each class center cluster; meanwhile, the standard deviation was used to define the class threshold. (**a**) Background; (**b**) plant; (**c**) soil; (**d**) stalk.

Depth Camera Accuracy

For every plant image collected, a manual measurement of the distance between the camera and plant stalk was conducted with a ruler, to the nearest half centimeter. In the depth-camera images, the pixel values represented the distance between the object and the camera. The depth-camera images included some erroneous pixel values, so the average of a group of pixels belonging to the plant stalk was used to minimize the effect of noise in the images. Two representative areas of the stalk in the depth image were selected by the user, and a 3×3 window was selected from each area. The average value from the 18 selected pixels was used as the depth-camera distance between camera and plant. Depth-camera distances were evaluated by comparing them to manual stalk-to-camera distance measurements by way of simple linear regression.

(a)

(b)

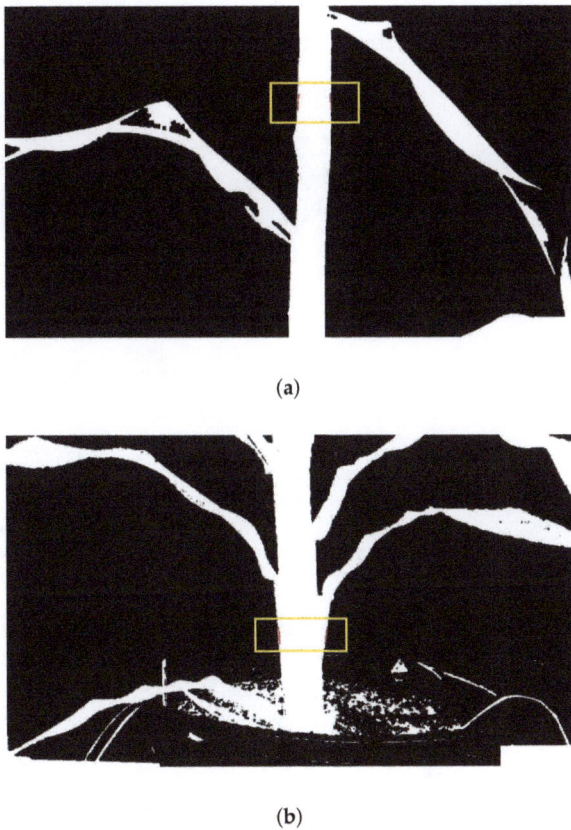

Figure 4. Results of image analysis. The golden box and red lines show the location where the stalk thickness was measured during this procedure: (**a**) K-means clustering, supervised stalk selection; (**b**) minimum-distance classification, supervised stalk selection.

3. Results and Discussion

The stalk thickness values determined through image analysis were compared to the manually-measured values and the percent error calculated for each plant and each date (Table 1). Averaged across all plants and dates, the average percent errors were 27.9 for K-means with unsupervised stalk measurement (KMU), 19.5 for minimum distance with unsupervised stalk measurement (MDU), 16.0 for K-means with supervised stalk measurement (KMS) and 16.7 for minimum distance with supervised stalk measurement (MDS). All KMS stalk-thickness measurements are plotted against the caliper values in Figure 5a. Linear regression of these data points resulted in an R^2 value of 0.28, indicating that KMS measurements only account for 28% of the variability in the caliper measurements.

Table 1. Comparison of stalk thickness measurements using K-means unsupervised (KMU), K-means supervised (KMS), minimum-distance unsupervised (MDU) and minimum-distance supervised (MDS) modes. The plant ID shows the plant number and the variety. The measured value indicates the value obtained with the stalk thickness calculation algorithm. The points in bold were determined to be outliers and removed from the final analysis (see Figure 5).

Date	Plant ID	Caliper Reading (mm)	Measured Value (mm)				Percent Error (%)			
			KMU	MDU	KMS	MDS	KMU	MDU	KMS	MDS
30 June	1A	25.58	24.12	32.27	28.47	27.47	5.71	26.17	11.29	7.41
	2A	33.17	28.96	40.13	30.15	29.18	12.69	20.98	9.12	12.04
	3A	30.82	31.18	31.41	28.25	26.88	1.17	1.91	8.32	12.77
	4A	32.59	35.75	35.11	32.85	27.31	9.70	7.74	0.80	16.20
	1B	29.75	21.84	30.62	29.65	31.45	26.59	2.92	0.35	5.70
	2B	28.61	27.57	37.41	34.60	32.96	3.64	30.78	20.95	15.19
	3B	27.76	21.65	26.22	27.49	25.36	22.01	5.56	0.96	8.65
	4B	25.62	60.49	39.14	32.77	26.98	136.10	52.79	27.90	5.29
7 July	**1A**	40.81	19.07	58.85	45.05	49.09	53.27	44.20	10.38	20.28
	2A	48.14	34.41	41.70	36.72	38.83	28.52	13.38	23.73	19.34
	3A	34.91	20.73	35.05	26.69	26.85	40.62	0.40	23.55	23.08
	4A	44.34	29.50	36.88	33.76	34.52	33.47	16.82	23.86	22.15
	1B	38.35	15.44	31.84	26.24	29.48	59.74	16.97	31.58	23.13
	2B	32.73	21.52	32.23	31.33	29.43	34.25	1.51	4.29	10.07
	3B	29.09	25.07	30.90	27.86	30.43	13.82	6.22	4.23	4.61
	4B	47.65	20.16	28.48	28.80	29.31	57.69	40.23	39.56	38.49
14 July	1A	36.54	35.44	34.25	38.05	32.10	3.01	6.27	4.13	12.16
	2A	36.04	40.80	40.50	34.42	32.92	13.21	12.38	4.51	8.64
	3A	35.38	41.10	41.98	37.50	36.87	16.17	18.65	5.99	4.22
	4A	33.68	41.59	35.09	37.14	36.47	23.49	4.20	10.27	8.29
	1B	36.68	35.40	36.81	32.76	31.55	3.49	0.37	10.69	14.00
	2B	34.51	31.20	36.08	34.98	35.55	9.59	4.54	1.36	3.00
	3B	22.22	35.39	38.57	29.83	30.71	59.27	73.59	34.23	38.23
	4B	37.28	44.04	34.63	34.32	31.37	18.13	7.11	7.94	15.85
21 July	**1A**	20.45	28.60	31.55	32.77	31.88	39.85	54.29	60.24	55.91
	2A	25.51	22.77	25.98	27.00	26.77	10.74	1.84	5.82	4.94
	3A	25.17	24.36	26.34	24.41	24.77	3.22	4.65	3.04	1.59
	4A	34.26	31.41	32.74	32.10	32.20	8.32	4.43	6.32	6.01
	1B	21.01	16.10	57.74	18.67	17.11	23.37	174.81	11.16	18.58
	2B	24.75	37.99	25.20	26.23	27.25	53.49	1.83	5.98	10.11
	3B	25.70	43.96	39.39	44.28	44.75	71.05	53.28	72.30	74.13
	4B	22.89	37.24	36.41	38.16	38.99	62.69	59.07	66.71	70.32
4 August	1A	21.01	24.12	23.31	24.66	23.14	14.80	10.97	17.35	10.12
	2A	25.22	28.96	29.13	32.09	32.09	14.83	15.50	27.24	27.24
	3A	25.58	31.18	26.33	27.66	26.99	21.89	2.94	8.13	5.51
	4A	27.03	35.75	24.83	29.06	29.70	32.26	8.15	7.49	9.88
	1B	20.45	21.84	21.55	23.79	23.54	6.80	5.39	16.31	15.11
	2B	23.02	27.57	27.43	30.18	30.48	19.77	19.17	31.08	32.41
	3B	22.26	21.65	21.28	24.46	25.90	2.74	4.42	9.88	16.33
	4B	26.22	60.49	22.49	24.16	24.31	130.70	14.21	7.88	7.29
11 August	1A	24.23	19.07	23.31	23.55	23.64	21.30	3.78	2.83	2.42
	2A	25.23	34.41	29.13	29.12	27.96	36.39	15.46	15.42	10.81
	3A	24.25	20.73	26.33	22.87	21.86	14.52	8.59	5.69	9.86
	4A	26.13	29.50	24.83	31.99	30.26	12.90	4.99	22.43	15.80
	1B	19.14	15.44	21.55	17.32	17.40	19.33	12.60	9.51	9.07
	2B	20.31	21.52	27.43	20.13	20.32	5.96	35.07	0.91	0.02
	3B	21.19	25.07	21.28	27.78	28.26	18.31	0.41	31.08	33.38
	4B	21.68	20.16	22.49	20.80	20.96	7.01	3.76	4.06	3.34

Close inspection of Figure 5a indicated that several of the data points may be outliers, so the following procedure (adapted from [30,31]) was used in an effort to identify actual outliers that should be removed from consideration. (1) Residuals were calculated and plotted against the caliper measurements, indicating heteroscedasticity (*i.e.*, residual values were lower at low stalk thickness values and higher at high stalk thickness values). (2) Based on the fact that residuals were heteroscedastic, both KMS and caliper measurements were transformed by taking their logarithms. Log(caliper) was regressed against log(KMS), and residuals were again plotted, indicating that heteroscedasticity had been mitigated. (3) Studentized residuals of the log(caliper) *vs.* log(KMS) relationship were calculated, and the distribution of the studentized residuals was determined. At the positive and negative tails of the distribution, seven data points were identified as being significantly disconnected from the rest of the data. This evaluation provided the basis for identifying suspected possible outliers. (4) The original images associated with the suspected outliers were observed to determine whether anomalies that could produce major errors could be identified. With each of the suspected outliers, the images suggested that significant error-causing anomalies were present, so these data points were removed, and simple linear regression was conducted on the resulting dataset (Figure 5b). Results of the analysis after removing outliers provided an R^2 value of 0.70, indicating that KMS measurements accounted for 70% of the variability in the caliper measurements. Furthermore, with the outliers (identified in Table 1) removed, the resulting average error for KMS measurements was reduced to 10.7%.

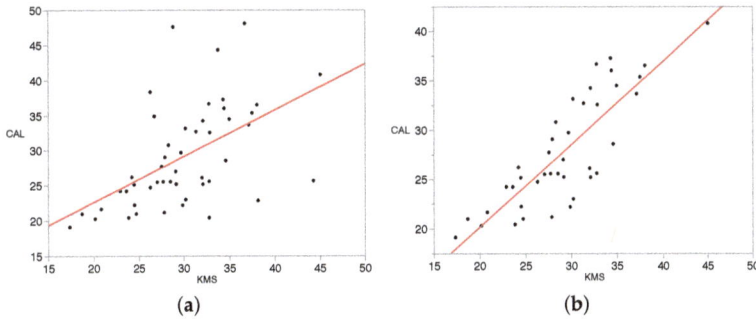

(a)   (b)

Figure 5. Linear regressions for caliper measurements *vs.* K-means supervised measurements: (a) the regression before removing outliers ($R^2 = 0.28, RMSE$ 6.27 mm); (b) after removal of outliers ($R^2 = 0.70, RMSE$ 3.19 mm). Plot units are in mm.

As a representation of the image classification errors, Figure 6 shows the original and two classified images of Plant 4B on 30 June, which had the highest error of all plants measured on that date (except with MDS). It can be observed that this plant had leaves emerging from the sides of the stalk over most of the stalk area. This problem made it difficult for the stalk measurement algorithm to identify the actual edges of the stalk, contributing error to the measurement. On 7 July, the plant with the highest percent error had non-plant pixels in the binary image of the stalk, indicating poor segmentation that led the analysis to measure only part of the stalk. These are the types of errors common to the high error image-based measurements.

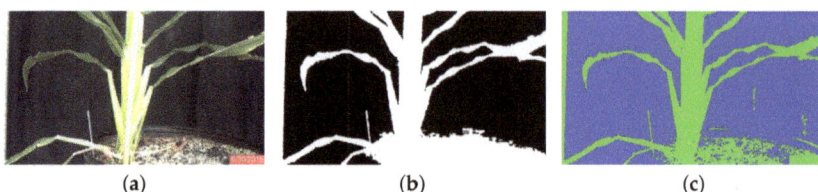

Figure 6. The images presented are examples of anomalies in the experiment. K-means, unsupervised (KMU), minimum-distance, unsupervised (MDU) and K-means, supervised (KMS) were three out of the four different methods in this experiment. (**a**) 30 June, 4B; (**b**) 30 June, 4B, KMU (136.10); (**c**) 30 June, 4B, MDU (52.79). In parenthesis, the error between the manual measurement and algorithm measurement is shown (see also Table 1).

In terms of the plant growth stage at which measurements were most accurate, later-stage plants had lower error than early-stage plants. By the seven-week point, the plants had developed reasonably well-defined stalks that were easier to segment and measure. The relatively low percent error at later growth stages showed the techniques to be most accurate at the time phenotyping would be most useful. In terms of which methods were most accurate, K-means clustering was more accurate with supervised stalk measurement, but minimum-distance classification was more accurate with unsupervised stalk measurement. Supervised stalk measurement had lower error than unsupervised when using either K-means or minimum distance. This result suggests that the position where stalk thickness is measured is critical to the accuracy of the measurement, and future research should focus on how the image-processing algorithm determines the best measurement location. As previously mentioned, the most obvious errors in measuring stalk thickness occurred when leaves protruding from the stalk were included in the stalk-thickness measurement. Additional image-processing techniques—e.g., to consider whether the left and right stalk edges are parallel—could better isolate the stalk, enabling the largest stalk thickness measurement errors to be mitigated. In other cases, the source of major error was either unknown or related to poor image segmentation, probably due to lack of contrast at particular points on the plant. To mitigate these errors, more research would be required, as solutions are not obvious.

The results of comparing depth-camera distance measurements to manual camera-to-stalk distance measurements are presented in Figure 7. The plot of data points indicates a high level of linearity, and the R^2 value of 0.96 indicates that the depth camera accounted for 96% of the variability in actual camera-to-stalk distance. The average error for camera-to-stalk distance on a given day ranged from about 5% to 20%, while the root mean square error ($RMSE$) was 2.5 cm. It is clear from the plot that at camera-to stalk distances greater than 65 cm, the percent error is low. Stalk thickness error associated with a depth-camera distance $RMSE$ of 2.5 cm is approximately 3%, meaning that the roughly 10% average stalk thickness error resulting from image processing is large compared to depth-camera error. The depth-camera feature of this combined camera thus appears to be acceptably accurate to provide distance data that would enable calculation of pixel-to-length conversion factors. If this capability could be combined with the image-processing techniques to segment the plant and calculate the number of pixels associated with the thickness of a stalk, then stalk thickness measurement for energy sorghum plants could be fully automated. The issue of pixel-to-pixel correspondence between depth images and RGB images is an issue remaining to be solved.

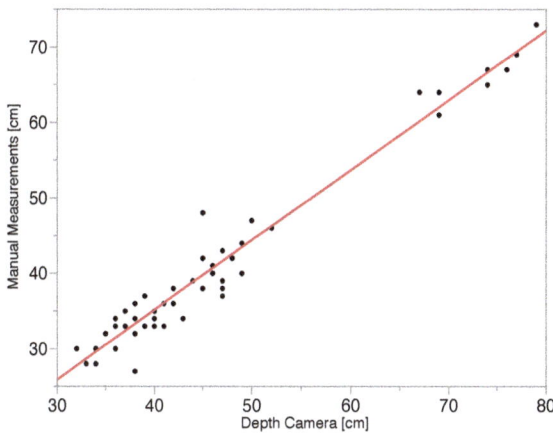

Figure 7. Linear regression of distance measurements from the camera to the plant, measured manually and measured with the depth camera images. The R^2 value calculated for this relationship is 0.96, and the $RMSE$ is 2.5 cm.

4. Conclusions

This study showed that an inexpensive combined (RGB plus depth) camera and common image analysis techniques (K-means clustering and minimum-distance classification) have potential for measuring stalk thickness in a high-throughput phenotyping system for energy sorghum plants. Furthermore, the integrated near-infrared time-of-flight depth camera showed excellent potential for providing distance measurements that would be required if stalk thickness were to be measured in an automated way. Incidentally, these techniques could be applied to other monocots, such as maize and rice. If successfully implemented, such a combined system could greatly reduce the labor and time required to make such measurements in experiments involving phenotyping, ultimately enabling many more plants to be measured than currently possible and potentially accelerating crop improvement.

Acknowledgments: The authors acknowledge the contributions of Brock Weers, Assistant Research Scientist, and John Mullet, Professor, both in the Department of Biochemistry and Biophysics at Texas A&M University. Weers planted and maintained the energy sorghum plants, and Mullet provided materials for the plant growth, space in the greenhouse and advice on sorghum phenotyping.

Author Contributions: Jose Batz developed procedures for data collection, unsupervised K-means clustering and stalk diameter measurement. He also collected and analyzed data and wrote significant portions of the manuscript. Mario A. Méndez compiled a review of the literature, developed procedures for minimum distance classification and wrote portions of and compiled the manuscript. J. Alex Thomasson oversaw the research and reviewed and approved the methods used, as well as the results obtained. Additionally, he edited and added final interpretive comments to the manuscript.

Conflicts of Interest: The authors declare no conflict of interest.

References

1. Davies, J. Better distribution holds the key to future food security for all. *Poult. World* **2015**, *170*, 1–19.
2. Koçar, G.; Civas, N. An overview of biofuels from energy crops: Current status and future prospects. *Renew. Sustain. Energy Rev.* **2013**, *28*, 900–916.
3. Chapman, S.C.; Chakraborty, S.; Dreccer, M.F.; Howden, S.M. Plant adaptation to climate change—Opportunities and priorities in breeding. *Crop Pasture Sci.* **2012**, *63*, 251–268.
4. Calviño, M.; Messing, J. Sweet sorghum as a model system for bioenergy crops. *Curr. Opin. Biotechnol.* **2012**, *23*, 323–329.

5. Cobb, J.N.; DeClerck, G.; Greenberg, A.; Clark, R.; McCouch, S. Next-generation phenotyping: Requirements and strategies for enhancing our understanding of genotype–phenotype relationships and its relevance to crop improvement. *Theor. Appl. Genet.* **2013**, *126*, 867–887.

6. Regassa, T.H.; Wortmann, C.S. Sweet sorghum as a bioenergy crop: Literature review. *Biomass Bioenergy* **2014**, *64*, 348–355.

7. Busemeyer, L.; Mentrup, D.; Möller, K.; Wunder, E.; Alheit, K.; Hahn, V.; Maurer, H.P.; Reif, J.C.; Würschum, T.; Müller, J.; *et al.* BreedVision—A Multi-Sensor Platform for Non-Destructive Field-Based Phenotyping in Plant Breeding. *Sensors* **2013**, *13*, 2830–2847.

8. Kicherer, A.; Herzog, K.; Pflanz, M.; Wieland, M.; Rüger, P.; Kecke, S.; Kuhlmann, H.; Töpfer, R. An Automated Field Phenotyping Pipeline for Application in Grapevine Research. *Sensors* **2015**, *15*, 4823–4836.

9. Klodt, M.; Herzog, K.; Töpfer, R.; Cremers, D. Field phenotyping of grapevine growth using dense stereo reconstruction. *BMC Bioinf.* **2015**, *16*, 1–11.

10. Upadhyaya, H.; Wang, Y.H.; Sharma, S.; Singh, S.; Hasenstein, K. SSR markers linked to kernel weight and tiller number in sorghum identified by association mapping. *Euphytica* **2012**, *187*, 401–410.

11. Mullet, J.; Morishige, D.; McCormick, R.; Truong, S.; Hilley, J.; McKinley, B.; Anderson, R.; Olson, S.; Rooney, W. Energy Sorghum—A genetic model for the design of C4 grass bioenergy crops. *J. Exp. Bot.* **2014**, *65*, 1–11.

12. van der Weijde, T.; Alvim Kamei, C.L.; Torres, A.F.; Wilfred, V.; Oene, D.; Visser, R.G.F.; Trindade, L.M. The potential of C4 grasses for cellulosic biofuel production. *Front. Plant Sci.* **2013**, *4*, 1–18.

13. Trouche, G.; Bastianelli, D.; Cao Hamadou, T.V.; Chantereau, J.; Rami, J.F.; Pot, D. Exploring the variability of a photoperiod-insensitive sorghum genetic panel for stem composition and related traits in temperate environments. *Field Crops Res.* **2014**, *166*, 72–81.

14. Chéné, Y.; Rousseau, D.; Lucidarme, P.; Bertheloot, J.; Caffier, V.; Morel, P.; Belin, É.; Chapeau-Blondeau, F. On the use of depth camera for 3D phenotyping of entire plants. *Comput. Electron. Agric.* **2012**, *82*, 122–127.

15. Vollmann, J.; Walter, H.; Sato, T.; Schweiger, P. Digital image analysis and chlorophyll metering for phenotyping the effects of nodulation in soybean. *Comput. Electron. Agric.* **2011**, *75*, 190–195.

16. Chaivivatrakul, S.; Tang, L.; Dailey, M.N.; Nakarmi, A.D. Automatic morphological trait characterization for corn plants via 3D holographic reconstruction. *Comput. Electron. Agric.* **2014**, *109*, 109–123.

17. Roscher, R.; Herzog, K.; Kunkel, A.; Kicherer, A.; Töpfer, R.; Förstner, W. Automated image analysis framework for high-throughput determination of grapevine berry sizes using conditional random fields. *Comput. Electron. Agric.* **2014**, *100*, 148–158.

18. Aksoy, E.E.; Abramov, A.; Wörgötter, F.; Scharr, H.; Fischbach, A.; Dellen, B. Modeling leaf growth of rosette plants using infrared stereo image sequences. *Comput. Electron. Agric.* **2015**, *110*, 78–90.

19. Fahlgren, N.; Gehan, M.A.; Baxter, I. Lights, camera, action: High-throughput plant phenotyping is ready for a close-up. *Curr. Opin. Plant Biol.* **2015**, *24*, 93–99.

20. Kolukisaoglu, Ü.; Thurow, K. Future and frontiers of automated screening in plant sciences. *Plant Sci.* **2010**, *178*, 476–484.

21. Lobet, G.; Draye, X.; Périlleux, C. An online database for plant image analysis software tools. *Plant Methods* **2013**, *9*, doi:10.1186/1746-4811-9-38.

22. Colombi, T.; Kirchgessner, N.; Le Marié, C.; York, L.; Lynch, J.; Hund, A. Next generation shovelomics: Set up a tent and REST. *Plant Soil* **2015**, *388*, 1–20.

23. Bucksch, A.; Burridge, J.; York, L.M.; Das, A.; Nord, E.; Weitz, J.S.; Lynch, J. Image-Based High-Throughput Field Phenotyping of Crop Roots. *Plant Physiol.* **2014**, *166*, 470–486.

24. Rooney, W.L. (Soil and Crop Science Professor, Texas A&M University, College Station, TX,USA) Stalk diameter measure location in energy sorghum. Personal communication, 2015.

25. Corke, P. *Robotics, Vision and Control: Fundamental Algorithms in MATLAB*; Springer Science & Business Media: Berlin, Germany, 2011; Volume 73.

26. Ghimire, S.; Wang, H. Classification of image pixels based on minimum distance and hypothesis testing. *Comput. Stat. Data Anal.* **2012**, *56*, 2273–2287.

27. Liang, B.; Zhang, J. KmsGC: An Unsupervised Color Image Segmentation Algorithm Based on K-Means Clustering and Graph Cut. *Math. Probl. Eng.* **2014**, *2014*, 464875.

28. Sert, E.; Okumus, I.T. Segmentation of mushroom and cap width measurement using modified K-means clustering algorithm. *Adv. Electr. Electron. Eng.* **2014**, *12*, 354–360.

29. Yu, W.; Wang, J.; Ye, L. An Improved Normalized Cut Image Segmentation Algorithm with K-means Cluster. *Appl. Mech. Mater.* **2014**, *548–549*, 1179–1184.

30. Neter, J.; Kutner, M.H.; Nachtsheim, C.J.; Wasserman, W. *Applied Linear Statistical Models*, 4th ed.; McGraw-Hill: Boston, MA, USA, 1996.

31. Cline, D. *Statistical Analysis*; STAT 601 Class Notes; Texas A&M University: College Station, TX, USA, 2015.

Journal of
Imaging

MDPI

Article

Viewing Geometry Sensitivity of Commonly Used Vegetation Indices towards the Estimation of Biophysical Variables in Orchards

Jonathan Van Beek [1,*], Laurent Tits [1,2], Ben Somers [3], Tom Deckers [4], Pieter Janssens [5] and Pol Coppin [1]

[1] KU Leuven, Department of Biosystems, Division of Crop Biotechnics, Willem de Croylaan 34, BE-3001 Leuven, Belgium; pol.coppin@kuleuven.be
[2] Flemish Institute for Technological Research (VITO), Remote Sensing Unit, Boeretang 200, BE-2400 Mol, Belgium; laurent.tits@vito.be
[3] KU Leuven, Department of Earth and Environmental Sciences, Division Forest, Nature and Landscape, Celestijnenlaan 200E, BE-3001 Leuven, Belgium; ben.somers@kuleuven.be
[4] Pcfruit research station, Fruittuinweg 1, BE-3800 Sint-Truiden, Belgium; tom.deckers@pcfruit.be
[5] Soil Service of Belgium, Willem de Croylaan 48, BE-3001 Leuven, Belgium; pjanssens@bdb.be
* Correspondence: jonathan.vanbeek@biw.kuleuven.be; Tel.: +32-16-328-146; Fax: +32-16-322-966

Academic Editor: Gonzalo Pajares Martinsanz
Received: 9 February 2016; Accepted: 26 April 2016; Published: 9 May 2016

Abstract: Stress-related biophysical variables of capital intensive orchard crops can be estimated with proxies via spectral vegetation indices from off-nadir viewing satellite imagery. However, variable viewing compositions affect the relationship between spectral vegetation indices and stress-related variables (*i.e.*, chlorophyll content, water content and Leaf Area Index (LAI)) and could obstruct change detection. A sensitivity analysis was performed on the estimation of biophysical variables via vegetation indices for a wide range of viewing geometries. Subsequently, off-nadir viewing satellite imagery of an experimental orchard was analyzed, while all influences of background admixture were minimized through vegetation index normalization. Results indicated significant differences between nadir and off-nadir viewing scenes ($\Delta R^2 > 0.4$). The Photochemical Reflectance Index (PRI), Normalized Difference Infrared Index (NDII) and Simple Ratio Pigment Index (SRPI) showed increased R^2 values for off-nadir scenes taken perpendicular compared to parallel to row orientation. Other indices, such as Normalized Difference Vegetation Index (NDVI), Gitelson and Merzlyak (GM) and Structure Insensitive Pigment Index (SIPI), showed a significant decrease in R^2 values from nadir to off-nadir viewing scenes. These results show the necessity of vegetation index selection for variable viewing applications to obtain an optimal derivation of biophysical variables in all circumstances.

Keywords: biophysical variables; orchards; vegetation indices; view angle sensitivity

1. Introduction

Monitoring and managing of capital intensive orchard crops through precision agriculture is centered around the estimation of stress-related biophysical variables such as chlorophyll content [1,2], water content [2–4] and Leaf Area Index (LAI) [5]. Time consuming, labor intensive and destructive *in situ* measurements of biophysical variables could be circumvented through remotely sensed imagery and spectral vegetation indices [6,7]. Examples include the Photochemical Reflectance Index (PRI) [8] for chlorophyll content [1,3], the Water Index (WI) [9] for water content [2] and the standardized LAI-Determining Index (sLAIDI) for LAI [5]. On the one hand, practical use of remotely sensed spectral imagery in temperate climates requires near-to-daily revisit times because of the high cloud cover [10]. On the other hand, heterogeneous orchards require high spatial resolution imagery to provide frequent and accurate information on field and plant conditions [11]. Currently, this combination of both high spatial and temporal resolutions is feasible with high spatial resolution satellite sensors capable of off-nadir viewing, such as GeoEye-1, Quickbird, Pleaides, WorldView-1, WorldView-2 and WorldView-3.

The main downside of using these agile satellites is the influence of variable viewing angles on the relationship between biophysical variables and spectral measurements or vegetation indices [4,12,13]. Although information from multiple view angles is considerably greater than from a single angle and could provide information regarding structural variables (*i.e.*, plant cover, canopy height and biomass) [14,15], varying viewing angles within and between time series could obstruct change detection due to the confusion between genuine changes and viewing geometry influences. In the past, several methods were constructed to remove the influence of variable viewing and illumination angles. The most common is the use of Bidirectional Reflectance Distribution Function (BRDF) models to standardize and produce nadir equivalent reflectance values [16–21]. However, this could remove useful information regarding vegetation structure [19,21,22] or plant stress [4] and requires a wide range of viewing angles over a short period of time. Therefore, existing research focused mostly on imagery of low spatial resolution images (*i.e.*, > 20 m spatial resolution e.g., Moderate Resolution Imaging Spectroradiometer (MODIS)) [17,19,21,23] and/or at fixed view angles (e.g., $+/- 55°$, $+/- 36°$ and $0°$ for the pushbroom CHRIS (Compact High Resolution Imaging Spectrometer) sensor mounted on the PROBA (Project for On-Board Autonomy) platform) [24].

In addition, the research on the effects of variable viewing conditions focused mostly on vegetative systems with a continuous canopy cover, such as forests [18,24,25], grasslands [24] and soybeans [26]. However, high spatial resolution satellite imagery over orchard cropping systems will always contain mixtures of canopies and background (*i.e.*, soil, grass and shadow) [27,28]. This mixture effects can be further aggravated by the variable viewing conditions. Recent studies have shown the necessity and usefulness of the removal or reduction of background effects through either unmixing algorithms [29] or vegetation index corrections [28]. However, both methods were developed for nadir viewing imagery and assumed the presence of the full range of canopy fractions.

The goal of this study was to investigate the view-angle sensitivity of common spectral vegetation indices on the estimation of biophysical variables—*i.e.*, chlorophyll content, water content and LAI—in orchards. In high spatial resolution imagery, changing view angles causes both BRDF- and mixture-related differences. As the latter influence can be minimized through a correction method, the view-angle sensitivity of common vegetation indices for high spatial resolution imagery of hedgerow cropping systems can be assessed and minimize the effects of varying viewing angles on change detection within and between time series. Synthetic imagery of a virtual orchard was used to include variable orchard conditions, as well as different background scenarios—*i.e.*, non-vegetated, vegetated and partially vegetated. Finally, imagery from a satellite with off-nadir viewing capacities over an experimental orchard was investigated to highlight the importance of changing viewing geometry towards the estimation of biophysical variables.

2. Materials

2.1. Synthetic Imagery

Synthetic imagery was used in this study to improve our understanding of the effects of a changing viewing geometry on the remotely sensed data and the information derived from it [28,30,31]. The synthetic data provided exact cover fractions, spectral signatures and biophysical variables of the target crop for a range of viewing compositions. Additionally, the simulations were used to investigate the overall effect of viewing composition on the estimation of biophysical variables via vegetation indices.

In compliance with [28,31], the virtual orchard, developed by [32], consisted of virtual citrus trees (*Citrus sinensis* L.). The trees were arranged in a 3.5 by 2 m grid (row orientation of $7°$; *i.e.*, north-south direction). Within each orchard, several sections of the virtual orchard were modified to mimic stressed conditions for chlorophyll content, water content and LAI. Stressed leaf spectra were modified through PROSPECT [33]. Chlorophyll contents were extracted from *in situ*-measured unmodified leaf spectra and reduced by 25% and 50% and the water content by 15% and 30% [32]. LAI stress was introduced by randomly removing or adding leaves to the virtual trees representing 56% and 125% of the original LAI [32]. The spatial distribution and range of each variable is shown in Figure 1.

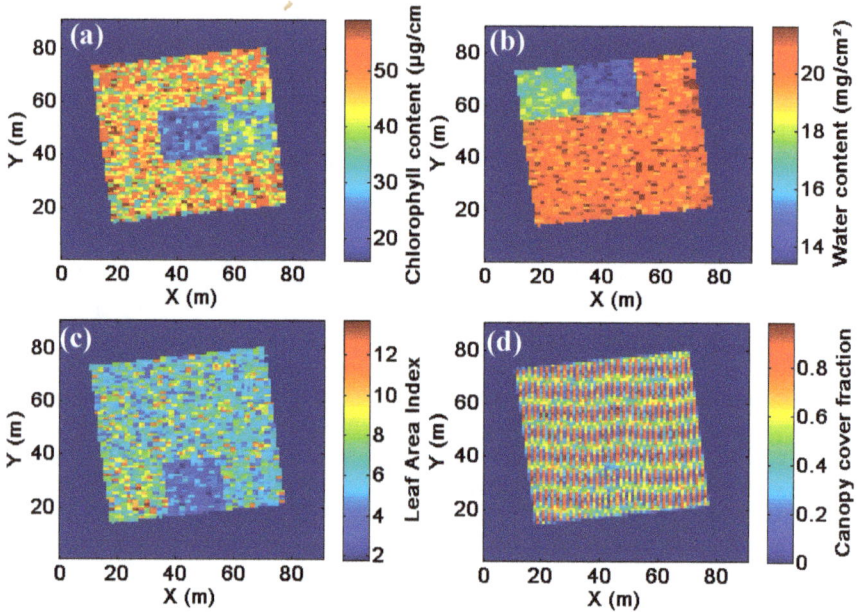

Figure 1. Spatial distribution of biophysical and structural variables in the virtual orchard for (**a**) chlorophyll content, (**b**) water content, (**c**) leaf area index and (**d**) canopy cover fractions.

From these virtual orchards, different images were rendered in a physically-based ray-tracer (PBRT) [34]. The model has previously been calibrated and validated with the Radiation Transfer Model Intercomparison (RAMI) online model checker [35,36] and field data obtained in a citrus orchard in Wellington, South Africa [32,37]. Images were rendered with a direct and diffuse light source (solar elevation of 79.2° and solar azimuth of 339.6°) mimicking South African winter solstice [32]. The spectral range of the synthetic images was 350-2500 nm with a spectral resolution of 10 nm, while the spatial resolution of the sensor was fixed at 2 m. The resulting canopy cover fractions within the 2 m pixels are shown in Figure 1d. The influence of different viewing compositions was investigated by adjusting the position of the sensor. The sensor's azimuth angles were varied between 0 and 315° with 45° increments and the sensor's view angle was varied between 0 and 60° with 15° increments. As a result, each orchard composition was rendered from 33 different sensor positions.

The influence of different orchard conditions was investigated by varying the orchard floor or background through four different scenarios, while the position of the virtual trees remained identical for each scenario. Through these modifications, the sensitivity of commonly used vegetation indices to orchard- and viewing parameters could be investigated.

- Scenario 1 (S1), a uniform soil background, consisting of an Albic Leptic Luvisol soil [38] measured *in situ* [39], shown in Figure 2a. Soil reflectance was assumed Lambertian.
- Scenario 2 (S2), a uniform weed background (Figure 2b), consisting of *Phleum pratense* L. modeled with leaf reflectance obtained from the Leaf Optical Experiment database [32,40].
- Scenario 3 (S3), a variable weed and soil background, consisting of a weed background with random and irregular soil patches (65/35% cover fraction distribution), depicted in Figure 2c.
- Scenario 4 (S4), a variable weed background with a chlorophyll gradient. The weed background was modified similar to the leaf reflectances, increasing the chlorophyll content from 75% to the reference value (*i.e.*, uniform weed background). A true color representation of the variable weed background is shown in Figure 2d.

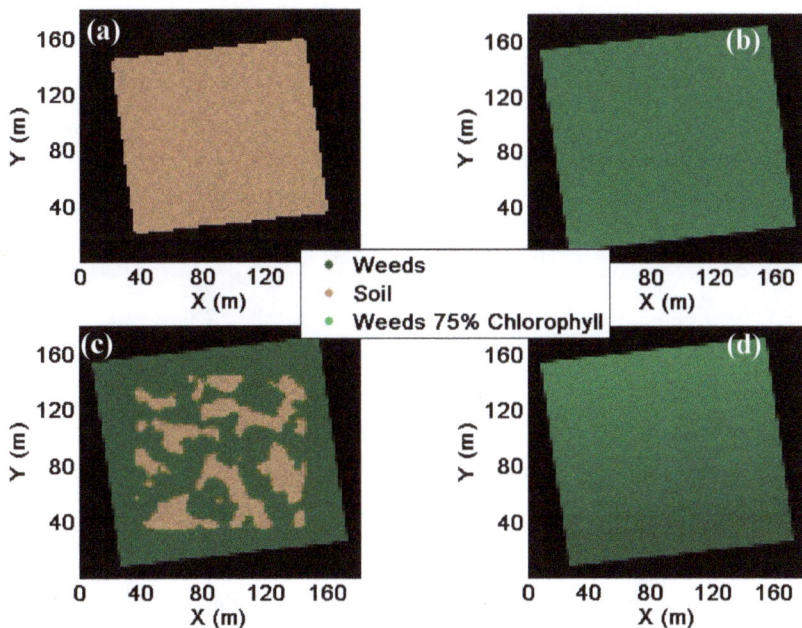

Figure 2. Spatial distribution of background scenarios (a) uniform soil background (S1); (b) uniform weed background (S2); (c) variable weed and soil background (S3); (d) variable weed background with a chlorophyll gradient.

2.2. Real Imagery

2.2.1. Study Area

The irrigated orchard, planted with Conference pear trees (Pyrus communis L. cv. "Conference") on Quince C rootstock, was situated in Bierbeek, Belgium (50°49′34.59″N, 4°47′42.83″E). The 2.5 m high trees were planted in 2000 in a 3.5 by 1 meter grid. They were trained in a V-system with four fruiting branches on one central stem [41]. The orchard was situated on a south-east facing slope (3.5°) with a loamy soil and two dominant row azimuths, namely 41 and 131°. The trees received 100% of the crop evapotranspiration (ETc) [42,43], throughout most of the growing season.

The non-irrigated or rainfed orchard, situated in Kerkom, Belgium (50°46′24.25″N, 5°09′27.05″E), was planted in 2000 with Conference pear trees on Quince A rootstock. The 3.5 m high trees, planted in a 3.75 by 1.75 m grid, were trained in a Spindle bush system [41]. The orchard was situated on a south facing slope (1.1°) with a loamy soil and a row azimuth of 197°. The trees were rainfed and received no additional water input.

In both orchards, the soil under the trees was kept weed free for about 0.3 m from the trunk and grass was sown in between the tree rows.

2.2.2. Satellite Imagery

WorldView-2 multispectral images were acquired under different off-nadir viewing angles, with a ground sampling distance of 2.0 m and a spectral resolution complying eight bands: Coastal (400–450 nm), Blue (450–510 nm), Green (510–580 nm), Yellow (585–625 nm), Red (630–690 nm), Red Edge (705–745 nm), NIR1 (Near InfraRed 1; 770–895 nm) and NIR2 (860–1,040 nm). The acquisition details for the WorldView-2 images are shown in Table 1. All images were radiometrically [44], atmospherically [45] and geometrically corrected [46].

Table 1. Metadata of WorldView-2 acquisitions used in this study.

Location	Year	Day Of Year (DOY)	Off-Nadir Viewing Angle (°)	Satellite Azimuth (°)	Satellite Elevation (°)
Irrigated Orchard	2011	214	10.8	45.9	78
	2012	148	2.7	181.1	86.7
		232	18.9	209.8	68.6
	2013	189	26.1	14.7	60.7
		214	25.6	107.9	61
Rainfed Orchard	2011	196	43.3	116.1	39.7
		214	4.8	68.6	84.7
	2012	148	15	199.8	72.9
		232	23.7	211.1	62.9
	2013	187	28	99.1	58.2
		214	27.4	133.5	58.7

2.2.3. Reference Plots

During each satellite acquisition, ground reference plots were monitored, each plot consisting of four adjacent trees. In the rainfed orchard, 16 plots were selected on fixed intervals within two adjacent rows of similar structure and age. In the irrigated orchard, 32 plots were selected with variable age, soil type and row orientation. The position of each plot was determined with a differential global positioning system (Trimble RTK 58000). Spatial variation was introduced by the application of one-sided root-pruning (rainfed orchard) and deficit irrigation (irrigated orchard). The location of the ground measurements and both root-pruning and deficit irrigation treatments is shown in Figure 3.

3. Methods

3.1. Vegetation Indices

The spectral vegetation indices used in this study were divided into categories based on their proven or expected relationship with biophysical and structural variables (*i.e.*, chlorophyll content, water content and LAI). The used vegetation indices are listed in Table 2 [5,8,9,47–63]. The indices were chosen because of a proven link with biophysical variables in fruit orchards [1,2,64].

Table 2. Overview of the biophysical and structural vegetation indices used in this study, their formulation and reference. The distinction was made between chlorophyll content, water content and Leaf Area Index (LAI)—related indices. Vegetation indices were approximated by WorldView-2 bands.

Index	Formulation	Reference	WorldView-2 Band Combination
Chlorophyll content related indices			
NDVI (Normalized Difference Vegetation Index)	$(R_{NIR} - R_{Red})/(R_{NIR} + R_{Red})$	[47]	$(R_{NIR1} - R_{Red})/(R_{NIR1} + R_{Red})$
OSAVI (Optimization SAVI [a])	$(1 + 0.16) * (R_{800} - R_{670})/(R_{800} - R_{670} + 0.16)$	[48]	$(1 + 0.16) * (R_{NIR1} - R_{Red})/(R_{NIR1} - R_{Red} + 0.16)$
MCARI (Modified CARI [b])	$[(R_{700} - R_{670}) - 0.2 * (R_{700} - R_{550})] * (R_{700}/R_{670})$	[49]	$[(R_{Red-edge} - R_{Red}) - 0.2 * (R_{Red-edge} - R_{Green})] * (R_{Red-edge}/R_{Red})$
TCARI (Transformed CARI [b])	$3 * (R_{700} - R_{670}) - 0.2 * (R_{700} - R_{550}) *$ (R_{700}/R_{670})	[50]	$3 * (R_{Red-edge} - R_{Red}) - 0.2 * (R_{Red-edge} - R_{Green}) * (R_{Red-edge}/R_{Red})$
ZM (Zarco and Miller)	R_{750}/R_{710}	[51]	-
SRPI (Simple Ratio Pigment Index)	R_{430}/R_{680}	[52]	$R_{Coastal}/R_{Red}$
PRI (Photochemical Reflectance Index)	$(R_{531} - R_{570})/(R_{531} + R_{570})$	[8]	$(R_{Green} - R_{Blue})/(R_{Green} + R_{Blue})$
NPCI (Normalized Pigment Chlorophyll Index)	$(R_{680} - R_{430})/(R_{680} + R_{430})$	[53]	$(R_{Red} - R_{Coastal})/(R_{Red} + R_{Coastal})$
CTR1 (Carter Index)	R_{695}/R_{420}	[54]	$(R_{Red-edge} + R_{Red})/(2*R_{Coastal})$
CTR2 (Carter Index)	R_{695}/R_{760}	[55]	$(R_{Red-edge} + R_{Red})/(2*R_{NIR1})$
SIPI (Structure Insensitive Pigment Index)	$(R_{800} - R_{450})/(R_{800} + R_{650})$	[52]	$(R_{NIR1} - R_{Coastal})/(R_{NIR1} + R_{Red})$
GM (Gitelson and Merzlyak Index)	R_{750}/R_{550}	[56]	$R_{Red-edge}/R_{Green}$
Water content related indices			
WI (Water Index)	R_{900}/R_{970}	[9]	R_{NIR1}/R_{NIR2}
MSI (Moisture Stress Index)	R_{1600}/R_{820}	[57,58]	-
CAI (Cellulose Absorption Index)	$0.5 * (R_{2000} + R_{2200}) - R_{2100}$	[59]	-
NDWI (Normalized Difference Water Index)	$(R_{850} - R_{1240})/(R_{850} + R_{1240})$	[60]	-
LAI related indices			
RDVI (Renormalized Difference Vegetation Index)	$(R_{800} - R_{670})/\sqrt{(R_{800} + R_{670})}$	[61]	$(R_{NIR1} - R_{Red})/\sqrt{(R_{NIR1} + R_{Red})}$
TVI (Triangular Vegetation Index)	$0.5 * [120 * (R_{750} - R_{550}) - 200 * (R_{670} - R_{550})]$	[62]	$0.5 * [120 * (R_{Red} - R_{Green}) - 200 * (R_{Red} - R_{Green})]$
NDII (Normalized Difference Infrared Index)	$(R_{850} - R_{1650})/(R_{850} + R_{1650})$	[63]	-
sLAIDI (standardized LAI Determining Index)	$5 * [(R_{1050} - R_{1250})/(R_{1050} + R_{1250})]$	[5]	-

With R_x the reflectance at band or wavelength x; Near Infrared (NIR); [a] Soil-Adjusted Vegetation Index (SAVI); [b] Chlorophyll Absorption in Reflectance Index (CARI).

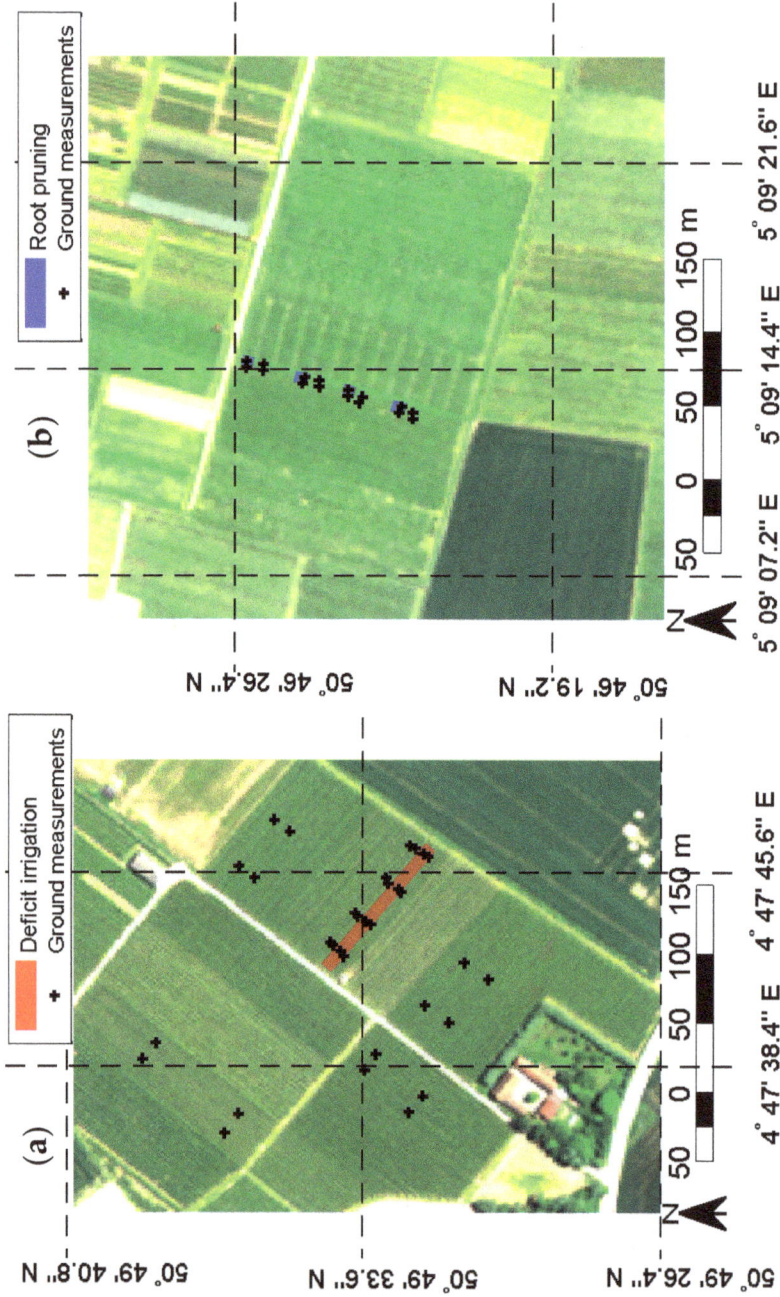

Figure 3. RGB WorldView-2 satellite imagery taken over (**a**) the irrigated pear orchard in Bierbeek, Belgium on Day of Year (DOY) 232 in 2012 and (**b**) the rainfed orchard in Kerkom, Belgium on DOY 148 in 2012. The highlighted areas indicate the deficit irrigation treatment (red) and the root-pruning treatment (blue), while the symbols represent the location of the ground measurement plots.

Vegetation indices for which representative WorldView-2 band combinations were available, are also listed in Table 2. The conversion from narrowband to broadband vegetation indices should be viewed as an approximation as the exact wavelengths from the narrowband vegetation indices overlap in two or more satellite bands, e.g., WI (Water Index [9]).

3.2. Vegetation Index Correction

To remove canopy fraction differences for all high spatial resolution orchard images, a vegetation index correction was applied [28]. This correction algorithm rescaled the range of index values for all pixels to the range of index values for the pure canopy pixels (*i.e.*, pixels with a canopy fractions over 0.8). The correction was applied in a moving window of size 7 by 7 pixels. Within each moving window, (*i*) background was assumed uniform, (*ii*) the pure canopy pixels were assumed to represent the true vegetation index range in the pure canopy pixels for that moving window and (*iii*) the presence of the full range of canopy fractions was assumed. Before the correction was applied, outliers were detected and removed, based on the threshold of 1.5 times the inter-quartile range from the upper or lower quartile [65]. Afterwards, the index values for each pixel were averaged from all the moving windows that included that pixel. In general, this correction assumed that the variability along the y-axis (*i.e.*, vegetation index value) was only caused by the variability in tree conditions, while the variability along the x-axis (*i.e.*, canopy fraction) was caused by the admixture of the background component. More information on this vegetation index correction can be found in [28].

For the synthetic images, the exact tree cover was known for each pixel. For the real satellite imagery, the canopy cover fraction was estimated through a Gram-Schmidt pan-sharpening [66] to produce multispectral bands with a panchromatic resolution (0.5 m). This pan-sharpened image was classified through unsupervised classification [67] and resampled for each 2 m multispectral pixel to provide an estimation of the canopy fractions, similar to Hamada *et al.* [68].

3.3. Determination of in situ Measured Biophysical and Structural Variables

During satellite acquisitions, biophysical variables were determined based on the inversion of leaf spectral measurements on 20 samples per plot. A validation experiment was acquired with lab measured chlorophyll and water content which achieved R^2 values of 0.79 and 0.85 between measured and modeled water and chlorophyll content respectively (results not shown here). In addition, hemispherical photographs were collected for each plot using a Kodak Professional DCS 660 digital camera (6 mega pixels) with a Sigma 8mm Circular Fisheye lens (Sigma Corporation, Tokyo, Japan). Each plot was sampled 5 times at fixed positions [69] and processed with the CAN-EYE software to determine LAI based on canopy gap fraction [70]. An overview of the mean and standard deviations of for both orchards is shown in Table 3.

Table 3. The mean (± standard deviation) of leaf chlorophyll, leaf water content and Leaf Area Index (LAI) for the reference plots in both the irrigated (*n* = 32) and rainfed orchard (*n* = 16).

Location	Chlorophyll Content (µg/cm²)	Water Content (mg/cm²)	LAI
Irrigated Orchard	82.9 (±14.1)	19.1 (±2.0)	2.5 (±0.6)
Rainfed Orchard	81.3 (±10.8)	17.8 (±3.2)	1.5 (±0.5)

4. Results

4.1. Synthetic Imagery

The synthetic imagery in this study provided exact cover fractions, spectral signatures and biophysical variables of the target crop for a range of view angles. The simulations were used to investigate the overall effect of viewing geometry on the estimation of biophysical variables through vegetation indices without the influence of sensor/target anomalies.

In a first step the ground references or tree endmembers—pure canopy pixels collected from scenes of individual trees without background—were used to illustrate optimal coefficient of determination values (R^2) between biophysical variables and vegetation indices. The variation of R^2 values between vegetation indices and chlorophyll content, water content and LAI for different viewing compositions is shown in Figure 4. As only pure canopy pixels were used, view angle effects regarding variable canopy fraction distributions could be assumed absent and resulting effects attributed to BRDF.

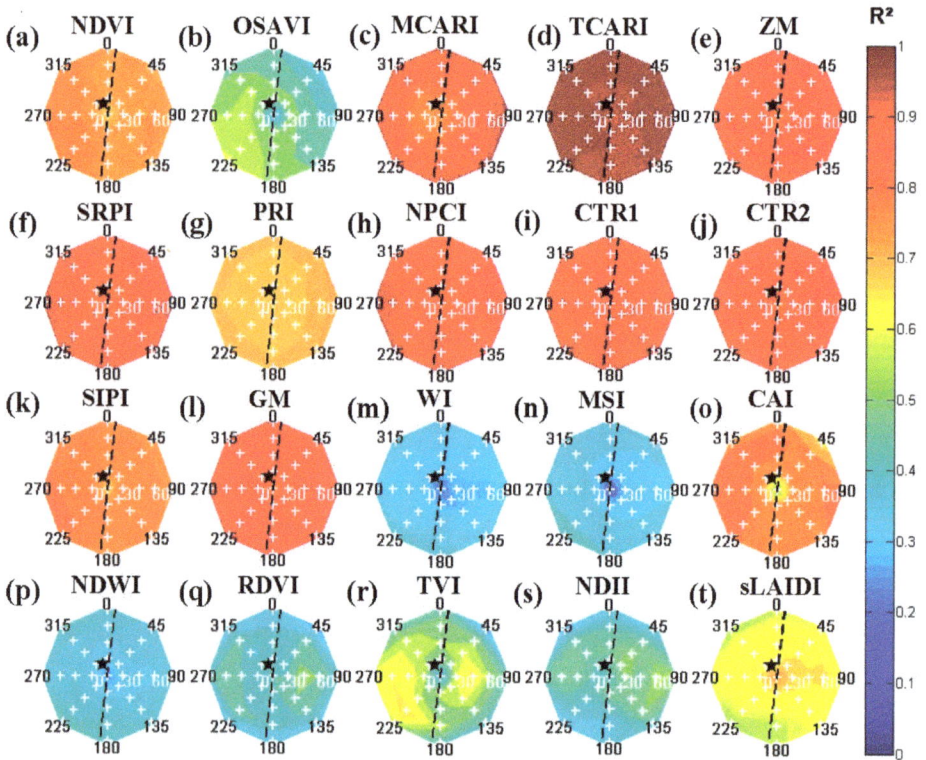

Figure 4. Viewing composition distribution of coefficient of determination (R^2) values between (**a**) Normalized Difference Vegetation Index (NDVI), (**b**) Optimization Soil-Adjusted Vegetation Index (OSAVI), (**c**) Modified Chlorophyll Absorption in Reflectance Index (MCARI), (**d**) Transformed Chlorophyll Absorption in Reflectance Index (TCARI), (**e**) Zarco and Miller (ZM), (**f**) Simple Ratio Pigment Index (SRPI), (**g**) Photochemical Reflectance Index (PRI), (**h**) Normalized Pigment Chlorophyll Index (NPCI), (**i**) Carter Index (CTR1), (**j**) Carter Index (CTR2), (**k**) Structure Insensitive Pigment Index (SIPI), (**l**) Gitelson and Merzlyak Index (GM), (**m**) Water Index (WI), (**n**) Moisture Stress Index (MSI), (**o**) Cellulose Absorption Index (CAI), (**p**) Normalized Difference Water Index (NDWI), (**q**) Renormalized Difference Vegetation Index (RDVI), (**r**) Triangular Vegetation Index (TVI), (**s**) Normalized Difference Infrared Index (NDII), (**t**) standardized Leaf Area Index Determining Index (sLAIDI) and (**a–l**) chlorophyll-, (**m–p**) water content and (**q–t**) Leaf Area Index (LAI) respectively. Star symbols denote illumination source, dotted lines denote tree row orientation and white crosses indicate different scenes. Index values were obtained from scenes without background to represent tree endmembers.

Figure 4 illustrates the complexity of viewing composition influences on biophysical and structural variables estimation through vegetation indices. On the one hand, several indices—such as the NDVI,

TCARI, PRI and CAI indices—showed an increase of R^2 values with increased sensor zenith angles. On the other hand, the sLAIDI showed higher R^2 values with LAI at nadir compared to off-nadir viewing angles. Figure 4 was only illustrative for imagery under ideal circumstances—*i.e.*, pure canopy scenes without the influence of background. Therefore, the viewing composition distribution of R^2 values for the uniform weed background (S2) are shown in Figure 5 prior to the vegetation index correction (Section 3.2). Figure 5 was limited to results from the S2 background scenario, as other scenarios presented similar distributions.

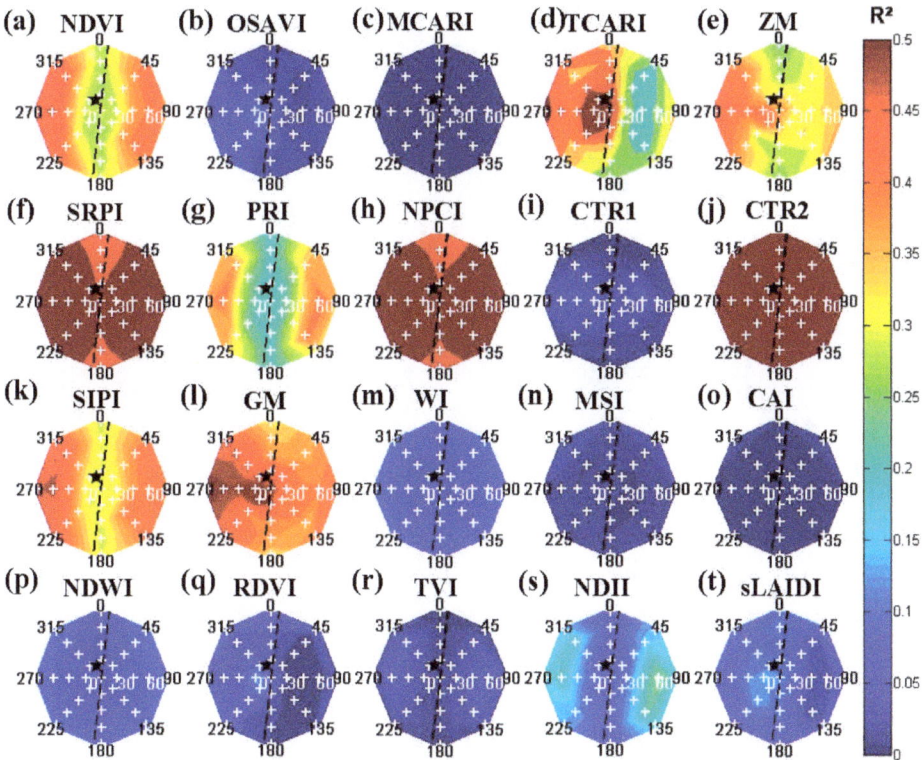

Figure 5. Viewing composition distribution of coefficient of determination (R^2) values between (a) NDVI, (b) OSAVI, (c) MCARI, (d) TCARI, (e) ZM, (f) SRPI, (g) PRI, (h) NPCI, (i) CTR1, (j) CTR2, (k) SIPI, (l) GM, (m) WI, (n) MSI, (o) CAI, (p) NDWI, (q) RDVI, (r) TVI, (s) NDII, (t) sLAIDI and (a–l) chlorophyll-, (m–p) water content and (q–t) LAI respectively. Star symbols denote illumination source, dotted lines denote tree row orientation and white crosses indicate different scenes. Index values were obtained from scenes with a uniform weed background (S2) before a vegetation index correction.

The distribution of R^2 values in Figure 5 illustrated the influence of variable viewing angles on vegetation indices in a realistic scenario (*i.e.*, uniform background of vegetation). For several indices, this distribution was mirrored around the row orientation (*i.e.*, NDVI, SRPI, PRI, SIPI and NDII). This effect could most likely be attributed to variable canopy fraction distribution, as off-nadir viewing imagery perpendicular to the row orientation would represent a decreasingly smaller fraction of background. To circumvent this problem, a vegetation index correction algorithm (Section 3.2) was applied. The distribution of R^2 values for S2 after the correction is shown in Figure 6. In addition, viewing geometry differences for R^2 values between index values and biophysical variables are listed in Table 4. For Table 4, the differences between viewing compositions were quantified through the range of R^2 values for each index and all viewing compositions in each scenario (Section 2.1).

Despite the correction algorithm removing the effects of variable canopy distribution in Figure 6 and Table 4, several indices were still significantly dependent on viewing composition. Only the distribution of R^2 values for S2 was shown in Figure 6, as other background scenarios showed similar distributions, although the scale varied depending on either vegetated or non-vegetated background scenarios.

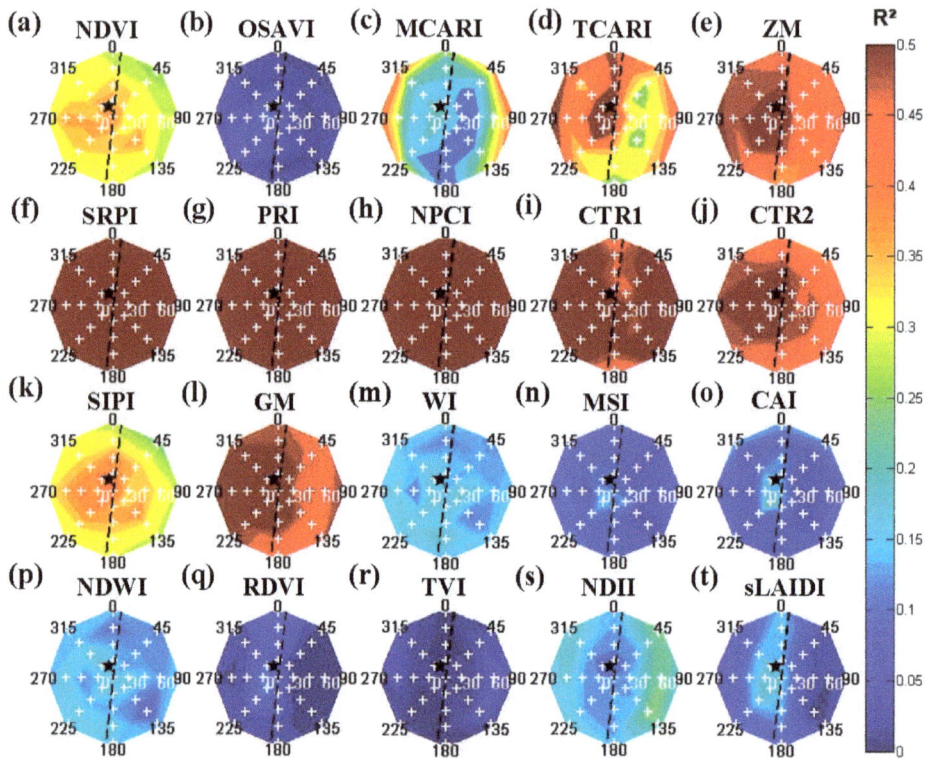

Figure 6. Viewing composition distribution of coefficient of determination (R^2) values between (a) NDVI, (b) OSAVI, (c) MCARI, (d) TCARI, (e) ZM, (f) SRPI, (g) PRI, (h) NPCI, (i) CTR1, (j) CTR2, (k) SIPI, (l) GM, (m) WI, (n) MSI, (o) CAI, (p) NDWI, (q) RDVI, (r) TVI, (s) NDII, (t) sLAIDI and (a–l) chlorophyll-, (m–p) water content and (q–t) LAI respectively. Star symbols denote illumination source, dotted lines denote tree row orientation and white crosses indicate different scenes. Index values were obtained from scenes with a uniform weed background (S2) after a vegetation index correction.

Figure 6 shows the dependency of viewing compositions for several vegetation indices while estimating biophysical and structural variables. Some indices showed higher R^2 values for nadir viewing scenes and a decrease towards more off-nadir viewing scenes, for example the ZM (Figure 6e), SIPI (Figure 6k) and GM1 indices (Figure 6l). SRPI (Figure 6f), PRI (Figure 6g) and NDII (Figure 6s) showed increased R^2 values for off-nadir scenes perpendicular to the row orientation, while R^2 values decreased for scenes parallel to row orientation for higher sensor zenith angles.

Table 4. Overview of the ranges for coefficient of determination values (R^2) between biophysical or structural variables and vegetation indices over different viewing compositions (Section 2.1). The distinction was made between chlorophyll content, water content and Leaf Area Index (LAI)—related indices. Results were shown for reference (without background) and a uniform soil background (S1), a uniform weed background (S2) background, a variable weed and soil background (S3) and a variable chlorophyll weed background (S4) after the use of a vegetation index correction (Section 3.2).

Index	Reference R^2 Range	S1 R^2 Range	S2 R^2 Range	S3 R^2 Range	S4 R^2 Range
	Chlorophyll content related indices				
NDVI	0.67 *–0.77 *	0.03 *–0.20 *	0.26 *–0.39 *	0.06 *–0.21 *	0.30 *–0.43 *
OSAVI	0.31 *–0.61 *	0.01 *–0.06 *	0.01 *–0.11 *	0.01 *–0.05 *	0.02 *–0.12 *
MCARI	0.77 *–0.95 *	0.42 *–0.53 *	0.09 *–0.46 *	0.09 *–0.33 *	0.11 *–0.49 *
TCARI	0.93 *–0.97 *	0.40 *–0.65 *	0.25 *–0.67 *	0.29 *–0.55 *	0.29 *–0.65 *
ZM	0.78 *–0.90 *	0.31 *–0.50 *	0.37 *–0.57 *	0.31 *–0.50 *	0.43 *–0.59 *
SRPI	0.81 *–0.94 *	0.41 *–0.57 *	0.59 *–0.67 *	0.45 *–0.61 *	0.59 *–0.70 *
PRI	0.68 *–0.71 *	0.56 *–0.62 *	0.52 *–0.60 *	0.34 *–0.58 *	0.55 *–0.65 *
NPCI	0.83 *–0.93 *	0.40 *–0.57 *	0.57 *–0.67 *	0.44 *–0.61 *	0.58 *–0.69 *
CTR1	0.87 *–0.91 *	0.42 *–0.58 *	0.40 *–0.63 *	0.22 *–0.48 *	0.40 *–0.66 *
CTR2	0.80 *–0.89 *	0.19 *–0.42 *	0.39 *–0.58 *	0.18 *–0.39 *	0.45 *–0.61 *
SIPI	0.69 *–0.81 *	0.04 *–0.22 *	0.24 *–0.43 *	0.07 *–0.21 *	0.28 *–0.48 *
GM	0.79 *–0.90 *	0.31 *–0.51 *	0.40 *–0.61 *	0.30 *–0.50 *	0.44 *–0.61 *
	Water content related indices				
WI	0.22 *–0.41 *	0.07 *–0.21 *	0.10 *–0.20 *	0.07 *–0.16 *	0.08 *–0.21 *
MSI	0.18 *–0.45 *	0.02 *–0.12 *	0.04 *–0.17 *	0.01 *–0.08 *	0.04 *–0.17 *
CAI	0.50 *–0.84 *	0.01 *–0.13 *	0.02 *–0.25 *	0.00–0.12 *	0.02 *–0.21 *
NDWI	0.24 *–0.45 *	0.05 *–0.15 *	0.08 *–0.18 *	0.03 *–0.08 *	0.08 *–0.20 *
	LAI related indices				
RDVI	0.26 *–0.55 *	0.01 *–0.10 *	0.00–0.13 *	0.01 *–0.08 *	0.00–0.15 *
TVI	0.23 *–0.71 *	0.00–0.01 *	0.00–0.06 *	0.00–0.01 *	0.00–0.02 *
NDII	0.30 *–0.53 *	0.07 *–0.29 *	0.09 *–0.26 *	0.05 *–0.22 *	0.09 *–0.26 *
SLAIDI	0.53 *–0.75 *	0.04 *–0.21 *	0.00 *–0.21 *	0.03 *–0.17 *	0.02 *–0.20 *

* *p*-value < 0.01.

4.2. Real Imagery

Similarly to the synthetic imagery, the satellite imagery was corrected for canopy cover fraction distribution based on the vegetation index correction algorithm described in [28]. Afterwards, index values for each plot were extracted. The R^2 values between vegetation indices and measured biophysical variables are presented in Table 5.

The results depicted in Table 5 showed significant correlations between biophysical variables and vegetation indices when combining both orchards. Significant R^2 values were found between chlorophyll content and the PRI, SRPI, NPCI and CTR1 indices. However, after processing imagery for both orchards separately, significantly higher R^2 values were found in the rainfed orchard for chlorophyll related indices. For example, R^2 values of 0.64, 0.59 and 0.58 for OSAVI, NDVI and CTR2 index values respectively. On the other hand, indices in the irrigated orchard showed no significant correlations. To visualize the influence of variable viewing geometry on the relationship between index values and biophysical variables, R^2 values were calculated for each available viewing composition. In order to compare different row orientations, only the relative angle between the row azimuth and view azimuth was used. The results are shown in Figure 7.

Table 5. Coefficient of determination (R^2) values between biophysical and structural variables and vegetation indices for satellite imagery (Section 2.2.2) after the use of a vegetation index correction (Section 3.2). The distinction was made between chlorophyll content, water content and Leaf Area Index (LAI)—related indices for both orchards combined and for each orchard separately. Indices for which appropriate bands were not available were omitted (Table 2).

Index	R^2 Values both Orchards (n = 232)	R^2 Values Irrigated Orchard (n = 144)	R^2 Values Rainfed Orchard (n = 88)	Index	R^2 Values both Orchards (n = 232)	R^2 Values Irrigated Orchard (n = 144)	R^2 Values Rainfed Orchard (n = 88)
Chlorophyll content related indices				*Water content related indices*			
NDVI	0.03	0.00	0.59 *	WI	0.00	0.01	0.16 *
OSAVI	0.01	0.06 *	0.64 *	MSI	-	-	-
MCARI	0.03	0.12 *	0.48 *	CAI	-	-	-
TCARI	0.01	0.00	0.02	NDWI	-	-	-
ZM	-	-	-				
SRPI	0.14 *	0.04	0.30 *	*LAI related indices*			
PRI	0.25 *	0.06 *	0.16 *	RDVI	0.00	0.00	0.01
NPCI	0.18 *	0.03	0.40 *	TVI	0.01	0.01	0.02
CTR1	0.37 *	0.01	0.27 *	NDII	-	-	-
CTR2	0.12 *	0.04 *	0.58 *	SLAIDI	-	-	-
SIPI	0.02	0.02	0.38 *				
GM	0.05 *	0.02	0.43 *				

* *p*-value < 0.01.

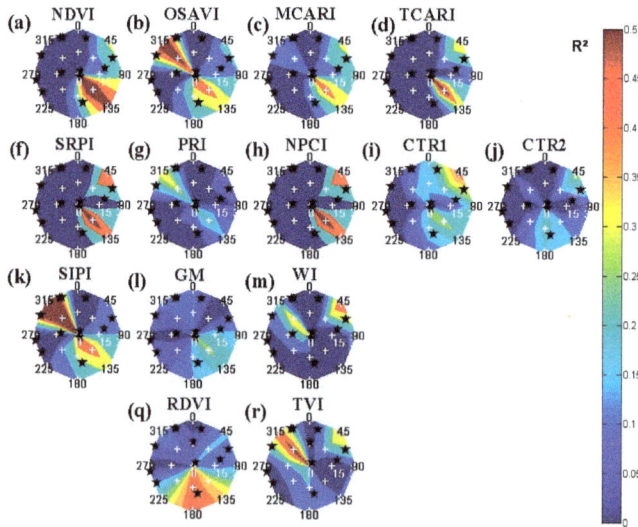

Figure 7. Viewing composition distribution of coefficient of determination (R^2) values between (**a**) Normalized Difference Vegetation Index (NDVI), (**b**) Optimization Soil-Adjusted Vegetation Index (OSAVI), (**c**) Modified Chlorophyll Absorption in Reflectance Index (MCARI), (**d**) Transformed Chlorophyll Absorption in Reflectance Index (TCARI), (**f**) Simple Ratio Pigment Index (SRPI), (**g**) Photochemical Reflectance Index (PRI), (**h**) Normalized Pigment Chlorophyll Index (NPCI), (**i**) Carter Index (CTR1), (**j**) Carter Index (CTR2), (**k**) Structure Insensitive Pigment Index (SIPI), (**l**) Gitelson and Merzlyak Index (GM), (**m**) Water Index (WI), (**q**) Renormalized Difference Vegetation Index (RDVI), (**r**) Triangular Vegetation Index (TVI) and (**a**–**l**) chlorophyll-, (**m**) water content and (**q**–**r**) Leaf Area Index (LAI) respectively. Star symbols viewing geometries. Index values were obtained after the use of a vegetation index correction (Section 3.2) from imagery at irrigated and rainfed orchard (Section 2.2).

Figure 7 showed the influence of viewing geometry on the estimation of biophysical and structural variables through vegetation indices. For several indices, the shaded side of the canopy presented higher R^2 values compared to the sunlit side—*i.e.*, NDVI (Figure 7a), MCARI (Figure 7c), TCARI (Figure 7d), CTR2 (Figure 7j) and RDVI (Figure 7q). Conversely, several indices were presenting higher R^2 values for the sunlit side compared to the shaded side—*i.e.*, CTR1 (Figure 7i) and TVI (Figure 7r). Finally, some indices presented high R^2 values for both sunlit and shaded sides—*i.e.*, OSAVI (Figure 7b), SIPI (Figure 7k) and WI (Figure 7m)—further illustrating that variable viewing geometry should be avoided or accounted for.

5. Discussion

The use of remote sensing within capital intensive orchard crops provides alternatives to time consuming, labor intensive and destructive *in situ* measurements. Practical use in hedgerow orchards under temperate climate conditions requires the use of high spatial resolution satellite sensors with off-nadir viewing capabilities [10,11]. However, this presence of variable viewing angles could affect the relationship between biophysical variables and spectral measurements [4,12,13]. Moreover, variable viewing angles within and between time series could obstruct the distinction between genuine changes and viewing geometry influences. In this study, the distribution of R^2 values for different viewing compositions (Figures 4–7) and the ranges for different background scenarios (Table 4) showed the necessity of careful selection of vegetation indices for applications using spectral imagery acquired under or subject to multiple viewing geometries.

For high spatial resolution imagery, changing viewing geometries would cause both differences related to Bidirectional Reflectance Distribution Function (BRDF) influences [21] and to variable canopy fraction mixtures. The former was visible for the tree endmembers, rendered without background and extracted from pure canopy pixels (Figure 4). For these reference images, several indices showed a variable distribution favoring more nadir viewing—*i.e.*, sLAIDI (Figure 4t)—or off-nadir viewing scenes—*i.e.*, OSAVI (Figure 4b), WI (Figure 4m), CAI (Figure 4o) and TVI (Figure 4r). These effect could be attributed to the inherent BRDF effects and should be avoided or corrected for [16–21]. However, other indices showed a less significant change in R^2 values and were less affected for these ideal circumstances.

The synthetic imagery also demonstrated a high dependence of several indices towards canopy fraction distribution (Figure 5), which was removed through a vegetation index correction (Figure 6). After the correction, the lowest R^2 values from the synthetic imagery were found for the most variable background scenario—*i.e.*, the soil and weed background or S3 (Table 4). Generally, the differences between different background scenarios were relatively small. This was most likely the result of the vegetation index correction [28], which removed most of the influence of background mixtures and variable canopy cover fractions. Exceptions hereof were indices that were related to vegetative cover fraction, e.g., NDVI [71], which showed significantly higher R^2 values for vegetated backgrounds (S2 and S4, Table 4) compared to non-vegetated backgrounds (S1 and S3, Table 4). This could be the result of the assumption of Lambertian behavior for the soil backgrounds. Although the range of R^2 values will be similar, the R^2 values could be lower for non-Lambertian backgrounds. Overall, the R^2 values were significantly higher between vegetation indices and chlorophyll content compared to LAI and water content, because of larger differences between treatments and the inherent relationship between leaf area and the amount of canopy water content [5].

For the synthetic imagery, several commonly used vegetation indices presented an influence towards variable viewing geometries. PRI, NDII and SRPI values showed higher R^2 values perpendicular to the row orientation compared to parallel to rows for higher off-nadir view angles (Figure 6). This might be attributed to the inclusion of background with parallel viewing to the rows, as imagery perpendicular to the rows would not present any background. Overall, differences between off-nadir viewing parallel or perpendicular to the row orientation might be greater without the use of the vegetation index correction to normalize for different canopy cover fractions. On the other hand, this view angle sensitivity of PRI was similar to [24,25,72], finding a large difference between PRI values from forward (shaded) and backward (sunlit) scattered scenes. Other indices, such as NDVI, GM, ZM and SIPI, showed a decrease in R^2 values from nadir to off-nadir viewing scenes (Figure 6). These results were similar to studies indicating a high dependence of NDVI towards variable viewing angles [24,73,74] and a degraded correlation with NDVI-derived products [18].

The R^2 values of satellite derived vegetation indices (Table 5) were significantly lower compared to the synthetic images (Table 4). This was most likely by the restricted amount of point measurements for the real imagery and the use of different row orientations within and between the orchards. On the other hand, also a suboptimal vegetation index correction could cause a decrease in R^2 values between indices and variables. For the rainfed orchard, R^2 increased significantly because of the similar row azimuth for all ground control plots. For the irrigated orchard, a significant decrease in R^2 values was the result of both sunlit and shaded canopies within one scene, as both 41 and 131° row orientations were monitored. The increased influence of shadow compared to the synthetic imagery was caused by a decrease in illumination elevation and the presence of a hedgerow cropping system causing one predominantly shaded and sunlit side [41,75]. Several studies have shown the negative effect of shaded canopy parts on the correlation between vegetation indices and biophysical variables [3,4,76]. As the virtual orchard trees consisted of spherical canopies, shaded canopy sections and the effects of shadow on the vegetation indices were of less importance [3].

The vegetation index correction algorithm could not mitigate the inclusion of shadow on the variation of index values, as it is not directly related to canopy cover fraction. Conversely, problems could be avoided by segmenting or classifying the canopies into sunlit and shaded areas—e.g., Stagakis *et al.* [2,3]. However, for hedgerow systems problems with canopy anisotropy would be enlarged and cause significant differences between both faces of the canopy [13], obstructing derivation of structural variables—*i.e.*, plant cover, canopy height and biomass—from multiple viewing angles. Another solution would be the minimization of shaded pixels in the analysis [3,76,77]. To illustrate the usefulness of this approach, a selection was made for all satellite images (Section 2.2) based on row orientation, viewing and illumination angles. Similarly to [4], scenes with high off-nadir viewing angles—*i.e.*, off-nadir viewing angles over 20°—and viewing geometries opposite to the sun were removed as they consisted mostly of shaded canopies. Off-nadir viewing angles under 20° would result in partially sunlit scenes and were shown less affected [4]. The average R^2 values for all indices after the selection of sunlit image pixels are shown in Table 6.

Table 6 illustrated the usefulness of pixel selection based on scene illumination for the irrigated orchard, as a significant increase in R^2 values was found between biophysical variables and index values compared to Table 5. The indices showing a significant increase also showed higher R^2 values for sunlit scene compositions in Figure 7. On the other hand, indices which presented higher R^2 values for shaded scene compositions—*i.e.*, NDVI, CTR2, MCARI, OSAVI and TCARI—showed similar or decreased R^2 values.

For the rainfed orchard, a significant decrease of R^2 values was present for almost all indices after the selection of sunlit pixels compared to Table 5. This was the effect of the different growing systems within both orchards. On the one hand, the Spindle bush system (rainfed orchard) resulted in open tree canopies and more sunlit areas [41]. On the other hand, the hedgerow V-system (irrigated orchard) resulted in variable canopy faces—*i.e.*, one predominantly sunlit and shaded side [41].

With regards to the optimal vegetation indices for high spatial resolution imagery, PRI, CTR1, SIPI and GM provided more stable correlations with chlorophyll content for variable viewing geometries (Figure 6, Tables 5 and 6). However, these indices were highly affected by the canopy fraction distribution and should be corrected or normalized prior to analysis (Figures 5 and 6). With regards to water content and LAI, WI and sLAIDI provided good correlations for nadir viewing scenes, with a significant decrease towards off-nadir viewing angles. Other vegetation indices could be used in certain circumstances, but should be avoided in time series with variable viewing angles, e.g., NDVI, OSAVI, MCARI, TCARI, RDVI and TVI (Figure 6, Tables 5 and 6). Although the vegetation indices (Table 2) were selected based on previous studies in fruit orchards, the list was not complete [1,2,64]. Further research is required for vegetation indices not represented here prior to their use for applications with multiple viewing angles.

Table 6. Coefficient of determination (R^2) values between biophysical and structural variables and vegetation indices for satellite imagery (Section 2.2.2) after the use of a vegetation index correction (Section 3.2). The distinction was made between chlorophyll content, water content and Leaf Area Index (LAI)—related indices for both orchards combined and for each orchard separately. Pixels were selected based on the combination of large off-nadir viewing angles (*i.e.*, over 20°) and a viewing geometry on the opposite side of the rows with regards to the illumination source. Indices for which appropriate bands were not available were omitted (Table 2).

Index	R^2 Values both Orchards ($n = 160$)	R^2 Values Irrigated Orchard ($n = 112$)	R^2 Values Rainfed Orchard ($n = 48$)	Index	R^2 Values both Orchards ($n = 160$)	R^2 Values Irrigated Orchard ($n = 112$)	R^2 Values Rainfed Orchard ($n = 48$)
Chlorophyll content related indices				*Water content related indices*			
NDVI	0.01	0.01	0.01	WI	0.07 *	0.13 *	0.00
OSAVI	0.04	0.07 *	0.04	MSI	-	-	-
MCARI	0.14 *	0.22 *	0.00	CAI	-	-	-
TCARI	0.06 *	0.04	0.02	NDWI	-	-	-
ZM	-	-	-				
SRPI	0.07 *	0.09 *	0.13	*LAI related indices*			
PRI	0.24 *	0.28 *	0.20	RDVI	0.02	0.05	0.00
NPCI	0.09 *	0.11 *	0.14	TVI	0.00	0.17 *	0.01
CTR1	0.37 *	0.42 *	0.29 *	NDII	-	-	-
CTR2	0.02	0.06*	0.01	SLAIDI	-	-	-
SIPI	0.16*	0.24*	0.21				
GM	0.20*	0.23*	0.07				

* *p*-value < 0.01.

Overall, the results illustrated the necessity of a vegetation index selection or correction based on each specific circumstance and data set. However, through the careful selection of vegetation indices, an optimal derivation of biophysical variable should be plausible for applications with multiple viewing angles. Moreover, depending on the size of the available time series for a fixed target location, index values could be normalized with regards to view angle—e.g., Seaquist and Olsson [73]. However, the most important limitation to this approach is the high amount of images required within a relatively small time window to minimize influences of a changing solar angle.

6. Conclusions

Stress-related biophysical and structural variables of capital intensive orchard crops can be approximated with accurate and consistent estimations of spectral vegetation indices from off-nadir viewing satellite imagery. However, the variable viewing compositions of each image could affect this relationship. Most of the research studying this effect has focused on cropping systems with complete canopy cover, while imagery over orchard cropping systems could cause mixtures between canopies and backgrounds. This study investigated the sensitivity of common spectral vegetation indices on changing viewing geometries and its relationship with the estimation of biophysical variables—*i.e.*, chlorophyll content, water content and Leaf Area Index (LAI). Ultimately to minimize variable viewing influences within and between time series, which could obstruct change detection because of the confusion with genuine changes. This was achieved through the use of synthetic and satellite imagery over a virtual citrus orchard and an experimental hedgerow pear orchard respectively.

Results indicated significant differences between nadir and off-nadir viewing scenes for some indices ($\Delta R^2 > 0.4$). Several indices—such as the Photochemical Reflectance Index (PRI), Normalized

Difference Infrared Index (NDII) and Simple Ratio Pigment Index (SRPI)—showed higher coefficient of determination (R^2) values for higher off-nadir view angles perpendicular compared to parallel to the row orientation. On the other hand, indices—such as the Normalized Difference Vegetation Index (NDVI), Gitelson and Merzlyak (GM), Zarco and Miller (ZM) and Structure Insensitive Pigment Index (SIPI)—showed a decrease in R^2 values from nadir to off-nadir viewing scenes.

In general, this study showed the necessity for a careful selection of vegetation indices for estimating biophysical variables, especially for applications with multiple viewing angles.

Acknowledgments: This work was supported by the Agency for Innovation by Science and Technology in Flanders (IWT-Vlaanderen). The research was funded through a project in cooperation with the Soil Service of Belgium (BDB) and the Research Center for Fruit (Proefcentrum Fruitteelt in Sint-Truiden). The authors would like to thank fruit grower Jan Vandervelpen for allowing research in his orchard and making this work possible.

Author Contributions: The contribution of the authors to this work were: Jonathan Van Beek [1,2,3,4,5], Laurent Tits [2,3,4], Ben Somers [2,3,4], Tom Deckers [1,4,5], Pieter Janssens [1,4,5] and Pol Coppin [1,4,5]. [1] Experimental design; [2] analysis and interpretation; [3] manuscript preparation and writing; [4] paper review; [5] project coordination.

Conflicts of Interest: The authors declare no conflict of interest.

References

1. Zarco-Tejada, P.J.; Berjon, A.; Lopez-Lozano, R.; Miller, J.R.; Martin, P.; Cachorro, V.; Gonzalez, M.R.; de Frutos, A. Assessing vineyard condition with hyperspectral indices: Leaf and canopy reflectance simulation in a row-structured discontinuous canopy. *Remote Sens. Environ.* **2005**, *99*, 271–287. [CrossRef]
2. Stagakis, S.; Markos, N.; Sykioti, O.; Kyparissis, A. Monitoring canopy biophysical and biochemical parameters in ecosystem scale using satellite hyperspectral imagery: An application on a Phlomis fruticosa Mediterranean ecosystem using multiangular CHRIS/PROBA observations. *Remote Sens. Environ.* **2010**, *114*, 977–994. [CrossRef]
3. Stagakis, S.; González-Dugo, V.; Cid, P.; Guillén-Climent, M.L.; Zarco-Tejada, P.J. Monitoring water stress and fruit quality in an orange orchard under regulated deficit irrigation using narrow-band structural and physiological remote sensing indices. *ISPRS J. Photogramm. Remote Sens.* **2012**, *71*, 47–61. [CrossRef]
4. Van Beek, J.; Tits, L.; Somers, B.; Coppin, P. Stem Water Potential Monitoring in Pear Orchards through WorldView-2 Multispectral Imagery. *Remote Sens.* **2013**, *5*, 6647–6666. [CrossRef]
5. Delalieux, S.; Somers, B.; Hereijgers, S.; Verstraeten, W.; Keulemans, W.; Coppin, P. A near-infrared narrow-waveband ratio to determine Leaf Area Index in orchards. *Remote Sens. Environ.* **2008**, *112*, 3762–3772. [CrossRef]
6. Pinter, P.J.; Hatfield, J.L.; Schepers, J.S.; Barnes, E.M.; Moran, M.S.; Daughtry, C.S.T.; Upchurch, D.R. Remote Sensing for Crop Management. *Photogramm. Eng. Remote Sens.* **2003**, *69*, 647–664. [CrossRef]
7. Dorigo, W.A.; Zurita-Milla, R.; de Wit, A.J.W.; Brazile, J.; Singh, R.; Schaepman, M.E. A review on reflective remote sensing and data assimilation techniques for enhanced agroecosystem modeling. *Int. J. Appl. Earth Obs. Geoinf.* **2007**, *9*, 165–193. [CrossRef]
8. Gamon, J.A.; Penuelas, J.; Field, C.B. A narrow-waveband spectral index that tracks diurnal changes in photosynthetic efficiency. *Remote Sens. Environ.* **1992**, *41*, 35–44.
9. Penuelas, J.; Filella, I.; Biel, C.; Serrano, L.; Savé, R. The reflectance at the 950–970 nm region as an indicator of plant water status. *Int. J. Remote Sens.* **1993**, *14*, 1887–1905. [CrossRef]
10. Moran, S.; Fitzgerald, G.; Rango, A.; Walthall, C.; Barnes, E.; Bausch, W.; Clarke, T.; Daughtry, C.; Everitt, J.; Escobar, D.; *et al.* Sensor Development and Radiometric Correction for Agricultural Applications. *Photogramm. Eng. Remote Sens.* **2003**, *69*, 705–718. [CrossRef]
11. Perry, E.M.; Dezzani, R.J.; Seavert, C.F.; Pierce, F.J. Spatial variation in tree characteristics and yield in a pear orchard. *Precis. Agric.* **2009**, *11*, 42–60. [CrossRef]
12. Stuckens, J.; Dzikiti, S.; Verstraeten, W.W.; Verreynne, S.; Swennen, R.; Coppin, P. Physiological interpretation of a hyperspectral time series in a citrus orchard. *Agric. For. Meteorol.* **2011**, *151*, 1002–1015. [CrossRef]
13. Meggio, F.; Zarco-Tejada, P.J.; Miller, J.R.; Martín, P.; González, M.R.; Berjón, A. Row orientation and viewing geometry effects on row-structured vine crops for chlorophyll content estimation. *Can. J. Remote Sens.* **2008**, *34*, 220–234.

14. Chopping, M.; Moisen, G.G.; Su, L.; Laliberte, A.; Rango, A.; Martonchik, J.V.; Peters, D.P.C. Large area mapping of southwestern forest crown cover, canopy height, and biomass using the NASA Multiangle Imaging Spectro-Radiometer. *Remote Sens. Environ.* **2008**, *112*, 2051–2063. [CrossRef]
15. Wolf, A.; Berry, J.A.; Asner, G.P. Allometric constraints on sources of variability in multi-angle reflectance measurements. *Remote Sens. Environ.* **2010**, *114*, 1205–1219. [CrossRef]
16. Shepherd, J.D.; Dymond, J.R. BRDF Correction of Vegetation in AVHRR Imagery. *Remote Sens. Environ.* **2000**, *74*, 397–408. [CrossRef]
17. Schaaf, C.B.; Gao, F.; Strahler, A.H.; Lucht, W.; Li, X.; Tsang, T.; Strugnell, N.C.; Zhang, X.; Jin, Y.; Muller, J.-P.; *et al.* First operational BRDF, albedo nadir reflectance products from MODIS. *Remote Sens. Environ.* **2002**, *83*, 135–148. [CrossRef]
18. Huete, A.; Didan, K.; Miura, T.; Rodriguez, E.; Gao, X.; Ferreira, L. Overview of the radiometric and biophysical performance of the MODIS vegetation indices. *Remote Sens. Environ.* **2002**, *83*, 195–213. [CrossRef]
19. Vermote, E.; Justice, C.O.; Breon, F.-M. Towards a Generalized Approach for Correction of the BRDF Effect in MODIS Directional Reflectances. *IEEE Trans. Geosci. Remote Sens.* **2009**, *47*, 898–908. [CrossRef]
20. Bréon, F.-M.; Vermote, E. Correction of MODIS surface reflectance time series for BRDF effects. *Remote Sens. Environ.* **2012**, *125*, 1–9. [CrossRef]
21. De Abelleyra, D.; Verón, S.R. Comparison of different BRDF correction methods to generate daily normalized MODIS 250m time series. *Remote Sens. Environ.* **2014**, *140*, 46–59. [CrossRef]
22. Sandmeier, S.; Müller, C.; Hosgood, B.; Andreoli, G. Physical mechanisms in hyperspectral BRDF data of grass and watercress. *Remote Sens. Environ.* **1998**, *66*, 222–233. [CrossRef]
23. Okin, G.S.; Clarke, K.D.; Lewis, M.M. Comparison of methods for estimation of absolute vegetation and soil fractional cover using MODIS normalized BRDF-adjusted reflectance data. *Remote Sens. Environ.* **2013**, *130*, 266–279. [CrossRef]
24. Verrelst, J.; Schaepman, M.E.; Koetz, B.; Kneubühler, M. Angular sensitivity analysis of vegetation indices derived from CHRIS/PROBA data. *Remote Sens. Environ.* **2008**, *112*, 2341–2353. [CrossRef]
25. Hall, F.G.; Hilker, T.; Coops, N.C.; Lyapustin, A.; Huemmrich, K.F.; Middleton, E.; Margolis, H.; Drolet, G.; Black, T.A. Multi-angle remote sensing of forest light use efficiency by observing PRI variation with canopy shadow fraction. *Remote Sens. Environ.* **2008**, *112*, 3201–3211. [CrossRef]
26. Galvão, L.S.; Roberts, D.A.; Formaggio, A.R.; Numata, I.; Breunig, F.M. View angle effects on the discrimination of soybean varieties and on the relationships between vegetation indices and yield using off-nadir Hyperion data. *Remote Sens. Environ.* **2009**, *113*, 846–856. [CrossRef]
27. Stuckens, J.; Somers, B.; Albrigo, G.L.; Dzikiti, S.; Verstraeten, W.W. Off-nadir Viewing for Reducing Spectral Mixture Issues in Citrus Orchards. *Photogramm. Eng. Remote Sens.* **2010**, *76*, 1261–1274. [CrossRef]
28. Van Beek, J.; Tits, L.; Somers, B.; Deckers, T.; Janssens, P.; Coppin, P. Reducing background effects in orchards through spectral vegetation index correction. *Int. J. Appl. Earth Obs. Geoinf.* **2014**, *34*, 167–177. [CrossRef]
29. Tits, L.; Somers, B.; Saeys, W.; Coppin, P. Site-Specific Plant Condition Monitoring through Hyperspectral Alternating Least Squares Unmixing. *IEEE J. Sel. Top. Appl. Earth Obs. Remote Sens.* **2014**, *7*, 3606–3618. [CrossRef]
30. Tits, L.; De Keersmaecker, W.; Somers, B.; Asner, G.P.; Farifteh, J.; Coppin, P. Hyperspectral shape-based unmixing to improve intra- and interclass variability for forest and agro-ecosystem monitoring. *ISPRS J. Photogramm. Remote Sens.* **2012**, *74*, 163–174. [CrossRef]
31. Tits, L.; Somers, B.; Stuckens, J.; Farifteh, J.; Coppin, P. Integration of *in situ* measured soil status and remotely sensed hyperspectral data to improve plant production system monitoring: Concept, perspectives and limitations. *Remote Sens. Environ.* **2013**, *128*, 197–211. [CrossRef]
32. Stuckens, J.; Somers, B.; Delalieux, S.; Verstraeten, W.W.; Coppin, P. The impact of common assumptions on canopy radiative transfer simulations: A case study in Citrus orchards. *J. Quant. Spectrosc. Radiat. Transf.* **2009**, *110*, 1–21. [CrossRef]
33. Jacquemoud, S.; Baret, F. Prospect: A model of leaf optical properties spectra. *Remote Sens. Environ.* **1990**, *34*, 75–91. [CrossRef]
34. Pharr, M.; Humphreys, G. *Physically Based Rendering, From Theory to Implementation*; Morgan Kaufmann: San Fransisco, CA, USA, 2004; p. 1019.
35. Widlowski, J.-L.; Taberner, M.; Pinty, B.; Bruniquel-Pinel, V.; Disney, M.; Fernandes, R.; Gastellu-Etchegorry, J.-P.; Gobron, N.; Kuusk, A.; Lavergne, T.; *et al.* Third Radiation Transfer Model

Intercomparison (RAMI) exercise: Documenting progress in canopy reflectance models. *J. Geophys. Res.* **2007**, *112*, D09111. [CrossRef]

36. Widlowski, J.-L.; Mio, C.; Disney, M.; Adams, J.; Andredakis, I.; Atzberger, C.; Brennan, J.; Busetto, L.; Chelle, M.; Ceccherini, G.; *et al.* The fourth phase of the radiative transfer model intercomparison (RAMI) exercise: Actual canopy scenarios and conformity testing. *Remote Sens. Environ.* **2015**, *169*, 418–437. [CrossRef]

37. Somers, B.; Delalieux, S.; Verstraeten, W.W.; Coppin, P. A Conceptual Framework for the Simultaneous Extraction of Sub-pixel Spatial Extent and Spectral Characteristics of Crops. *Photogramm. Eng. Remote Sens.* **2009**, *75*, 57–68. [CrossRef]

38. FAO. *Soils Map of the World: Revised Legend*; Food and Agriculture Organization of the United Nations: Rome, Italy, 1988.

39. Somers, B.; Gysels, V.; Verstraeten, W.W.; Delalieux, S.; Coppin, P. Modelling moisture-induced soil reflectance changes in cultivated sandy soils: a case study in citrus orchards. *Eur. J. Soil Sci.* **2010**, *61*, 1091–1105. [CrossRef]

40. Hosgood, B.; Jacquemoud, S.; Andreoli, G.; Verdebout, J.; Pedrini, A.; Schmuck, G. *Leaf Optical Properties EXperiment 93 (LOPEX93)*; Report EUR 16095 EN; European Commission—Joint Research Centre: Ispra, Italy, 1994.

41. Sansavini, S.; Musacchi, S. Canopy architecture, training and pruning in the modern European pear orchards: An overview. *Acta Hortic.* **1994**, *367*, 152–172. [CrossRef]

42. Allen, R.G.; Pereira, L.S.; Raes, D.; Smith, M. *Crop Evapotranspiration—Guidelines for Computing Crop Water Requirements—FAO Irrigation and Drainage Paper 56*; FAO: Rome, Italy, 1998.

43. Janssens, P.; Deckers, T.; Elsen, F.; Elsen, A.; Schoofs, H.; Verjans, W.; Vandendriessche, H. Sensitivity of root pruned "Conference" pear to water deficit in a temperate climate. *Agric. Water Manag.* **2011**, *99*, 58–66. [CrossRef]

44. Updike, T.; Comp, C. *Radiometric Use of WorldView-2 Imagery Technical Note*; DigitalGlobe: Westminster, CO, USA, 2010; p. 16.

45. Adler-Golden, S.M.; Berk, A.; Bernstein, L.S.; Richtsmeier, S.; Acharya, P.K.; Matthew, M.W.; Anderson, G.P.; Allred, C.L.; Jeong, L.S.; Chetwynd, J.H. FLAASH, A MODTRAN4 Atmospheric correction package for Hyperspectral Data retrievals and simulations. In Proceedings of the Summaries of the Seventh JPL Airborne Earth Science Workshop, Pasadena, CA, USA, 12–16 January 1998; p. 442.

46. Grodecki, J.; Dial, G. Block Adjustment of High-Resolution Satellite Images Described by Rational Polynomials. *Photogramm. Eng. Remote Sens.* **2003**, *69*, 59–68. [CrossRef]

47. Rouse, J.W.; Haas, R.H.; Schell, J.A.; Deering, D.W.; Harlan, J.C. *Monitoring the Vernal Advancements and Retrogradation of Natural Vegetation*; NASA Technical Report 1974; NASA: College Station, TX, USA; p. 371.

48. Rondeaux, G.; Steven, M.; Baret, F. Optimization of Soil-Adjusted Vegetation Indices. *Remote Sens. Environ.* **1996**, *107*, 95–107. [CrossRef]

49. Daughtry, C.S.T.; Walthall, C.L.; Kim, M.S.; Brown de Colstoun, E.; McMurtrey, J.E., III. Estimating Corn Leaf Chlorophyll Concentration from Leaf and Canopy Reflectance. *Remote Sens. Environ.* **2000**, *74*, 229–239. [CrossRef]

50. Haboudane, D.; Miller, J.R.; Tremblay, N.; Zarco-Tejada, P.J.; Dextraze, L. Integrated narrow-band vegetation indices for prediction of crop chlorophyll content for application to precision agriculture. *Remote Sens. Environ.* **2002**, *81*, 416–426. [CrossRef]

51. Zarco-Tejada, P.J.; Miller, J.R.; Noland, T.L.; Mohammed, G.H.; Sampson, P.H. Scaling-up and model inversion methods with narrowband optical indices for chlorophyll content estimation in closed forest canopies with hyperspectral data. *IEEE Trans. Geosci. Remote Sens.* **2001**, *39*, 1491–1507. [CrossRef]

52. Penuelas, J.; Filella, I.; Lloret, P.; Munoz, F.; Vilajeliu, M. Reflectance assessment of mite effects on apple trees. *Int. J. Remote Sens.* **1995**, *16*, 2727–2733. [CrossRef]

53. Penuelas, J.; Gamon, J.A.; Freeden, A.L.; Merino, J.; Field, C.B. Reflectance Indices Associated with Physiological Changes in Nitrogen- and Water-Limited Sunflower Leaves. *Remote Sens. Environ.* **1994**, *48*, 135–146. [CrossRef]

54. Carter, G.A. Ratios of leaf reflectances in narrow wavebands as indicators of plant stress. *Int. J. Remote Sens.* **1994**, *15*, 697–703. [CrossRef]

55. Carter, G.A.; Cibula, W.G.; Dell, T.R. Spectral reflectance characteristics and digital imagery of a pine needle blight in the southeastern United States. *Can. J. For. Res.* **1996**, *26*, 402–407. [CrossRef]

56. Gitelson, A.A.; Merzlyak, M.N. Signature Analysis of Leaf Reflectance Spectra: Algorithm Development for Remote Sensing of Chlorophyll. *J. Plant Physiol.* **1996**, *148*, 494–500. [CrossRef]

57. Hunt, E.R.; Rock, B.N. Detection of changes in leaf water of, content using near- and middle-infrared reflectances. *Remote Sens. Environ.* **1989**, *30*, 43–54.
58. Ceccato, P.; Flasse, S.; Tarantola, S.; Jacquemoud, S.; Grégoire, J.-M. Detecting vegetation leaf water content using reflectance in the optical domain. *Remote Sens. Environ.* **2001**, *77*, 22–33. [CrossRef]
59. Nagler, P.L.; Daughtry, C.S.T.; Goward, S.N. Plant litter and soil reflectance. *Remote Sens. Environ.* **2000**, *71*, 207–215. [CrossRef]
60. Gao, B. NDWI A Normalized Difference Water Index for Remote Sensing of Vegetation Liquid Water From Space. *Remote Sens. Environ.* **1996**, *58*, 257–266. [CrossRef]
61. Roujean, J.-L.; Breon, F.-M. Estimating PAR absorbed by vegetation from bidirectional reflectance measurements. *Remote Sens. Environ.* **1995**, *51*, 375–384. [CrossRef]
62. Broge, N.H.; Leblanc, E. Comparing prediction power and stability of broadband and hyperspectral vegetation indices for estimation of green leaf area index and canopy chlorophyll density. *Remote Sens. Environ.* **2001**, *76*, 156–172. [CrossRef]
63. Yilmaz, M.T.; Hunt, E.R.; Jackson, T.J. Remote sensing of vegetation water content from equivalent water thickness using satellite imagery. *Remote Sens. Environ.* **2008**, *112*, 2514–2522. [CrossRef]
64. Rodríguez-pérez, J.R.; Riaño, D.; Carlisle, E.; Ustin, S.; Smart, D.R. Evaluation of Hyperspectral Reflectance Indexes to Detect Grapevine Water Status in Vineyards. *Am. J. Enol. Vitic.* **2007**, *58*, 302–317.
65. Laurikkala, J.; Juhola, M.; Kentala, E. Informal Identification of Outliers in Medical Data. In *Fifth International Workshop on Intelligent Data Analysis in Medicine and Pharmacology IDAMAP-2000 Berlin, 22 August*, Proceedings of the Organized as a workshop of the 14th European Conference on Artificial Intelligence ECAI-2000, Berlin, Germany, 20–25 August 2000.
66. Laben, C.A.; Brower, B.V. Process for Enhancing the Spatial Resolution of Multispectral Imagery Using Pan-Sharpening. U.S. Patent No. 6,011,875, 4 January 2000.
67. Tou, J.T.; Gonzalez, R.C. *Pattern Recognition Principles*; Addison-Wesley Publishing Company: Reading, MA, USA, 1974; p. 377.
68. Hamada, Y.; Stow, D.A.; Roberts, D.A. Estimating life-form cover fractions in California sage scrub communities using multispectral remote sensing. *Remote Sens. Environ.* **2011**, *115*, 3056–3068. [CrossRef]
69. López-Lozano, R.; Baret, F.; García de Cortázar-Atauri, I.; Bertrand, N.; Casterad, M.A. Optimal geometric configuration and algorithms for LAI indirect estimates under row canopies: The case of vineyards. *Agric. For. Meteorol.* **2009**, *149*, 1307–1316. [CrossRef]
70. Demarez, V.; Duthoit, S.; Baret, F.; Weiss, M.; Dedieu, G. Estimation of leaf area and clumping indexes of crops with hemispherical photographs. *Agric. For. Meteorol.* **2008**, *148*, 644–655. [CrossRef]
71. Jiang, Z.; Huete, A.R.; Chen, J.; Chen, Y.; Li, J.; Yan, G.; Zhang, X. Analysis of NDVI and scaled difference vegetation index retrievals of vegetation fraction. *Remote Sens. Environ.* **2006**, *101*, 366–378. [CrossRef]
72. Barton, C.V.M.; North, P.R.J. Remote sensing of canopy light use efficiency using the photochemical reflectance index: Model and sensitivity analysis. *Remote Sens. Environ.* **2001**, *78*, 264–273. [CrossRef]
73. Seaquist, J.W.; Olsson, L. A simple method to account for off-nadir-scattering in the NOAA/NASA Pathfinder AVHRR Land Data Set. *Int. J. Remote Sens.* **1998**, *19*, 1425–1431. [CrossRef]
74. Goodin, D.G.; Gao, J.; Henebry, G.M. The Effect of Solar Illumination Angle and Sensor View Angle on Observed Patterns of Spatial Structure in Tallgrass Prairie. *IEEE Trans. Geosci. Remote Sens.* **2004**, *42*, 154–165. [CrossRef]
75. Van Beek, J.; Tits, L.; Somers, B.; Deckers, T.; Verjans, W.; Bylemans, D.; Janssens, P.; Coppin, P. Temporal Dependency of Yield and Quality Estimation through Spectral Vegetation Indices in Pear Orchards. *Remote Sens.* **2015**, *7*, 9886–9903. [CrossRef]

76. Suárez, L.; Zarco-Tejada, P.J.; Sepulcre-Cantó, G.; Pérez-Priego, O.; Miller, J.R.; Jiménez-Muñoz, J.C.; Sobrino, J. Assessing canopy PRI for water stress detection with diurnal airborne imagery. *Remote Sens. Environ.* **2008**, *112*, 560–575. [CrossRef]

77. Zarco-Tejada, P.J.; Miller, J.R. Minimization of shadow effects in forest canopies for chlorophyll content estimation using red edge optical indices through radiative transfer: implications for MERIS. In Proceedings of the IEEE 2001 International Geoscience and Remote Sensing Symposium, 2001. (IGARSS '01), Sydney, NSW, Australia, 9–13 July 2001; Volume 2, pp. 736–738.

Journal of
Imaging

MDPI

Article

Estimating Mangrove Biophysical Variables Using WorldView-2 Satellite Data: Rapid Creek, Northern Territory, Australia

Muditha K. Heenkenda [1,*], Stefan W. Maier [2] and Karen E. Joyce [3]

1   Research Institute for the Environment and Livelihoods, Charles Darwin University, Ellengowan Drive, Casuarina, NT 0909, Australia
2   Maitec, P.O. Box U19, Charles Darwin University, NT 0815, Australia; stefan.maier@maitec.com.au
3   College of Science, Technology and Engineering, James Cook University, P.O. Box 6811, Cairns, QLD 4870, Australia; karen.joyce@jcu.edu.au
*   Correspondence: mudithakumari.heenkenda@cdu.edu.au or mheenkenda@selkirk.ca; Tel.: +1-403-970-2850

Academic Editors: Gonzalo Pajares Martinsanz and Francisco Rovira-Más
Received: 2 June 2016; Accepted: 5 September 2016; Published: 8 September 2016

Abstract: Mangroves are one of the most productive coastal communities in the world. Although we acknowledge the significance of ecosystems, mangroves are under natural and anthropogenic pressures at various scales. Therefore, understanding biophysical variations of mangrove forests is important. An extensive field survey is impossible within mangroves. WorldView-2 multi-spectral images having a 2-m spatial resolution were used to quantify above ground biomass (AGB) and leaf area index (LAI) in the Rapid Creek mangroves, Darwin, Australia. Field measurements, vegetation indices derived from WorldView-2 images and a partial least squares regression algorithm were incorporated to produce LAI and AGB maps. LAI maps with 2-m and 5-m spatial resolutions showed root mean square errors (RMSEs) of 0.75 and 0.78, respectively, compared to validation samples. Correlation coefficients between field samples and predicted maps were 0.7 and 0.8, respectively. RMSEs obtained for AGB maps were 2.2 kg/m^2 and 2.0 kg/m^2 for a 2-m and a 5-m spatial resolution, and the correlation coefficients were 0.4 and 0.8, respectively. We would suggest implementing the transects method for field sampling and establishing end points of these transects with a highly accurate positioning system. The study demonstrated the possibility of assessing biophysical variations of mangroves using remotely-sensed data.

Keywords: mangrove; above ground biomass; leaf area index; WorldView-2; partial least squares regression

1. Introduction

Mangrove forests are a dominant feature of many tropical and subtropical coastlines. They have a variety of growth forms, including intertidal trees, palms and shrubs, that often grow in dense stands [1]. Although mangroves form valuable ecosystems along sheltered coastal environments, on the global scale, they are disappearing at an alarming rate [2]. For instance, by 2000, the worldwide mangrove extent has fallen below 15 million ha, down from 19.8 million ha in 1980 [3]. The world has thus lost five million ha of mangroves over twenty years, or 25% of the extent found in 1980. The main reasons for rapid mangrove destruction and land clearings are urbanization, population growth, water diversion, aquaculture, agriculture and salt pond construction.

Land clearings throughout catchments and in urbanized areas can cause an increased volume of water entering watercourses, carrying substances, such as topsoil, chemicals, rubbish and nutrients. These substances deposit on sediments in which mangroves grow. The increased amount of water influences the rate of erosion or deposition of sediments, causing a significant problem for the health of aquatic habitats. Therefore, when sustainable developments are progressing through land clearing, it is necessary to ensure effective mangrove conservation.

To set the balance between mangrove conservation and developments, one of the vital roles of monitoring and ultimately managing mangroves is to create an accurate and up-to-date quantitative analysis of their baseline health parameters. However, assessing mangrove health is not a straightforward task. This is due to the complex structure of forests, their biophysical variations and the interaction between mangroves, soil, water and salinity. Therefore, indirect measures that correlate with the levels of vegetation stresses, such as above ground biomass (AGB), canopy nutrient levels, particularly canopy nitrogen level, and leaf area index (LAI), can eventually be considered.

Most of the conventional methods that have been developed for estimating AGB, LAI and canopy nutrient levels have limitations when extended over space and time. For example, estimating biomass using the allometric method is based on measureable canopy dynamics, such as tree height and diameter at breast height (DBH). Due to the within-stand heterogeneity of canopies, labour-intensive, site- and species-specific field measurements are crucial [4–6]. However, field sampling within mangrove forests is challenging. Many mangrove species have complex aerial root systems, which make sampling difficult. Furthermore, mangroves are dense and rather difficult to walk through. Remotely-sensed data addresses the major challenges (especially field sampling) identified with already developed conventional methods that estimate AGB and LAI. Most remote sensing-based approaches are capable of estimating plant biophysical characteristics by reducing the shortcomings of field observations.

There are numerous studies to estimate the biophysical characteristics of vegetation using remotely-sensed data. The key issue is to correlate the intensity of electromagnetic energy absorbed or reflected by the plant (spectral reflectance) with ground measurements of biophysical variables. This spectral reflectance is either measured in situ (using spectrometers) or via airborne or spaceborne sensors. Several studies have found strong correlations between forest biomass or LAI and spectral reflectance values at different wavelengths [7,8]. Eckert [9] and Ahamed et al. [10] summarized different studies that analysed the levels of vegetation greenness in terms of vegetation indices derived from remotely-sensed data for estimating biomass. For example, Anaya et al. [7] and Satyanarayana et al. [11] used the normalized difference vegetation index (NDVI) calculated from red and near infrared wavelengths of remotely-sensed data for estimating biomass. However, compared to other terrestrial vegetation types, little has been adapted to mangroves.

Compared to other terrestrial ecosystems, a few studies associated with field samples and remotely-sensed data for producing thematic maps over mangrove forests can be found [11–16]. Clough et al. [17], Green and Clark [18] and Green et al. [14] established a relationship between vegetation indices derived from remotely-sensed data and in situ LAI samples for mangroves. Kamal et al. [15] studied the spatial resolution of satellite images, spectral vegetation indices and different mapping approaches for LAI estimation at Moreton Bay, Australia, and Karimunjawa Island, Indonesia. The study confirmed that the LAI estimation accuracy using remotely-sensed data was site specific and varied across pixel sizes and image segmentation scales. Since there is no recorded study related to the Rapid Creek mangrove forest in Darwin, Australia, the demand still exists for estimating the LAI of mangroves by integrating in situ samples, remotely-sensed data and advanced statistical regression algorithms. To the same extent, we could find only a few studies that mapped AGB of mangroves over a large area. For example, recently, Zhu et al. [16] retrieved mangrove AGB from field data and WorldView-2 (WV2) satellite images. This study tested *Sonneratia apetala* and *Kandelia candel* mangrove species and a limited number of vegetation indices for AGB mapping over the study site. The study confirmed the importance of the red-edge band of WV2 satellite image and associated vegetation indices for AGB estimation over other spectral regions. Komiyama et al. [19], Komiyama et al. [5], Fu and Wu [4] and Perera and Amarasinghe [20] derived only a relationship between mangrove canopy dynamics, AGB and vegetation indices. Therefore, testing the possibility of mapping AGB of mangroves over a large area from remotely-sensed data is still required.

This study aimed to quantify the spatial distribution of LAI and AGB of mangroves using field measurements and WV2 satellite images. To achieve this aim, the study compared two different spatial resolutions of WV2 data (original spatial resolution of multispectral bands (2 m) with resampled multispectral bands (5 m)) and partial least squares regression algorithm for mapping LAI and AGB variations over a large area.

2. Data and Methods

2.1. Study Area

The Rapid Creek mangrove forest in Darwin, Northern Territory, Australia, was selected as the test site for this study (Figure 1). This area is dominated by five different mangrove species: *Avicennia marina, Ceriops tagal, Bruguiera exaristata, Lumnitzera racemosa* and *Rhizophora stylosa* [21]. Other species do not represent significant coverage to be considered separately.

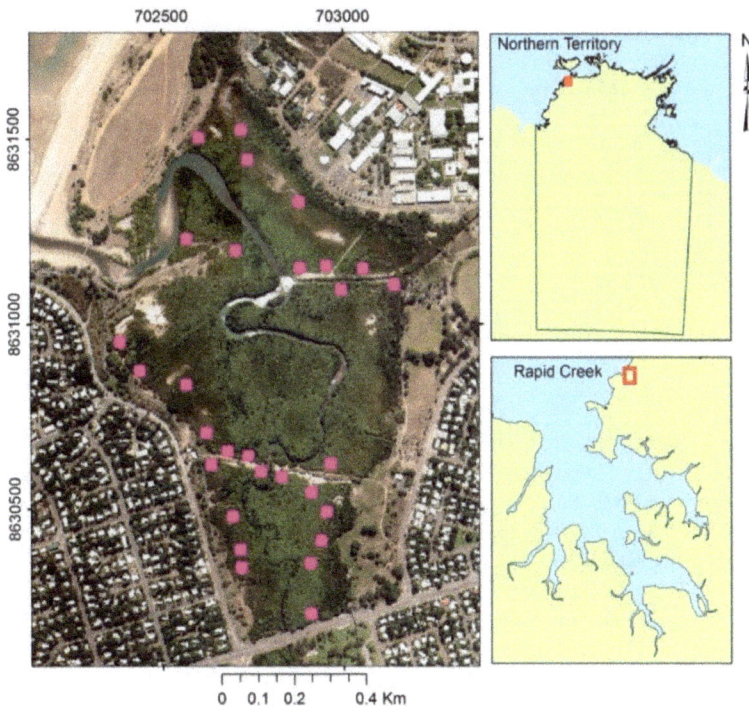

Figure 1. Rapid Creek coastal mangrove forest, Darwin, Northern Territory, Australia. The distribution of field sampling plots (5 m × 5 m) is shown in the map (the sizes of magenta squares are not to the scale). Aerial photographs © Northern Territory Government.

2.2. Field Sampling, Satellite Data and Predictor Variables

A total of 29 plots, 5 m × 5 m in extent, were identified in the field. These plots were selected with trees having similar characteristics (species, age, height and DBH) to represent calculated values accurately. Each of them was oriented in a north-south and east-west direction in order to locate them easily in the satellite image. Inside each plot, five trees were selected to take measurements based on their ability to be identified in satellite images; that is isolated and without clumping to neighbouring tree crowns. A sampling pattern was determined considering the accessibility, density of mangrove forest, mangrove greenness and species variation. Attention was also given to avoid areas close to water features due to the danger of salt water crocodiles that inhabit the region.

Tree height, diameter at breast height and species were recorded. Species were identified based on the guidelines provided by Duke [22] and Wightman [23]. The positions of each the tree and the four corners of each plot were recorded using the non-differentially-corrected Global Positioning System (GPS). Apart from the GPS measurements, we collected some additional measurements to support positioning field plots in the image. These measurements especially refer to distances to closest roads, water features and other features that can easily be identified in the image.

2.2.1. Remotely-Sensed Data and Predictor Variables

An overlapping pair of WV2 satellite images acquired on 26 July 2013 was used as the remotely-sensed data for this study. The main reason for having an overlapping pair of images was to prepare a stereo model of the area to manually extract individual tree crowns. The satellite image capture coincided with the field data acquisition. The satellite is in a nearly circular, Sun-synchronous orbit with a period of 100.2 min. The spatial resolution of multispectral bands was 2.0 m, and the panchromatic band was 0.5 m. Table 1 shows the detailed spectral band description of WV2 images.

Table 1. Spectral band information of WorldView-2 images.

Band	Spectral Range (nm)	Spatial Resolution (m)
Panchromatic	447–808	0.5
Coastal	396–458	
Blue	442–515	
Green	506–586	
Yellow	584–632	2
Red	624–694	
Red-edge	699–749	
NIR1	765–901	
NIR2	856–1043	

WV2 images were radiometrically corrected according to the method described by Heenkenda et al. [21]. First, digital numbers were converted to at-sensor-radiance values, and then, they were converted to top-of-atmosphere reflectance values. The dark pixel subtraction technique in ENVI 5.0 software was applied to remove the additive path radiance. Finally, images were geo-referenced using rational polynomial coefficients provided with the images and ground control points extracted from digital topographic maps of Darwin, Australia. The georeferenced surface reflectance values were used for further analysis.

Non-mangrove areas of images were masked as per Heenkenda et al. [21]. Class-specific rules were developed based on the contextual information from the WV2 images, geometry and neighbourhood characteristics of objects at different hierarchical levels to separate mangrove coverage only (see Heenkenda et al. [21] for a detailed description).

Plant growth is correlated with the expansion of leaf canopy and the increasing weights of woody elements. Hence, the AGB and LAI of mangroves can relate to vegetation indices derived from remotely-sensed data. To take the maximum advantage of the relatively narrow spectral bands of WV2 multispectral images, this study selected nine vegetation indices and the green, yellow, red, red-edge, NIR1 and NIR2 bands of the WV2 image to predict LAI and AGB over the study area (Table 2). These narrowband greenness indices use spectral reflectance values in the red, red-edge, NIR1 and NIR2 regions, and they produce a measure of the photosynthetic characteristics of vegetation. Hence, narrowband greenness indices are more suitable measures of the biophysical variations and vigour of green vegetation than broadband greenness vegetation indices. Vegetation indices shown in

Table 2 were calculated for predicting AGB and LAI from WV2 data. Then, spectral bands of the WV2 satellite image and vegetation indices were stacked together to form one image with multiple layers (15 layers) for further processing.

Table 2. Predictor variables generated from the WorldView-2 multispectral image for estimating above ground biomass and leaf area index. NIR1, NIR2, red edge, red and green: surface reflectance of Near Infrared 1, Near Infrared 2, red-edge, red and green wavelength regions, respectively.

Vegetation Index	Band Relationship	Source
Normalized Difference Vegetation Index (NDVI)	$(NIR1 - red) / (NIR1 + red)$	Rouse et al. [24] and Ahamed et al. [10]
Normalized Difference Red Edge index (NDRE)	$(NIR1 - Red\ Edge) / (NIR1 + Red\ Edge)$	Ahamed et al. [10] and Barnes et al. [25]
Green Normalized Difference Vegetation Index (GNDVI)	$(NIR1 - green) / (NIR1 + green)$	Ahamed et al. [10], Li et al. [26] and Gitelson et al. [27]
Green Normalized Difference Vegetation Index 2 (GNDVI2)	$(NIR2 - green) / (NIR2 + green)$	Mutanga et al. [28]
Normalized Difference Vegetation Index 2 (NDVI2)	$(NIR2 - red) / (NIR2 + red)$	Mutanga et al. [28]
Normalized Difference Red Edge index 2 (NDRE2)	$(NIR2 - Red\ Edge) / (NIR2 + Red\ Edge)$	Mutanga et al. [28]
Renormalized Vegetation Index (RDVI)	$(NIR1 - red) / \sqrt{NIR1 + red}$	Li et al. [26]
Ratio Vegetation Index (RVI)	$NIR1 / red$	Li et al. [26]
Modified Soil Adjusted Vegetation Index (MSAVI)	$(1 + 0.5)\,(NIR1 - red) / (NIR1 + red + 0.5)$	Qi et al. [29]

The WV2 multispectral bands were resampled to a 5-m spatial resolution using the cubic convolution resampling method. This was done to simulate remote sensing images from other satellite missions that provide multispectral images within the same spectral region, such as RapidEye. The number of band ratios and vegetation indices was calculated as explained in Table 2. The green (506 nm–586 nm), red (624–694 nm), red-edge (699 nm–749 nm), NIR1 (765 nm–901 nm) and NIR2 (856 nm–1043 nm) bands of the WV2 image and the calculated band ratios and indices were stacked together to form a single image with 15 bands of a 5-m spatial resolution for further processing.

An overlapping panchromatic image pair was oriented to ground coordinates following the digital photogrammetric image orientation steps in the Leica Photogrammetric Suite (LPS)/ERDAS IMAGINE software to obtain a stereo model of the area. The rational polynomial coefficients (RPC's) calculated during the image acquisition and ground control points extracted from the digital topographic map of Darwin, Australia (scale of 1: 10,000), were used as references. Further, the quality of the stereo model was assessed by comparison with ground control points. Then, outlines of the mangrove field plots that were assessed in the field were digitised using a stereo model. To identify the four corners of field plots in the image, the GPS locations, as well as the indirect measurements that were collected during field sampling were used. To calculate the corresponding predictor variable values of each plot, all pixels within a field plot polygon were considered. However, pixels that represented less than 70% in extent within a plot were discarded from further calculation.

The study selected the digital cover photography method developed by Macfarlane et al. [30] for calculating the LAI of field sampled mangrove trees. Since mangroves are densely clustered with muddy soil underneath the canopy cover, the digital cover photography method is more reliable than plant canopy analysers or area meters (LAI-2200C, LI-3100C, LI-COR 6400, etc.). Vertical photographs of the canopy, pointing the lens of the camera upwards, were taken using a Panasonic Lumix DMC-FT2 compact digital camera. Two levelling bubbles were attached to the camera to ensure that photographs were taken without a tilt. The camera was set to an aperture priority-automatic exposure mode. Due to the densely-clustered nature of mangroves, it is impossible to isolate one tree and take photographs that represent its canopy area. Therefore, 16 evenly-spaced photographs were taken within each plot. The number of trees within the plot was counted.

2.2.2. Estimating the Leaf Area Index

According to Macfarlane et al. [30], vertical photographs were used to estimate the large gaps between mangrove trees, small gaps within mangrove trees, the proportion of the ground area covered by the vertical projection of foliage and branches, the crown cover, the crown porosity and the woody-to-total area ratio of each sampling plot. A detailed description of the method to estimate LAI using digital photographs can be found in Macfarlane et al. [30], Pekin and Macfarlane [31] and Heenkenda et al. [32].

All digital photographs were classified using eCognition software for further calculations. By considering the contrast between vegetation and the background (the sky) of the photographs, areas with higher blue reflectance values were classified as sky. The ratio between the reflectance of blue and red bands was further considered to separate the background and the vegetation. The background was further classified into large gaps (g_L) between tree crowns and total gaps (g_T), considering the relation to neighbouring features. For example, if the area of the gap is less than 50 pixels and it is surrounded by vegetation, it is classified as a small gap within a tree. Finally, a total number of pixels for large gaps and small gaps was summed to obtain g_T (see Heenkenda et al. [32] for the detailed description).

Vegetation was further classified into leaf area and branches/stems. The greenness values were used to identify branches/stems from vegetation. Hence, the greenness was calculated from photographs using Equation (1) [33].

$$\text{Greenness} = 2 \times G - B - R \tag{1}$$

where G, B and R represent the reflectance (intensity levels) recorded with the green, blue and red bands of the digital camera.

The fraction of foliage cover (f_f) that is the proportion of the ground area covered by the vertical projection of foliage and branches [30,31,34] and the crown cover (f_c) were calculated as Equations (2) and (3).

$$f_c = 1 - g_L / \sum \text{Pixels} \tag{2}$$

$$f_f = 1 - g_T / \sum \text{Pixels} \tag{3}$$

where g_L is the total number of pixels of large gaps (gaps between tree crowns), g_T is the total number of pixels of gaps, f_c is the crown cover, f_f is the fraction of the foliage cover and $\sum \text{Pixels}$ is the total number of pixels of the photograph.

Considering these results, the crown porosity (ϕ), which is the proportion of the ground area covered by the vertical projection of foliage and branches within the perimeter of the crowns of individual plants, was calculated from the Equation (4):

$$\phi = 1 - \frac{f_f}{f_c} \tag{4}$$

where f_f is the fraction of the foliage coverage and f_c is the crown cover.

The effective plant area index (L_t) includes the contribution from woody elements to the total plant cover. Hence, it provides an overestimation for leaf area index [30]. The effective plant area index was estimated using the modified version of the Beer–Lambert law as specified in Equation (5).

$$L_t = -f_c \times \frac{\ln(\phi)}{k} \tag{5}$$

where f_c is the crown cover; ϕ is the crown porosity; and k is the canopy extinction coefficient.

As Perera et al. [35] suggested, the canopy extinction coefficient (k) was taken as 0.5 for this study. This represents the average value of already published k values for mangroves around the world. The factor k mainly depends on stand structure and canopy architecture; therefore, different vegetation types have different values. For example, Breda [36] reviewed k values for forest stands with broad leaves and found a range of values from 0.42 to 0.58. Macfarlane et al. [30] assumed k =0.5 for eucalypt forests. Finally, we calculated the woody-to-total area ratio (α) and the actual leaf area index (LAI) from Equations (6) and (7) [37,38].

$$\alpha = \frac{\sum (Area_B)}{\sum (Area_V)} \tag{6}$$

$$LAI = L_t \times (1 - \alpha) \tag{7}$$

where $\sum (Area_B)$ is the total area of branches, except leaves; $\sum (Area_V)$ is the total area of vegetation including branches; and L_t is the effective plant area index.

The final results represented an actual LAI of individual plots or a cluster of trees. This was then divided by the number of trees to get a value per tree.

2.2.3. Estimating the Above Ground Biomass

The study selected the allometric method for estimating AGB from field measurements. The basic theory of the allometric relationship is that the growth rate of one part of the organism is proportional to that of another [5], and therefore, it depends on measurable canopy dynamics, such as tree height and DBH. Once the regression relationship between canopy dynamics is established, the regression equation estimates the standing biomass of a tree. However, allometric relationships for mangroves are species and site specific [5]. Therefore, special attention was given to use allometric equations developed for mangrove species in Northern Australia or Southeast Asia. Since there was no equation derived for *Sonneratia alba* and *Excoecaria agallocha var. ovalis*, the common equation developed by Bai [39] for mangroves in the Northern Territory, Australia, was used (Table 3).

The logarithmic transformation of DBH values was calculated using Microsoft Excel software. Equations from Table 3 were used to calculate Log_{10} (Biomass). Finally, the calculated above ground biomass values were converted to AGB per square meter (unit = kg/m^2) considering the extent of each plot.

Table 3. Log_{10}-transformed allometric relationships used for different mangrove species. The equations are in the form of $log_{10} (Biomass) = B_0 + B_1 * log_{10} (DBH)$; where DBH is the diameter at breast height; B_0 and B_1 are regression coefficients. These equations are specific to Northern Australia [39–41], North-eastern Queensland, Australia [41], and Sri Lanka [20] (biomass in kg and DBH in cm).

Mangrove Species	B_0	B_1	Study
Avicennia marina	−0.511	2.113	Comley and McGuinness [40]
Bruguiera exaristata	−0.643	2.141	Comley and McGuinness [40]
Ceriops tagal	−0.7247	2.3379	Clough and Scott [41]
Lumnitzera racemosa	1.788	2.529	Perera and Amarasinghe [20]
Rhizophora stylosa	−0.696	2.465	Comley and McGuinness [40]
Sonneratia alba	−0.634	2.248	Bai [39]
Excoecaria agallocha var. ovalis	−0.634	2.248	Bai [39]

2.3. Predicting LAI and AGB

Multivariate regression methods are some of the most-widely used methods for estimating plant biophysical and biochemical variables, in particular for developing empirical models of variables of interest based on satellite image data. Among them, partial least squares regression (PLSR) receives much popularity, as it has been designed to outperform the problems of collinearity of prediction variables and over-fitting with relatively few samples. Hence, PLSR is suitable when the number of predictor variables is large and they are highly collinear [26,42,43]. This study selected the PLSR algorithm to model the relationship between field samples and satellite data. The "pls" package in R software [44] was used for the calculation.

In its simplest form, PLSR specifies a regression model by projecting the predictor variables and response variables to a new space. That is, data are first transformed into a different and non-orthogonal basis similar to a principal component analysis, and the most important partial least square components are considered for building a regression model. Hence, the PLSR procedure extracts successive linear combinations of partial least square factors or components. Once the optimal number of partial least square components is selected, the relationship can be predicted over a large area. A detailed description of the PLSR algorithm is available in Mevik and Wehrens [44], Hastie et al. [45] and Wang et al. [43].

The PLSR algorithm performs best with normalized data without outliers [45]. Therefore, the LAI sample data were normalized with the mean equal to zero and the variance equal to one and analysed for outliers. The mangrove field plot polygons were considered to extract predictor variable values from all pixels within plots, as described previously for a 2-m spatial resolution. Finally, these sampled pixels were randomly divided into two sets: training (70% of sample data) and validation (30% of sample data). By using the "pls" package, the optimal number of components for the prediction or predictive abilities of the model was assessed considering the prediction root mean square error (RMSEP). To obtain RMSEP, the model was internally cross-validated using the leave-one-out cross-validation method with training data. Once the optimal number of components was selected, the model was used to predict LAI over the study area using the "raster" package in R software [46]. Finally, we de-normalized the raster maps considering the mean and variance of the original field samples to obtain real LAI values. A low pass filter (3 × 3 kernel) was applied to smooth data by reducing local variation and removing noise. This process was repeated to analyse the spatial variation of LAI over the Rapid Creek mangrove forest with predictor variables of a 5-m spatial resolution.

Field samples of log-transformed AGB values (allometric equations directly calculated log-transformed AGB) were analysed for the normality and outliers. Then, the above process was repeated to predict AGB over the study area. The sampling plot polygons were considered to extract predictor variable values from all pixels within plots.

2.4. Accuracy Assessment

The accuracies of models were internally assessed using a leave-one-out cross-validation method with training samples at all instances. However, the accuracy of the predicted LAI and AGB maps was assessed using validation data. Approximately one third of field samples of LAI and AGB separately was considered as the validation data (randomly selecting 30% of the sampling pixels within field plots). Root mean squared errors (RMSE's) and correlation coefficients (r) between predicted values and field measurements were recorded.

3. Results

The locations of field plots are shown in Figure 1. Three field plots at the northeast corner of the study area have low biomass values. Their DBH values vary from 1.5 cm to 3.5 cm, and tree heights vary from 1.4 m to 3.6 m. Figure 2A–C shows mangrove trees in these areas. They are newly-regenerated, small trees. Some areas are covered with large, wide-spread, multi-stemmed mangrove trees (Figure 2D,E). Their DBH values vary from 10.0 cm to 19.0 cm, and heights range from 4.0 m to 8.0 m. There are some areas especially near water features covered by densely-clustered mangroves with different heights (Figure 2F,G).

Figure 2. Mangrove trees inside the Rapid Creek forest; (**A–C**) small, newly-regenerated mangrove trees; (**D,E**) large, multi-stemmed mangrove trees; (**F,G**) densely-clustered areas.

The normal score plot of field samples is shown in Figure 3. The normal score represents alternative values to data points within a dataset that would be expected from a normal distribution. Points on the reference line are closer to normality, and horizontal departures indicate departures from normality. Hence, few sampled mangrove trees were removed from further calculations, as they exhibited non-normality with extremely low or high values (Figure 3A,B). The normality was measured with respect to the coefficient of determination of regression (R^2). The R^2 of LAI equals 0.96, and the log-transformed AGB showed 0.9. Once outliers were removed, the remaining field samples were normally distributed.

(A)

(B)

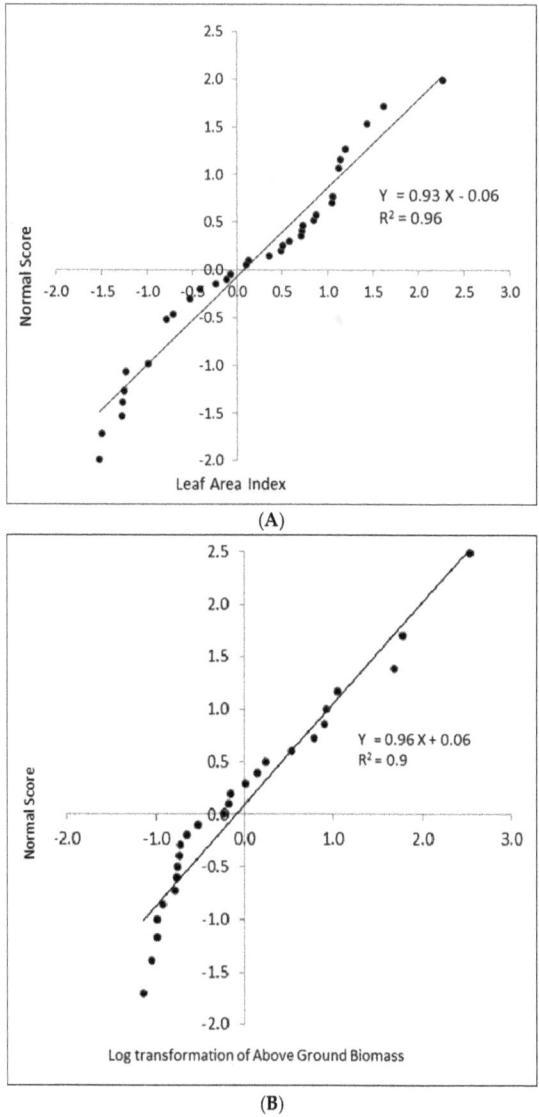

Figure 3. Normal score plots of field estimates of (**A**) leaf area index and (**B**) log-transformed above ground biomass.

3.1. Predicting AGB and LAI

A number of studies indicated PLSR as a powerful tool to extract spectral signatures and to create reliable models of LAI [25]. The performance of the PLSR model: a "goodness of fit" is represented by a root mean squared error of prediction (RMSEP). The RMSEP of the PLSR for LAI prediction was 0.69 with four components when considering a 2-m spatial resolution. This value confirmed a good relation between predictor variables and LAI field samples. Therefore, four components with all predictor variables were used for final mapping. The RMSEP of a 5-m spatial resolution was 0.78 with six components. Six components with all predictor variables were used for final mapping. If the prediction model is perfect, the RMSEP and RMSE (the root mean squared error for training and validation data) should be very similar. However, the RMSE values of LAI map with a 2-m and a 5-m spatial resolution were 0.75 and 0.78, respectively (Table 4). Although the slight difference with respect to a 2-m spatial resolution shows model overfitting, the model performance for 5-m spatial resolution data was good.

Figure 4A–D shows the cross-validated predictions with selected components versus measured values (or field samples). Most of the points (normalized LAI and AGB values) slightly deviate from the aspect ratio = 1 line. However, there is no indication of a curvature or other anomalies. The validation results (RMSEP) are shown in Figure 4.

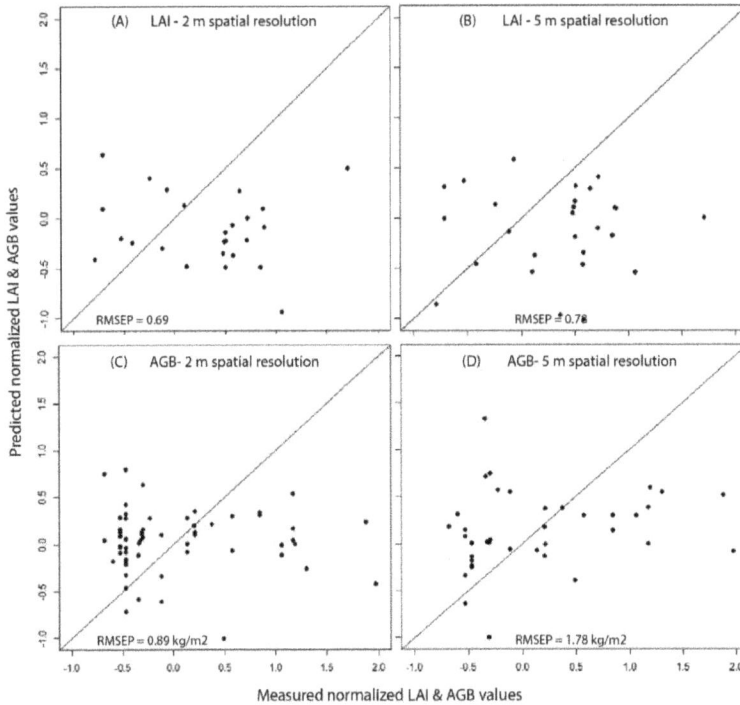

Figure 4. Cross-validated predictions for the normalized leaf area index (LAI) and above ground biomass (AGB): (**A**) predicted versus field measured normalized LAI with 2-m spatial resolution predictor variables; (**B**) predicted versus field measured normalized LAI with 5-m spatial resolution predictor variables; (**C**) predicted versus field measured normalized LogAGB with 2-m spatial resolution predictor variables; (**D**) predicted versus field measured normalized LogAGB with 5-m spatial resolution predictor variables. (RMSEP, root mean square error of prediction).

The predicted LAI map with a 2-m spatial resolution is shown in Figure 5A. The highest LAI value is 13.0, and the lowest one is 0.2. The mean LAI value is 3.9 with a standard deviation of 1.1. Approximately 70% of data ranged from 0.8 to 4.5. Although LAI values are normally distributed, the pattern has a high dispersion. The majority of data values are between 0.8 and 4.5. LAI values are high where closer to edges of water features. The upper right side of the map (Figure 5A) shows low LAI values, and this area is dominated by relatively small recently re-generated mangrove plants.

The predicted LAI map with a 5-m spatial resolution is shown in Figure 5B. The highest LAI value is 13.3, and the lowest one is 1.1. The mean LAI value is 4.2 with a standard deviation of 0.8. More than 70% of LAI values are between 3.2 and 4.8 showing a normal distribution. However, LAI values are high where closer to edges of water features. The upper right side of the map (Figure 5B) shows low LAI values, as shown in the map with a 5-m spatial resolution (Figure 5B).

When comparing Figure 5A,B at this scale, it can be seen that results obtained from 2-m spatial resolution images provide finer LAI spatial variations than results from 5-m spatial resolution images. However, at a larger scale, Figure 5A shows scattered LAI variation patterns. These patterns do not correctly represent the spatial variations of mangrove trees around those areas and, thus, the spatial distribution of LAI of the area.

The visual appearances of the AGB maps are in-line with the field observations. Most of the areas having low AGB values are dominated by relatively small and young mangrove trees. For instance, three field plots at the northeast corner of the study area (Figure 1) have low biomass values. The average measured DBH in this area was 1.8 cm, and the average height of mangrove trees was 1.6 m. Mangrove trees are dense and tall along the water features, and these areas showed high AGB values.

When considering the performance of the PLSR model for AGB, the RMSEP for a 2-m spatial resolution was 0.89 kg/m^2; however, once predicted AGB over the study area, the RMSE was 2.2 kg/m^2. The correlation coefficient obtained with respect to validation samples was 0.4 (Table 4). Hence, it can be concluded that the model was over fitted. The RMSEP for a 5-m spatial resolution was 1.7 kg/m^2, and the RMSE with respect to validation data was 2.0 kg/m^2 with strong linear correlation between predicted and sampled AGB values.

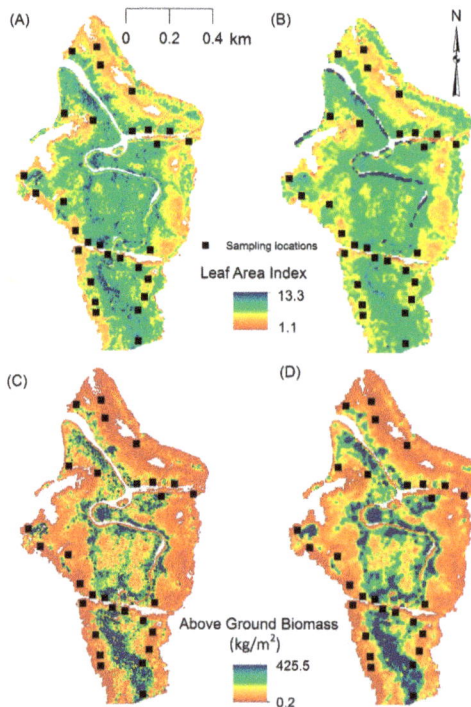

Figure 5. Predicted maps: (**A**) leaf area index with a 2-m spatial resolution; (**B**) leaf area index with a 5-m spatial resolution; (**C**) above ground biomass with a 2-m spatial resolution; and (**D**) above ground biomass with a 5-m spatial resolution; using the partial least squares regression algorithm. The distribution of field sampling plots is shown in the maps (the sizes of the black squares are not to scale).

The predicted AGB values with a 2-m spatial resolution ranged from 0.12 kg/m^2 to 425.5 kg/m^2 with a mean value of 22.5 kg/m^2 and a standard deviation of 8.1 kg/m^2 (Figure 5C). AGB values with a 5-m spatial resolution ranged from 0.12 kg/m^2 to 423.2 kg/m^2 with mean value of 18.4 kg/m^2 and a standard deviation of 7.2 kg/m^2 (Figure 5D). At both instances, the AGB values showed a skewed distribution rather than a normal distribution. The majority of the data (more than 70%) were in between 30 kg/m^2 and 267 kg/m^2. The main reason should be the extremely large, multi-stemmed *Avicenna marina* and *Rhizophora stylosa* mangrove trees in this forest. They were spread over a large area without secondary forest underneath. The highest measured, as well as highest predicted AGB values represent these areas (423.2 kg/m^2 and 425.5 kg/m^2).

3.2. Accuracy Assessment

The RMSE values with respect to the validation samples and correlation coefficients between predicted values and validation samples were recorded (Table 4). We used randomly-selected, approximately one third of field samples as the validation samples for both instances. Although the AGB map with a 2-m resolution showed low linear correlation between predicted values with validation samples, the LAI map showed a good correlation. AGB and LAI maps with a 5-m spatial resolution showed a strong linear correlation between predicted values and validation samples.

Table 4. Root mean square errors (RMSEs) and correlation coefficients (r) for above ground biomass and leaf area index maps with respect to the validation samples.

Biophysical Variable	RMSE		Correlation Coefficient	
Spatial resolution	2 m	5 m	2 m	5 m
Above ground biomass (AGB)	2.2 kg/m^2	2.0 kg/m^2	0.4	0.8
Leaf area index (LAI)	0.75	0.78	0.7	0.8

4. Discussion

The major advantage of assessing AGB and LAI from remotely-sensed data is the estimation of AGB and LAI over large areas without having extensive field campaigns. This is also a solution for many logistical and practical problems arising with field efforts. For example, access to interior extremely-dense mangrove patches is extremely difficult. In addition, the methods are non-destructive, relatively fast and economical with less labour force.

4.1. Predicting LAI

Mangrove canopies are densely clustered with overlaps. Extended root systems and muddy soil underneath the canopy cover make it difficult for sampling. Although the widely-accepted method for LAI estimation is the simulation of canopy light profiles using conventional instruments, such as portable plant canopy analysers or area meters, the digital cover photography method is more reliable than conventional instruments for mangrove forests. The digital cover photography approach is independent from the radiation condition of the forest and can be used for an extensive sampling. This method was later tested for types of cameras, digital file compression, image size and ISO equivalences, and little or no effect on estimating LAI was found [31]. Lui and Pattey [33] also recommended using the digital cover photography method by comparing its results with LAI estimation from conventional equipment.

The mean value of the predicted LAI map with a 2-m spatial resolution (3.9) is slightly lower than previously-recorded mangrove studies around the world. For instance, Clough et al. [17] estimated the LAI of *Rhizophora apiculata* mangrove forest in Malaysia using three different methods and obtained mean LAI values: 4.9, 4.4 and 5.1. LAI of *Avicennia marina* plantations in Thailand varied from 0.5 to 5.0. For the homogeneous mangrove stands at Moreton Bay, Australia, LAI values ranged from 0.26 to 3.23 with 1.97 as the mean value [15]. Kamal et al. [15] also assessed heterogeneous mangrove stands at Karimunjawa Island, Indonesia, and obtained LAI values from 0.88 to 5.33 (mean value = 2.98). A remote sensing study on LAI in the British West Indies showed a range from 0.8–7.0 with 3.96 as the mean. Statistics from the predicted LAI map of this study with a 5-m spatial resolution (4.2) was as close as previous mangrove studies. However, comparing results from this study with other studies is not a perfect approach to make a conclusion regarding the spatial distribution of LAI. They are site and species specific. When looking at cross-validated predictions, most of the predicted LAI values are lower than field sample values (Figure 4A,B). One consideration is the canopy extinction coefficient (k) used to calculate L_t. Although the used value (0.5) represents the average of already published k values for mangroves around the world, this might not be the correct value for this area. Further, LAI values vary with the spatial resolutions of remote sensing images used for data processing. Kamal et al. [15] recently confirmed that the accuracy of LAI estimation was site specific and depends on the pixel size of the remotely-sensed images, and their study even indicated two different LAI distribution patterns for homogeneous and heterogeneous mangrove forests.

The performance of the PLSR model: a "goodness of fit" is represented by a root mean squared error of prediction (RMSEP). The accuracy of the predicted maps was assessed using the root mean squared error (RMSE) compared to the validation samples. If the prediction model is perfect, RMSEP and RMSE (the root mean squared error for training and validation data) should be very similar. However, in this study, values related to a 2-m spatial resolution are slightly different, indicating a model overfitting. The best model performances were shown when using a 5-m spatial resolution. According to Zheng and Moskal [47], the accuracy of LAI estimation depends on two main reasons: overlapping and clumping between leaves within canopies due to the non-random distribution of foliage and light obstruction from canopy branches, trunks and stems. The first reason relates to this study, as well, due to mangrove canopy overlapping. The latter will not be a problem because of using vertical photographs rather than simulating light transmission. Another reason for obtaining low accuracies with a 2-m spatial resolution than a 5-m spatial resolution would be the spatial mismatch between field samples and predictor variables due to positional errors.

To establish the relationship between field samples and predictor variables, a corresponding predictor pixel value for each field sample should correctly be extracted. Hence, the positional accuracy of field samples and the pixel size is important. Laongmanee et al. [48] assessed the optimal spatial resolution for estimating the LAI of *Avicennia marina* plantations in Thailand. The best results were produced by satellite images with a 10-m spatial resolution. Although there is still room to confirm this finding, Green and Clark [18] suggested analysing the information of satellite data within a 5 m × 5 m block of pixels regardless of the spatial resolution for LAI estimation. Green and Clark [18] also argued that although high resolution satellite data provide significantly greater levels of accuracy, they were not capable of fixing positional errors. Hence, it is clear that the satellite data with a 2-m spatial resolution are not suitable to establish this match-minimizing effect from the positioning errors. According to this study, although 2-m or 5-m spatial resolutions can interchangeably be used for mapping LAI in the Rapid Creek mangrove forest, we would recommend testing a 5 m × 5 m block of pixels regardless of the spatial resolution for LAI estimation, as suggested by Green and Clark [18]. Further, we would recommend using the transect method for field sampling to minimize positional errors. Two ends of transects can be established outside the mangrove forest with DGPS (differential GPS). Then, all other measurements can be based on the established transect line with greater accuracy.

4.2. Predicating AGB

We used allometric equations for AGB calculation. Allometric relationships are highly species specific and less site specific [5,19]. Although special attention was given to using allometric relationships that were specifically developed for Northern Territory, Australia, mangrove species, we used some equations that were developed for Northern Queensland, Australia, and Sri Lanka. In this study, it was not possible to use the common equation proposed by Komiyama et al. [19], as it requires the vegetation density, which we did not measure in the first place. Therefore, for the mangrove species for which we could not find already developed allometric equations, we used a generic equation developed by Bai [39] for the Northern Territory, Australia, mangroves.

Some of the sampled *Avicennia marina* and *Rhizophora stylosa* species are multi-stemmed mangrove trees. As explained by Clough et al. [49], when there are multi-stemmed mangrove trees, each branch should be treated as a discrete tree, and the dry weight of the common butt should be included to obtain a robust AGB of the tree. However, this discrepancy is important if the consideration is to calculate the AGB of different woody parts of trees only, rather than total AGB [49]. Hence, in this study, we treated each branch of multi-stemmed mangrove trees as a separate tree. The dry weight of the common butt was neglected.

The predicted AGB maps are approximately in-line with visual field observations and AGB calculations. Along water features, mangroves exhibit high AGB values. Mangroves are known to have low productivity where closer to the landward margin [35]. These areas experience relatively less tidal inundation frequency and shorter duration together with a minimal freshwater influence [50]. The low productivity is also connected with higher salinity and less nutrients, and this is strongly evidenced in the Rapid Creek mangrove forest. Above ground biomass varies with the age of the tree, temperature, solar radiation and oxygen, as well [51]. In this study area, there is no doubt that all trees would experience the same temperature, solar radiation and oxygen, but the age of trees, salinity and nutrients levels are different. As some parts of the Rapid Creek mangrove forest are regenerated forests after clearings and natural disasters, there are trees with different ages. Especially closer to the northeast corner of the forest, trees are younger than the rest of the forest and small, and they showed low AGB values.

When considering the performance of the PLSR model for predicting AGB with a 2-m spatial resolution, it was found that the model was over fitted (Figure 4). The high spectral variation and shadows caused by canopy may create difficulty in developing PLSR model. However, the model performance for a 5-m spatial resolution was at an acceptable level. There was a strong linear correlation between predicted and field sample AGB values. The correlation coefficient is relatively high compared to the very limited number of mangrove studies that are available to-date without indicating any model overfitting. As we suspected earlier when estimating LAI from satellite data, one reason for obtaining low accuracies with a 2-m spatial resolution than a 5-m spatial resolution would be the spatial mismatch between field samples and predictor variables due to positional errors. When estimating AGB from satellite data, a 5-m spatial resolution would be the most suitable approach against a 2-m spatial resolution.

Although there are several studies investigating the relationship between the physical parameters of mangrove trees and AGB and developing allometric equations, there is a limited number of recorded studies mapping the spatial distribution of AGB over large areas around the world. For instance, in Qi'ao Island, Guangdong Province, China, predicted AGB of mature *Kandelia candel* mangroves ranged from 15.51 kg/m² to 40.66 kg/m² with the average value of 24.77 kg/m², and artificially-restored *Sonneratia apetala* mangroves ranged from 3.4 kg/m² to 23.42 kg/m² with the average value of 11.38 kg/m² [16]. However, in this study, we did not find *Kandelia candel* and *Sonneratia apetala* mangroves, and thus, this limits the comparison possibilities.

Apart from the statistical analysis of model accuracies, the authors are confident about the LAI and AGB model performances with respect to a 5-m spatial resolution at the northeast corner of the study area, southern areas and areas along water features. For instance, as shown in Figure 2A–C, northeast areas are dominated by regenerated small mangrove trees. Southern areas have relatively tall and dense mangroves. There is no recorded natural or anthropogenic disasters related to these areas, and mangrove trees are relatively mature. However, the model performances along the western edge of the mangrove forest are confusing. These edges are dominated by *Lumnitzera racemosa* mangrove trees and are tall and mature with relatively small DBH values (around 5 cm). On the other hand, these trees receive less amounts of water, thus nutrients, compared to the other areas due to ground height variations.

J. Imaging **2016**, *2*, 24

The canopy height model of the vegetation has a direct and increasingly well-understood relationship to above ground biomass, and therefore, one of the best predictors for biomass would be a canopy height model [8]. Simard et al. [12] also used elevation data derived from the Shuttle Radar Topography Mission (SRTM) for estimating mangrove heights and above ground biomass. They used airborne laser scanning data and extensive field samples to calibrate these elevation data. We created an accurate digital surface model of the area using a pair of WV2 panchromatic images. However, due to the lack of elevation data with reliable accuracy, we were not able to create a canopy height model of the area. Hence, although we identified the possibility of integrating the canopy height model as a predictor variable, we did not use it.

In this study, we did not consider very small gaps (less than 2 m) between mangrove trees due to processing difficulties with the spatial resolutions with which we dealt. Rather than masking out large non-mangrove areas from satellite images, we would suggest identifying gaps between mangrove trees and excluding them from further analysis. This would remove the noise of predicted images. Additionally, the complexity of species composition, stand structures and the densely-clustered nature of mangrove forests might add some errors, forming mixed pixels of remotely-sensed data.

5. Conclusions and Recommendations

One of the key issues associated with effective planning and management of mangrove forests is up-to-date information about the ecosystem. To collect information, extensive field sampling is difficult and time consuming. Compared to other terrestrial ecosystems, a few studies that are associated with field samples and remotely-sensed data for producing thematic maps over mangrove forests can be found. This study investigated mapping above ground biomass (AGB) and leaf area index (LAI) from WorldView-2 satellite images and field samples. Site- and species-specific allometric relationships were used for calculating above ground biomass for sampled trees. The leaf area index was obtained from the digital cover photography method. Hence, the field sampling process was fast and economical.

The relationships between both biophysical variables, LAI and AGB, with predictor variables (2-m spatial resolution) were established separately using the partial least squares regression algorithm. Once these relationships were established, they were used to predict response variables over the study area. The accuracies of predicted maps were analysed compared to validation samples. The process was repeated for the predictor variables with a 5-m spatial resolution. The LAI map with a 2-m spatial resolution showed a root mean square error of 0.75, and the map with a 5-m spatial resolution showed a root mean square error of 0.78 compared to the validation samples. The correlation coefficients between field samples and predicted maps were 0.7 and 0.8, respectively. Root mean square errors obtained for AGB maps were 2.2 kg/m^2 and 2.0 kg/m^2 for a 2-m and a 5-m spatial resolution, and the correlation coefficients were 0.4 and 0.8, respectively.

The satellite data with higher spatial resolution should be associated with accurately-positioned field samples. Therefore, we would suggest implementing the transects method for field sampling and establishing end points of these transects outside the mangrove forest with a highly accurate positioning system, such as a differential GPS. Further, we would recommend analysing a 5 m × 5 m block of pixels regardless of the spatial resolution for LAI and AGB estimation.

In conclusion, the study demonstrated the possibility of assessing the biophysical variations of mangroves using WorldView-2 satellite data. This would lead to better mangrove conservation and management of the Rapid Creek area.

Acknowledgments: The authors thank Miguel Tovar Valencia, Evi Warintan Saragih, Silvia Gabrina Tonyes, Olukemi Ronke Alaba and Dinesh Gunawardena for their assistance in the field surveying.

Author Contributions: Muditha K. Heenkenda designed the research, processed the remote sensing data and drafted the manuscript with the co-authors providing supervision and mentorship throughout the process.

Conflicts of Interest: The authors declare no conflicts of interest.

References

1. Food and Agriculture Organisation of the United Nations (FAO). *The World's Mangroves 1980–2005;* Food and Agriculture Organisation of the United Nations: Rome, Italy, 2007.
2. Bouillon, S.; Rivera-Monroy, V.H.; Twilley, R.R.; Kairo, J.G. Mangroves. In *The Management of Natural Coastal Carbon Sinks;* Laffoley, D., Grimsditch, G., Eds.; International Union for Conservation of Nature and Natural Resources (IUCN): Gland, Switzerland, 2009.
3. Wilkie, M.L.; Fortuna, S.; Forestry Department; FAO. *Status and Trends in Mangrove Area Extent Worldwide;* FAO: Rome, Italy, 2003; Available online: www.fao.org/docrep/007/j1533e/J1533E02.htm (accessed on 17 August 2016).
4. Fu, W.; Wu, Y. Estimation of aboveground biomass of different mangrove trees based on canopy diameter and tree height. *Procedia Environ. Sci.* **2011**, *10*, 2189–2194. [CrossRef]
5. Komiyama, A.; Ong, J.E.; Poungparn, S. Allometry, biomass, and productivity of mangrove forests: A review. *Aquat. Bot.* **2008**, *89*, 128–137. [CrossRef]
6. Metcalfe, K.; Franklin, D.C.; McGuinness, K.A. Mangrove litter fall: Extrapolation from traps to a large tropical macrotidal harbour. *Estuar. Coast. Shelf Sci.* **2011**, *95*, 245–252. [CrossRef]
7. Anaya, J.A.; Chuvieco, E.; Palacios-Orueta, A. Aboveground biomass assessment in Colombia: A remote sensing approach. *For. Ecol. Manag.* **2009**, *257*, 1237–1246. [CrossRef]
8. Roy, P.S.; Ravan, S.A. Biomass estimation using satellite remote sensing data—An investigation on possible approaches for natural forest. *J. Biosci.* **1996**, *21*, 535–561. [CrossRef]
9. Eckert, S. Improved forest biomass and carbon estimations using texture measures from WorldView-2 satellite data. *Remote Sens.* **2012**, *4*, 810–829. [CrossRef]
10. Ahamed, T.; Tian, L.; Zhang, Y.; Ting, K.C. A review of remote sensing methods for biomass feedstock production. *Biomass Bioenergy* **2011**, *35*, 2455–2469. [CrossRef]
11. Satyanarayana, B.; Mohamad, K.A.; Idris, I.F.; Husain, M.-L.; Dahdouh-Guebas, F. Assessment of mangrove vegetation based on remote sensing and ground-truth measurements at Tumpat, Kelantan delta, east coast of Peninsular Malaysia. *Int. J. Remote Sens.* **2011**, *32*, 1635–1650. [CrossRef]
12. Simard, M.; Rivera-Monroy, V.H.; Mancera-Pineda, J.E.; Castañeda-Moya, E.; Twilley, R.R. A systematic method for 3d mapping of mangrove forests based on shuttle radar topography mission elevation data, ICEsat/GLAS waveforms and field data: Application to Ciénaga Grande de Santa Marta, Colombia. *Remote Sens. Environ.* **2008**, *112*, 2131–2144. [CrossRef]
13. Mitchard, E.T.A.; Saatchi, S.S.; White, L.J.T.; Abernethy, K.A.; Jeffery, K.J.; Lewis, S.L.; Collins, M.; Lefsky, M.A.; Leal, M.E.; Woodhouse, I.H.; et al. Mapping tropical forest biomass with radar and spaceborne LiDAR in Lope National Park, Gabon: Overcoming problems of high biomass and persistent cloud. *Biogeosciences* **2012**, *9*, 179–191. [CrossRef]
14. Green, E.P.; Mumby, P.J.; Edwards, A.J.; Clark, C.D.; Ellis, A.C. Estimating leaf area index of mangrove from satellite data. *Aquat. Bot.* **1997**, *58*, 11–19. [CrossRef]
15. Kamal, M.; Phinn, S.; Johansen, K. Assessment of multi-resolution image data for mangrove leaf area index mapping. *Remote Sens. Environ.* **2016**, *176*, 242–254. [CrossRef]
16. Zhu, Y.; Liu, K.; Liu, L.; Wang, S.; Liu, H. Retrieval of mangrove aboveground biomass at the individual species level with WorldView-2 images. *Remote Sens.* **2015**, *7*, 12192–12214. [CrossRef]
17. Clough, B.F.; Ong, J.E.; Gong, W.K. Estimating leaf area index and photosynthetic production in canopies of the mangrove *Rhizophora apiculata. Mar. Ecol. Prog. Ser.* **1997**, *159*, 285–292. [CrossRef]
18. Green, E.; Clark, C. Assessing mangrove leaf area index and canopy closure. In *Remote Sensing Handbook for Tropical Coastal Management (Extracts);* Edwards, A.J., Ed.; UNESCO: Paris, France, 2000.
19. Komiyama, A.; Poungparn, S.; Kato, S. Common allometric equations for estimating the tree weight of mangroves. *J. Trop. Ecol.* **2005**, *21*, 471–477. [CrossRef]
20. Perera, K.A.R.S.; Amarasinghe, M.D. Carbon partitioning and allometric relashionships between stem diameter and total organic carbon (TOC) in plant components of *Bruguiera gymnorrhiza* (L.) Lamk. and *Lumnitzera racemosa* willd. in a Microtidal Basin Estuary in Sri Lanka. *Int. J. Mar. Sci.* **2013**, *3*, 72–78.
21. Heenkenda, M.K.; Joyce, K.E.; Maier, S.W.; Bartolo, R. Mangrove species identification: Comparing WorldView-2 with aerial photographs. *Remote Sens.* **2014**, *6*, 6064–6088. [CrossRef]

22. Duke, N.C. *Australia's Mangroves—The Authoritative Guide to Australia's Mangrove Plants*; University of Queensland: Brisbane, Australia, 2006; p. 200.

23. Wightman, G. *Mangrove Plant Identikit for North Australia's Top End*; Greening Australia: Darwin, Australia, 2006; p. 64.

24. Rouse, J.W., Jr.; Haas, R.H.; Schell, J.A.; Deering, D.W. Monitoring vegetation systems in the great plains with Erts. *Third Earth Resources Technology Satellite-1 Symposium*; NASA: Washington, DC, USA, 1974; pp. 309–317.

25. Barnes, E.M.; Clarke, T.R.; Richards, S.E.; Colaizzi, P.D.; Haberland, J.; Kostrzewski, M.; Waller, P.; Choi, C.; Riley, E.; Thompson, T.; et al. Coincident detection of crop water stress, nitrogen status and canopy density using ground-based multispectral data. In Proceedings of the Fifth International Conference on Precision Agriculture, Bloomington, MN, USA, 16–19 July 2000.

26. Li, X.; Zhang, Y.; Bao, Y.; Luo, J.; Jin, X.; Xu, X.; Song, X.; Yang, G. Exploring the best hyperspectral features for lai estimation using partial least squares regression. *Remote Sens.* **2014**, *6*, 6221–6241. [CrossRef]

27. Gitelson, A.A.; Kaufman, Y.J.; Merzlyak, M.N. Use of a green channel in remote sensing of global vegetation from EOS-MODIS. *Remote Sens. Environ.* **1996**, *58*, 289–298. [CrossRef]

28. Mutanga, O.; Adam, E.; Cho, M.A. High density biomass estimation for wetland vegetation using WorldView-2 imagery and random forest regression algorithm. *Int. J. Appl. Earth Obs. Geoinf.* **2012**, *18*, 399–406. [CrossRef]

29. Qi, J.; Chehbouni, A.; Huete, A.R.; Kerr, Y.H.; Sorooshian, S. A modified soil adjusted vegetation index. *Remote Sens. Environ.* **1994**, *48*, 119–126. [CrossRef]

30. Macfarlane, C.; Hoffman, M.; Eamus, D.; Kerp, N.; Higginson, S.; McMurtrie, R.; Adams, M. Estimation of leaf area index in eucalypt forest using digital photography. *Agric. For. Meteorol.* **2007**, *143*, 176–188. [CrossRef]

31. Pekin, B.; Macfarlane, C. Measurement of crown cover and leaf area index using digital cover photography and its application to remote sensing. *Remote Sens.* **2009**, *1*, 1298–1320. [CrossRef]

32. Heenkenda, M.K.; Joyce, K.E.; Maier, S.W.; Bruin, S.D. Quantifying mangrove chlorophyll from high spatial resolution imagery. *ISPRS J. Photogramm. Remote Sens.* **2015**, *108*, 234–244. [CrossRef]

33. Lui, J.; Pattey, E. Retrieval of leaf area index from top-of-canopy digital photogrpahy over agricultural crops. *Agric. For. Meteorol.* **2010**, *150*, 1485–1490.

34. Walker, J.; Tunstall, B.R. *Field Estimation of Foliage cover in Australian Woody Vegetation*; CSIRO Institute of Biological Resources: Canberra, Australia, 1981; p. 18.

35. Perera, K.A.R.S.; Amarasinghe, M.D.; Somaratna, S. Vegetation structure and species distribution of mangroves along a soil salinity gradient in a micro tidal estuary on the North-Werstern coast of Sri Lanka. *Am. J. Mar. Sci.* **2013**, *1*, 7–15.

36. Breda, N.J.J. Ground-based measurements of leaf area index: A review of methods, instruments and current controversies. *J. Exp. Bot.* **2003**, *54*, 2403–2417. [CrossRef] [PubMed]

37. Chen, J.M. Optically-based methods for measuring seasonal variation of leaf area index in boreal conifer stands. *Agric. For. Meteorol.* **1996**, *80*, 135–163. [CrossRef]

38. Chen, J.M.; Rich, P.M.; Gower, S.T.; Norman, J.M.; Plummer, S. Leaf area index of boreal forests: Theory, techniques, and measurements. *J. Geophys. Res.* **1997**, *102*, 29429–29443. [CrossRef]

39. Bai, L. *The Colour of Mud: Blue Carbon Storage in Darwin Harbour*; Charles Darwin University: Darwin, Australia, 2012.

40. Comley, B.W.T.; McGuinness, K.A. Above-and below-ground biomass, and allometry, of four common Northern Australian mangroves. *Aust. J. Bot.* **2005**, *53*, 431–436. [CrossRef]

41. Clough, B.F.; Scott, K. Allometric relationships for etimating above-ground biomass in six mangroves species. *For. Ecol. Manag.* **1989**, *27*, 117–127. [CrossRef]

42. Carrascal, L.M.; Galvan, I.; Gordo, O. Partial least squares regression as an alternative to current regression methods used in ecology. *Oikos* **2009**, *118*, 681–690. [CrossRef]

43. Wang, F.; Huang, J.; Lou, Z. A comparison of three methods for estimating leaf area index of paddy rice from optimal hyperspectral bands. *Precis. Agric.* **2011**, *12*, 439–447. [CrossRef]

44. Mevik, B.; Wehrens, R. The pls pckage: Principal component and partial least squares regression in R. *J. Stat. Softw.* **2007**, *18*, 1–24. [CrossRef]

45. Hastie, T.; Tibshirani, R.; Friedman, J. Linear methods for regression. In *The Elements of Statistical Learning: Data Mining, Inference, and Prediction*; Springer: New York, NY, USA, 2001; p. 533.

46. Hijmans, R.J.; van Etten, J. Raster: Geographic Analysis and Modeling. R Package Version 2.2–31. Available online: http://CRAN.R-project.org/package=raster (accessed on 17 August 2016).
47. Zheng, G.; Moskal, L.M. Retrieving leaf area index (LAI) using remote sensing: Theories, methods and sensors. *Sensors* **2009**, *9*, 2719–2745. [CrossRef] [PubMed]
48. Laongmanee, W.; Vaiphasa, C.; Laongmanee, P. Assessment of spatial resolution in estimating leaf area index from satellite images: A case study with *Avicennia marina* plantations in Thaliland. *Int. J. Geomat.* **2013**, *9*, 69–77.
49. Clough, B.F.; Dixon, P.; Dalhaus, O. Allometric relationships for estimating biomass in multi-stemmed mangrove trees. *Aust. J. Bot.* **1997**, *45*, 1023–1031. [CrossRef]
50. Clough, B.F.; Andrews, T.J. Some ecophysiological aspects of primary production by mangroves in North Queensland. *Wetlands* **1981**, *1*, 6–7.
51. Alongi, D.M. Present state and future of the world's mangrove forests. *Environ. Conserv.* **2002**, *29*, 331–349. [CrossRef]

Journal of
Imaging

MDPI

Article

Land Cover Change Image Analysis for Assateague Island National Seashore Following Hurricane Sandy

Heather Grybas * and Russell G. Congalton

Department of Natural Resources and the Environment, University of New Hampshire, 114 James Hall, 56 College Road , Durham, NH 03824, USA; russ.congalton@unh.edu

* Author to whom correspondence should be addressed; hmg10@wildcats.unh.edu; Tel.: +1-603-862-4644.

Academic Editors: Gonzalo Pajares Martinsanz and Francisco Rovira-Más

Received: 1 September 2015; Accepted: 30 September 2015; Published: 5 October 2015

Abstract: The assessment of storm damages is critically important if resource managers are to understand the impacts of weather pattern changes and sea level rise on their lands and develop management strategies to mitigate its effects. This study was performed to detect land cover change on Assateague Island as a result of Hurricane Sandy. Several single-date classifications were performed on the pre and post hurricane imagery utilized using both a pixel-based and object-based approach with the Random Forest classifier. Univariate image differencing and a post classification comparison were used to conduct the change detection. This study found that the addition of the coastal blue band to the Landsat 8 sensor did not improve classification accuracy and there was also no statistically significant improvement in classification accuracy using Landsat 8 compared to Landsat 5. Furthermore, there was no significant difference found between object-based and pixel-based classification. Change totals were estimated on Assateague Island following Hurricane Sandy and were found to be minimal, occurring predominately in the most active sections of the island in terms of land cover change, however, the post classification detected significantly more change, mainly due to classification errors in the single-date maps used.

Keywords: Assateague Island; Hurricane Sandy; change detection; Landsat 8; object-based classification

1. Introduction

Hurricanes and other large coastal storms have significant impacts on human and vegetation communities but they are also important agents of ecological succession in coastal ecosystems [1,2]. There are direct impacts to coastal vegetation such as damages caused by wind and storm surge [3,4] or from increased soil salinity following a storm surge [2,5,6]. These impacts can alter vegetation patterns within coastal habitats causing minor to irreversible changes to the landscape which in turn affects the wildlife communities that utilize these unique and rare coastal habitats. It is expected that there will be an increase in the frequency and intensity of hurricanes and other extreme storm events due to the effects of changing global weather patterns [7,8]. Sea level is also expected to continue rising due to glacial melting resulting in larger storm surges, greater flood damage and shoreline erosion [7,9,10]. These factors make the accurate assessment of hurricane damages to coastal ecosystems critically important if resource managers are to understand how these changes may impact their lands so improved management strategies can be developed [2,6,9].

These kinds of assessments need to be rapid and be easily repeatable so that potential problem areas can be identified and mitigation strategies developed in a timely manner [11]. Remote sensing provides a means by which hurricane impacts can be assessed over large areas quickly, accurately, and repeatedly [2,5,6]. Because of its repetitive data collection, synoptic view, and digital format, remotely sensed data has become an important resource for monitoring land cover change resulting from human and natural processes [12] and has been extensively used in a number of other studies assessing hurricane damage to coastal ecosystems [2–6,9,13–15]

There is a variety of satellite remotely sensed imagery that can be utilized for studies involving land cover classification and change detection. For over 40 years the Landsat program has been providing moderate spatial resolution imagery that has proven to be invaluable for a large number of studies in a wide variety of different disciplines [16,17]. While there are a number of sources for higher spatial resolution imagery (e.g., Ikonos, Quickbird, Worldview), Landsat imagery has higher spectral resolution (*i.e.*, senses in more wavelengths) than most high spatial resolution data, covers a larger extent, and since 2008 is freely available, making repeat land cover analysis possible. The latest generation of Landsat satellites, Landsat 8, was launched 11 February 2013 and brings with it several improvements. Previous Landsat sensors had a radiometric resolution of 8 bits or 256 possible values, but the new Operational Land Imager (OLI) sensor onboard Landsat 8 collects 12 bit data which increases the number of potential values to 4096. This increase in radiometric resolution enhances the sensor's ability to detect small changes in the amount of energy reflected from objects on the Earth surface and can lead to an improved ability to distinguish between objects with similar spectral patterns [18]. The OLI also has three new bands, a blue band for coastal and aerosol studies, a shortwave infrared band for cloud detection, and finally a quality assessment band used to indicate possible bad pixels [19]. Furthermore, the spectral range of most of the original bands, especially the NIR and SWIR bands, were narrowed to reduce the effects of water absorption [20,21]. Because of these improvements, Landsat 8 is expected to produce superior analysis results compared to previous Landsat sensors [22–24]. Given the recent availability of this imagery, little work has been formally published testing these enhancements; nevertheless, a few have shown improved results [20,25].

Land cover classification has typically been accomplished using a pixel-based approach, where by each pixel is independently classified using only the spectral reflectance values for that pixel [26–28]. Recently, object-based image analysis (OBIA) has been utilized for image classification and change detection and has resulted in a significant improvement in accuracy especially for high spatial resolution imagery [29–32]. Unlike the traditional pixel-based approach, OBIA starts by first segmenting an image into spatially and/or spectrally homogenous, non-overlapping, groupings of pixels known as image objects. The objects are then assigned a land cover class much like pixels by a classification algorithm. Objects have an advantage over pixels in that they contain spatial and spectral information that single pixels do not such as mean band values, mean ratios, size, shape, and topological relationships that can all be utilized during the classification process [33,34].

OBIA is frequently used to extract information from high resolution imagery; however, it is not limited to such imagery. While it can be used with moderate resolution data like Landsat, it has been unclear whether there is an improvement in accuracy when doing so. Studies that have made comparisons between pixel-based (PBC) and object-based (OBC) classification with moderate resolution data have reached differing conclusions; some finding OBC outperformed PBC [35–37] while other have found the opposite to be true [38]. Several studies found no significant difference between methods [39,40].

Detecting and measuring change is important when studying natural resources. It can be accomplished using two broad approaches. The first approach involves the comparison of two individually classified maps. Post classification comparison (PCC) is a very popular method within this approach and is perhaps the most commonly used methodology for change detection in general [41]. It involves overlaying two land cover maps from different dates and identifying the differences between the classifications. The second approach involves analyzing multi-temporal data concurrently. While the second approach is comprised of numerous methodologies, univariate image differencing (UID) is often cited as the preferred method [42]. Furthermore, it is also the recommended change detection method of the National Oceanic and Atmospheric Administration (NOAA) Coastal Change Analysis Program (C-CAP) which, as part of the program, developed a comprehensive protocol for the implementation of land cover classification and change detection for coastal regions in the US from remotely sensed data [43].

The main objective of this study was to quantify and qualify the type and amount of land cover change that occurred on Assateague Island as a result of Hurricane Sandy. In addition to detecting land cover change, several additional objectives were established including, evaluating the contribution of the new Landsat 8 bands to classification accuracy, comparing the performance of Landsat 8 for land cover classification with Landsat 5, comparing OBC and PBC using moderate resolution data, and comparing the results of PCC to UID for change detection. Our results show that given the level of classification detail (*i.e.*, map classes) in this study, the new Landsat 8 coastal blue band did not improve classification accuracy. Also, there was no statistically significant difference in classification accuracy between using Landsat 8 compared to Landsat 5, again at the level of land cover detail utilized. Furthermore, there was no significant difference between OBC and PBC. Finally, while land cover change totals were found to be minimal on Assateague Island, issues of thematic error associated with the PCC methodology results in significantly higher amounts of change identified compared to the UID.

2. Methods

2.1. Study Area

Assateague Island is an undeveloped barrier island and part of the Delmarva (Delaware-Maryland-Virginia) Peninsula on the Mid-Atlantic Coast (Figure 1). Assateague is managed by three government agencies. The National Park Service (NPS) manages most of the land within the Maryland portion of the island as well as some within Virginia. The U.S Fish and Wildlife Service (USFWS) manage the lands within the rest of Virginia known as the Chincoteague National Wildlife Refuge. The third agency is the Maryland Park Service (MPS) who manages a small, 3 km long portion of the island within Maryland as a state park. The island is mostly undeveloped with the exception of a few roads and parking lots, as well as campgrounds [44]. Coastal storms are responsible for creating and maintaining a number unique habitat types that support a variety of important species such as the threatened piping plover (*Charadrius melodus*) as well as numerous migratory birds, deer, and the famous wild horses [45].

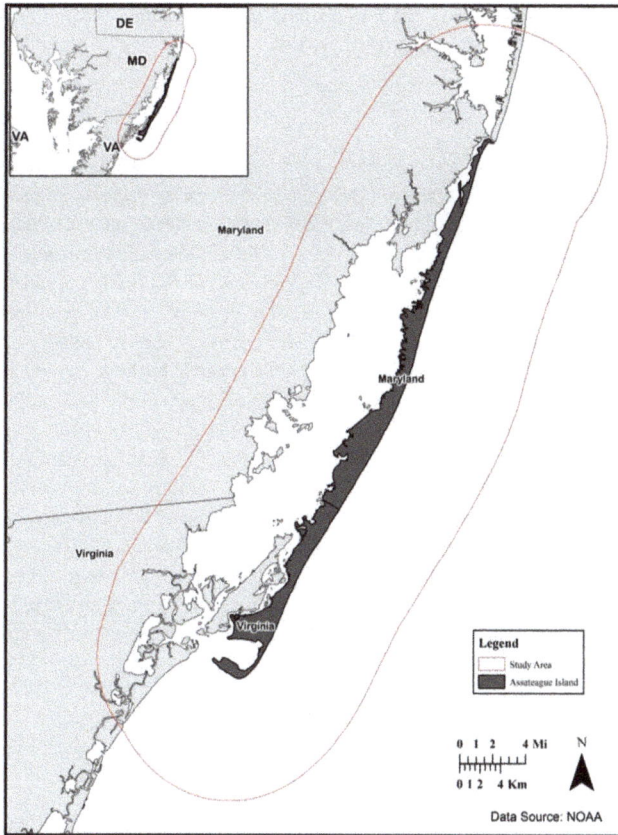

Figure 1. States shown include Delaware (DE), Maryland (MD), and Virginia (VA), located in the United States. The study area boundary shown in red includes both Assateague Island (dark grey) and a portion of the surrounding mainland (light gray).

The island can be broken into three sections. The northern 10 km of the island, also known as the North End, is a dynamic, storm structured environment that is much lower and narrower relative to the rest of the island. The North End has seen significant changes as a result of anthropogenic modifications. A jetty built in 1934 to hold open the Ocean City inlet has disrupted longshore transport from the north and thus starves the North End of sand leading to increased erosion [46,47]. The low and narrow characteristics of the North End make it vulnerable to storm waves and overwash and thus erosion [44]. The middle of the island is widest section and the oldest. It includes a substantial amount of maritime forest and is the most stable portion of the island. The southern portion of the island is known as Tom's Cove Hook and is an accretionary spit that has grown 6 km since the 1800s and continues to develop southward.

The study area for this project, depicted by the red outline Figure 1, encompasses not only Assateague Island but Sinepuxuent and Chincoteague Bay to the west as well as a portion of the surrounding mainland. While the main focus of this study was assessing the impact of Hurricane Sandy on Assateague Island, in order to expand the usefulness of this dataset, the surrounding main lands were included as well since they can have ecological impacts on the resources within Assateague [45]. The mainland is a mixture of both natural and manmade cover types, including forest, shrubland, agriculture and development. Significantly more anthropogenic modifications to the landscape occur on the mainland compared to Assateague Island. While the entire study area was classified and assessed for change, the change detection results will be elaborated on in terms of changes on Assateague Island.

2.2. Imagery

For this study, data from two Landsat satellites were used. The pre-Hurricane Sandy imagery was collected by the Landsat 5 Thematic Mapper (TM) sensor while the post-Hurricane Sandy imagery was collected by the Landsat 8 OLI sensor. Hurricane Sandy struck the east coast in October 2012, almost a year after Landsat 5 was decommissioned (November 2011). While Landsat 7 was functional during the time of the hurricane, the failure in the scan line corrector made the data unusable for this study.

Two multi-temporal layer stacks were created, one for before the hurricane and the other for after. Two images were picked for each stack, an early spring, leaf-off image and a late summer image to capture the full phenological differences between classes. All imagery was downloaded from USGS EarthExplorer (http://earthexplorer.usgs.gov). In addition to the seasonal requirements, all imagery had to be cloud free and have near anniversary collection dates within each season in order to avoid substantial differences in phenology between dates. Table 1 shows the collection dates for the imagery used. The summer images were collected within 5 days of each other, however, there was more than a month between the pre and post hurricane early spring imagery. Because Landsat 8 was launched February 2013, the earliest images available are in April 2013. There were no cloud free Landsat 5 images available until 8 March 2011. Ultimately, the summer images were more important for classification and change detection thus it was more important that these images be collected as close to each other as possible. Also, despite the month or so difference in the spring images, the phenology of the vegetation was consistent between the images.

Table 1. Imagery collection dates.

Layer Stack	Sensor	Spring Image	Summer Image
Pre-Hurricane Sandy	Landsat 5 TM	8 March 2011	31 August 2011
Post-Hurricane Sandy	Landsat 8 OLI	14 April 2013	5 September 2013

2.3. Image Processing

All images were pre-processed prior to performing single-date classifications and change detection using the ERDAS Imagine 2014 software [48]. The imagery for each date and season was first layer stacked and clipped to the study area. The COST correction method [49] was used to radiometrically and atmospherically correct all the images for increased comparability. The original COST correction Equation (1) is shown below along with a table describing the equation variables (Table 2).

$$\rho = \frac{\left[\pi d^2 \left(L_{min} + \frac{DN_i(L_{max} - L_{min})}{DN_{max}} \right) - \left(L_{min} + \frac{DN_{min}(L_{max} - L_{min})}{DN_{max}} \right) - \left(\frac{.01 d^2 \cos^2 \theta_z}{\pi E_{sun}} \right) \right]}{E_{sun} \cos^2 \theta_z} \tag{1}$$

Table 2. Description of COST parameters.

Variable	Description
d	The sun-earth distance at time of collection
L_{min} and L_{max}	Spectral radiance calibration factors
DN_i	The DN value at pixel i
DN_{min}	Band specific minimum DN value as determined by user
DN_{max}	Maximum possible DN value for the data (ex. 255 for 8 bit)
E_{sun}	Solar spectral irradiance
θ_z	Local solar zenith angle (90°- local solar elevation angle)

L_{min} and L_{max} can be found in the image metadata along with the solar elevation angle needed to calculate θ_z. E_{sun} and d can be found in Chander and Markham [50]. The band specific DN_{min} value is found through histogram exploration of each band. The above equation was adjusted in order to perform the correction of the Landsat 8 imagery. Landsat 8 no longer requires that the imagery be processed to radiance before converting to Top-of-Atmosphere reflectance (TOA). Instead, each band can be processed directly to TOA using a constant multiplicative and additive rescaling factor provided in the image metadata. Equation (2) presents the updated COST correction formula for Landsat 8.

$$\rho = \left(\frac{(DN_i * 0.00002 - 0.1)}{cos^2\theta_z}\right) - \left(\frac{(DN_{min} * 0.00002 - 0.1)}{cos^2\theta_z}\right) - 0.01 \tag{2}$$

Once corrected, derivative layers were generated in order to provide additional information along with the original spectral bands and to improve classification accuracy. Two vegetation index layers were generated for each image, a normalized vegetation index (NDVI) layer and a moisture stress index (MSI) layer. In addition to the vegetation index layers, three tasseled cap (TC) transformation feature layers (brightness, greenness, wetness) were also generated using the transformation coefficients for Landsat 5 and Landsat 8 taken from Crist *et al.* [51] and Baig *et al.* [52] respectively. The tasseled cap transformations for both sensors make use of the visible band (R, G, and B) as well as the Infrared bands (NIR, SWIR 1, SWIR 2). The Landsat 8 transformations do not utilize the coastal blue band. The corrected imagery and the five derivative layers were then stacked to create a single-date image stack. For each time period, pre-hurricane and post-hurricane, the early spring and late summer images were then stacked together to form the final multi-temporal image for classification.

2.4. Land Cover Classification

Land cover classification was carried out in three basic steps. First, reference data were collected for classification training and accuracy assessment. Next a series of classifications were completed. First the new coastal blue band was tested to assess its contribution to the overall classification accuracy. Then, the pre and post hurricane imagery were classified using both the PBC and OBC approaches. In the last step, the accuracy of all maps was calculated and statistically compared to fulfill several of the research objectives.

2.4.1. Reference Data Collection

A modified version of the NOAA C-CAP classification scheme [43] was chosen for all the land cover mapping in this study (Figure 2). The original 25 classes were condensed to 10 based upon *a priori* knowledge of the study area and an examination of previously generated NOAA C-CAP land cover maps. The classes in bold are the final land cover classes used. To improve classification accuracy, four temporary classes were added in order to decrease the spectral variability of the developed and unconsolidated shore classes.

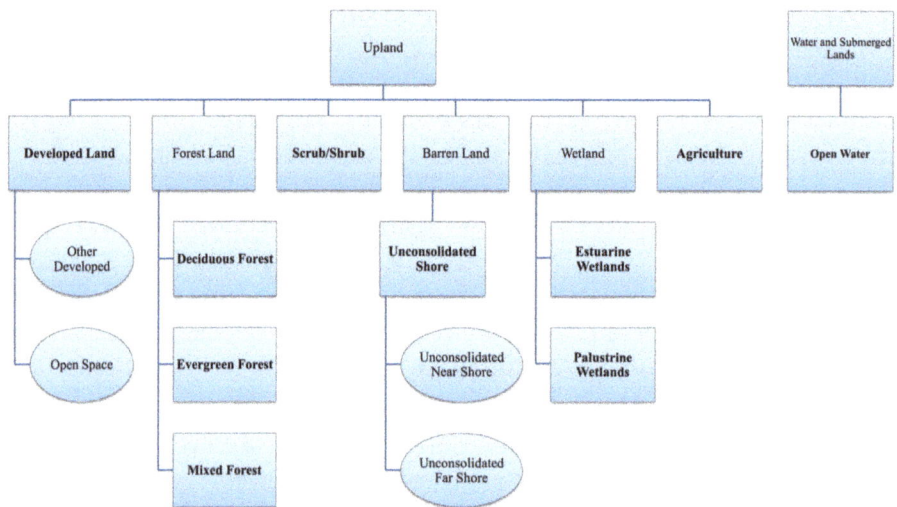

Figure 2. Modified National Oceanic and Atmospheric Administration (NOAA) Coastal Change Analysis Program (C-CAP) classification scheme organized hierarchically. Land cover types in bold, are the cover types of interest in this study. Land cover types in circles represent temporary classes that were aggregated for the final maps.

Reference data are samples of the land cover classes defined by the classification scheme and are used independently for developing the classification model (*i.e.* training) and assessing map accuracy. Units that were selected for inclusion as reference data had to meet several requirements outlined in Congalton and Green [53]: (1) the sample had to at least 90 m × 90 m in size, or 3 × 3 Landsat pixels ; (2) the area within each sample had to be spectrally homogenous (only one cover type); (3) a high degree of variation had to be captured for each class; and (4) the samples had to well distributed across the study area in order to avoid spatial autocorrelation.

The goal was to collect 100 samples per land cover class [53]. The reference samples were first carefully inspected using high resolution imagery (2011 and 2012 National Agricultural Inventory Program (NAIP) imagery and Google Earth). Sample unit labels were adjusted as needed. Field reconnaissance was performed between July and August 2014 to confirm the land cover labels of the selected reference data. Following the field reconnaissance, further photo interpretation was conducted to carefully edit the remainder of the reference samples. The final sample totals can be seen in Table 3. The goal of at least 100 samples was reached for all classes but scrub/shrub and unconsolidated shore. These two classes encompassed a small percentage of the study area and/or occurred in small patches that did not meet the minimum mapping unit. In order to avoid spatial autocorrelation and spectral redundancy, fewer samples were collected [54]. The samples were divided to achieve 50/50split between those samples used for training and those used for accuracy assessment. In the case of scrub/shrub and unconsolidated shore, at least 50 samples were needed to ensure the accuracy assessment was statistically valid for those classes and the remainders were used for training.

Table 3. Number of reference data samples collected for each land cover class.

Land Cover Class	Total Samples	Training	Accuracy
Agriculture	106	53	53
Deciduous Forest	102	52	50
Developed	121	71	50
Estuarine	100	50	50
Evergreen Forest	100	50	50
Mixed Forest	100	50	50
Open Water	121	61	60
Palustrine	105	55	50
Scrub/Shrub	85	35	50
Unconsolidated Shore	85	35	50

2.4.2. Land Cover Classifications

All classification processes were performed using Trimble eCognition Developer 9.0 [55]. Prior to classification, an analysis was performed to determine the best parameters for segmentation using the multi-resolution segmentation algorithm. The multiresolution segmentation algorithm requires three user defined parameters, scale, shape, and compactness. While all three play a role in the general shape and size of the polygons, the scale parameter has been found to have a significantly greater control since it ultimately controls the amount of heterogeneity allowed in each segment [56]. Scale values from 2 to 40 were tested to determine the optimal scale parameter while the shape and compactness parameters were held at 0.2 and 0.5 respectively. Three methods were used to determine the optimal scale, autocorrelation [57], average local variance [58,59], and a measure that combines both autocorrelation and segment variance known as the objective function [60]. The multiresolution segmentation algorithm was used to perform a series of segmentations with equal weight assigned to all bands in the image. The mean and standard deviation of the summer NIR band was exported along with the final vector for all segmentations. A visual assessment of the "optimal" scale value determined by each method was performed and a final scale was chosen. The same process was used to determine the optimal scale parameter for both the pre and post-hurricane images separately. Due to differences in the sensors that collected each image, it was not appropriate to use the same parameters.

An initial classification was performed on the post-hurricane summer image to assess the contribution of the new coastal blue band to overall classification accuracy. An object based classification was performed using the Random Forest (RF) classifier [61] (called Random Trees in eCognition). After segmenting the image using the optimal parameters, the image segments were intersected with the training data in order to select the training sample segments for the RF classifier. The RF classifier is simple to run and only requires two parameters, the number of trees and the number of randomly selected explanatory variables to select at each split. The number of trees has to be high enough to ensure the out of box (OOB) error, or accuracy of the model, converges (once converged the additional trees neither increases or decreases the OOB error) but the number of selected variables can have an effect on the strength of the individual trees and the correlation between trees [61]. Rodriguez-Galiano *et al.* [62] found that once convergence is reached, the number of randomly selected variables has very little effect on classification accuracy allowing it to run with little guidance. For this study, each forest was grown using 500 trees and the square root of the number of available features which is the default value for the number of randomly selected variables. While in an object-based environment the model can be trained using a number of object features such as spectral and shape, for this test only the mean spectral values were used to ensure that the RF classifier was only able to choose from the spectral bands while developing the classification model. The classification was performed twice, once with the coastal blue band and once without and the accuracy of each map was calculated and compared. The results determined whether the coastal blue band would remain included in the post-hurricane image.

The post-hurricane and pre-hurricane imagery were then fully classified using both an OBC and PBC with the RF classifier. After segmenting the image using the optimal parameters, the training segments were again selected by intersecting the sample units with the image segments. With a PBC, the model is trained solely on the mean and variance of all the bands and then applied to the unclassified pixels, not the remaining image objects. With an OBC, the model can be trained using a number of object features in addition to the mean band values and variance which is then applied to unclassified segments. The input features used to train the classifier in the object-based environment are given in Table 4.

The open space, other developed, unconsolidated far and near shore classes that were included in the classification as temporary classes in order to break down the spectral variance of the developed and unconsolidated shore classes were aggregated up one level (to unconsolidated shore and developed) following the classification.

2.4.3. Accuracy Assessment

All land cover classifications underwent an accuracy assessment using the independent reference data set. A traditional tally-based error matrix [63] was generated for each classification. The overall accuracy as well as the kappa were all calculated from the error matrix [63]. For each of the object-based classifications, an area-based error matrix [64] was also generated and the overall and kappa were calculated. Several pairwise kappa comparisons took place between the numerous classification results. First, the error matrices for the post-hurricane band testing were compared to assess whether there was a significant difference in classification accuracy when the coastal blue band was included in the classification. Next, a comparison took place between the OBC and PBC for both the pre and post-hurricane images to determine whether the classification method produced significantly different accuracies. Finally the matrices of the pre and post-hurricane maps were directly compared for both classification methods (OLI-PBC *vs.* TM-PBC and OLI-OBC *vs.* TM-OBC) to determine if there was a significant difference in accuracy using Landsat 8 compared to Landsat 5.

Table 4. Input features for object-based classification.

Spectral Feature	Spatial Features	Thematic Features
Mean Layer Values		
Standard Deviation		
Minimum Pixel Value		
Maximum Pixel Value	Area	Min and Maximum % overlap
Mean Difference to Neighbors	Border Length	with the National Wetland
Mean Difference To Super-Objects	Length	Inventory data layer
Ratio To Super-Objects	Width	
Difference in NDVI and NLWM	Length/Width	
Difference in NIR, SWIR1, SWIR2		

2.5. Land Cover Change

2.5.1. Univariate Image Differencing

A univariate image differencing was first carried out using the protocols outlined in Dobson *et al.* [43]. A pixel-by-pixel subtraction, also known as image differencing, was performed using the pre and post-hurricane NDVI layers and added into eCognition along with the vector boundaries of the object-based post-hurricane classification. The multi-resolution segmentation algorithm was used to segment the NDVI difference layer. During the segmentation, the polygons from the post-hurricane classification acted as boundaries and the segments were then assigned a class according to the thematic polygon it fell within [38] so that class specific change thresholds could be determined as explained in Xian *et al.* [65]. Change thresholds were determined using a visual assessment with the goal of reducing omission as much as possible especially on the island. Change thresholds were determined for all classes except for estuarine wetland to avoid errors associated with the higher tide in the pre-hurricane summer image. Using the different change thresholds, the segments were classified as either change or no-change based on their mean NDVI difference.

The next step in the process was to assign the pre-hurricane land cover class to each change area. Prior to this, reference samples that fell within the segments labeled as change were removed and new training samples selected using photointerpretation. The pre-hurricane image and updated training data were then used to classify only those areas that were labeled as change. The classified change areas were merged with the post-hurricane image to form a new complete pre-hurricane image.

2.5.2. Post Classification Comparison

In addition, a PCC between the two single-date maps was conducted to compare the results of the univariate image differencing as described above against the results of the PCC. The two single-date maps were overlaid. Areas that exhibited a change in land cover type between dates were considered changed. Since these changed areas were already classified for both dates, no further classification was necessary.

3. Results

3.1. Land Cover Classification

An initial classification was performed on the post-hurricane summer image to assess the contribution of the new coastal blue band to overall classification accuracy. The RF classifier calculates the importance of each variable in terms of model accuracy automatically while generating the classification model. Table 5 gives the importance value of each band after running the RF classifier on all bands in the summer image. Importance values greater than 0, indicate that the band was used at a node within the model to generate a decision rule. Since all bands have a value greater than 0, all bands were used at some point in the development of all the individual decision trees. Higher values indicate that the band has a greater impact on classification accuracy; so for example, the NIR band had the greatest impact on the accuracy of the model. The coastal blue band is at the very bottom of the list indicating that its impact on accuracy of model was the least significant. To test this, two classifications were performed, one with the coastal blue band and the other without. The accuracy of the two maps was calculated and compared using a pairwise kappa comparison. The results are presented in Table 6.

Table 5. Variable importance.

Band	Importance
NIR	0.108458534
SWIR 1	0.101598412
SWIR 2	0.101203956
Brightness	0.099140197
Greenness	0.092806019
NDVI	0.089701816
Blue	0.083893582
Red	0.076913401
Green	0.069847479
MSI	0.066674672
Wetness	0.062730104
Coastal Blue	0.047031853

Table 6. Overall accuracy and kappa of coastal blue band test.

	Overall Accuracy	Kappa
With Coastal Band	67.57%	0.63966
Without Coastal Band	67.57%	0.63964

The overall accuracy of the two maps was exactly the same. Likewise, the kappa values were almost identical (Table 6), and a pairwise kappa comparison (Z-score = 0.470), confirms that there was no significant difference at the 95% confidence interval (Z-score < 1.96). Since no significant difference was found, the coastal blue bands were removed from the post-hurricane image in order to reduce the number of unnecessary features during the single-date classifications.

The pre and post-hurricane imagery were then classified using both a pixel and object-based classification. Figures 3 and 4 are the results of the pixel-based and object-based classification for both of the single-date classifications respectively. There is strong agreement between the two maps created using both methods.

3.2. Accuracy Assessment

For each of the maps produced an accuracy assessment was performed. For the object-based maps, both a traditional tally-based error matrix and an area-based error matrix were generated. For the pixel-based classifications, only a tally-based error matrix is appropriate. For each matrix, the overall accuracy was calculated as well as the kappa. A pairwise kappa comparison was then run to compare several of the matrices in order to determine if: (1) there was a significant difference in the accuracy between the Landsat 8 and Landsat 5 maps; and (2) whether there was an improvement in accuracy using the object-based compared to pixel-based classification.

3.2.1. Single Date Classification Accuracy

The overall accuracies are summarized in Table 7 along with the kappa and Z statistic for each map. Remember that there is no area-based accuracy assessment for pixel-based classifications. All single-date classifications had overall accuracies greater than 75% and all kappa values indicate a moderate agreement between the map and the reference data. Additionally, the Z statistics indicate that all maps were better than a random classification at the 95% confidence interval (Z Statistic > 1.96). The overall accuracy and kappa statistics for the area-based maps attained better accuracies compared to the assessment using the tally-based matrix. These higher values for the area-based *vs.* traditional analysis were expected as shown in MacLean and Congalton [64] and Campbell *et al.* [38]. The overall accuracies between the pre and post hurricane maps were very similar for both methods. The object-based pre-hurricane map did have a slightly higher overall accuracy than the post-hurricane map. Secondly, the overall accuracies were similar for both the pixel-based and object-based classifications, with the object based obtaining a slightly higher accuracy for both dates.

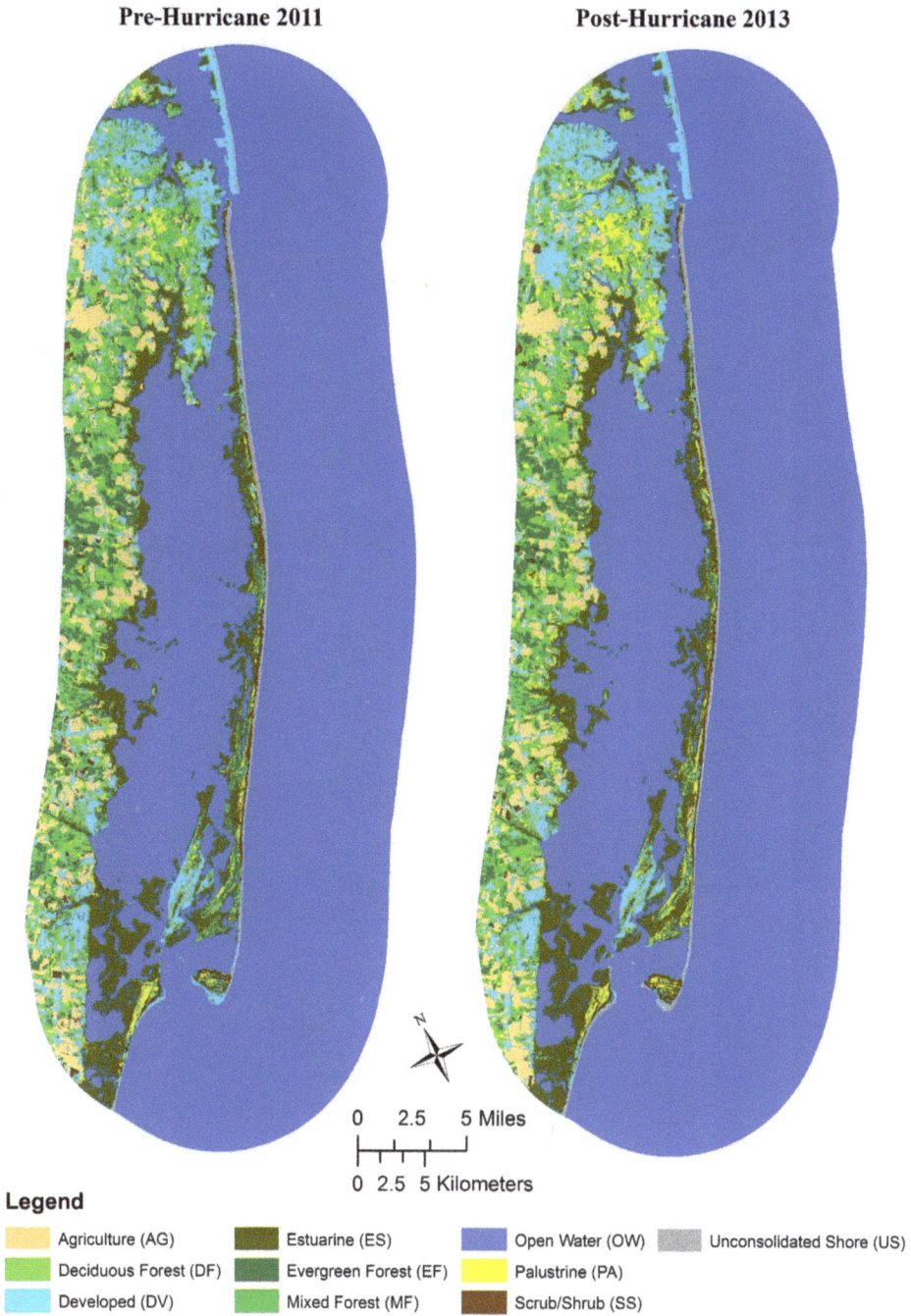

Figure 3. Results of the single-date pixel-based classification.

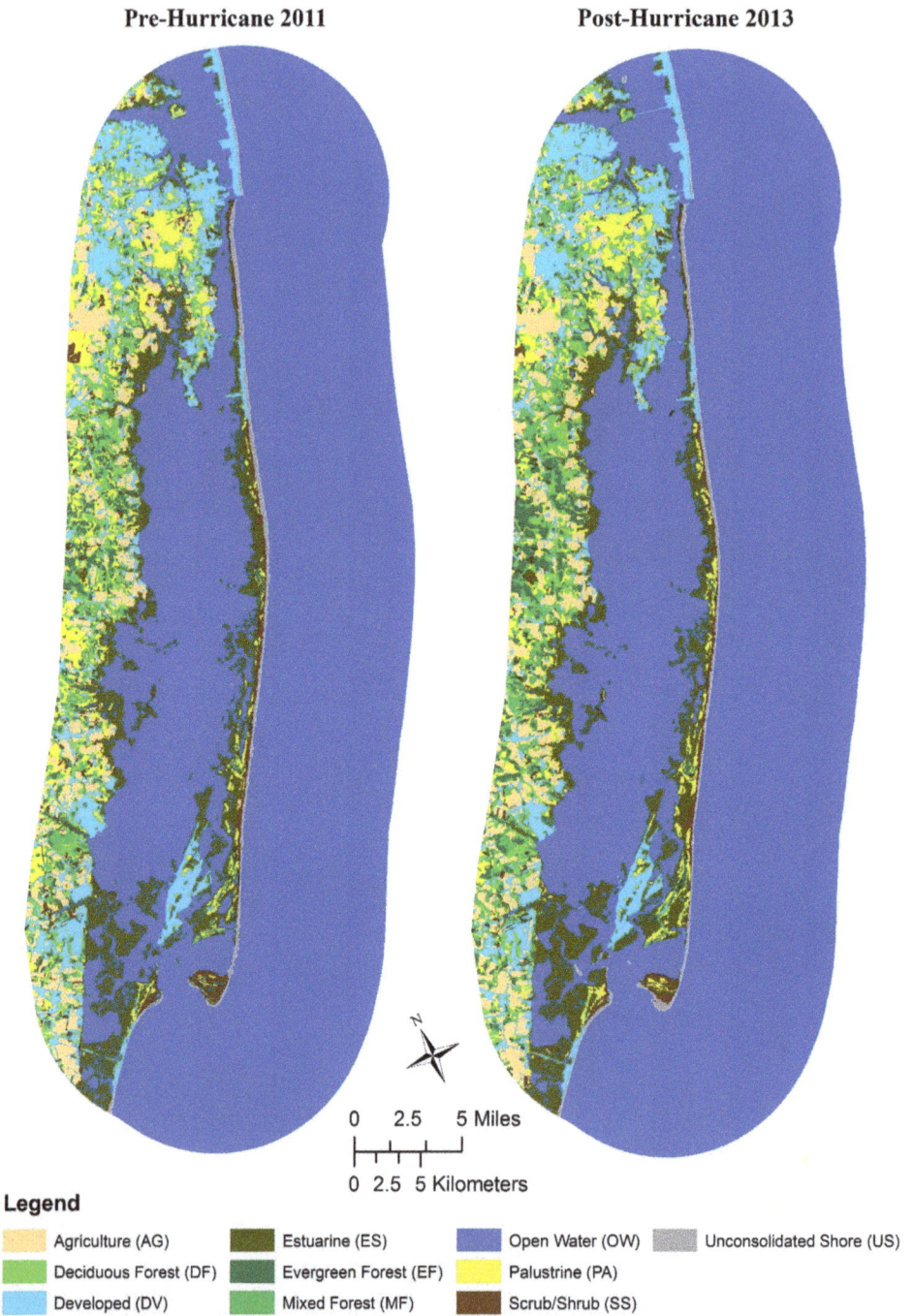

Figure 4. Results of the single-date object-based classification.

Table 7. Summary of single-date classification accuracies.

Date	Method	Tally-Based			Area-Based		
		Overall	Kappa	Z Statistic	Overall	Kappa	Z Statistic
Pre-Hurricane	Pixel	78.90%	0.765	38.325	*NA*	*NA*	*NA*
	Object	82.64%	0.807	43.255	89.59%	0.874	144.097
Post-Hurricane	Pixel	80.50%	0.783	40.567	*NA*	*NA*	*NA*
	Object	81.66%	0.796	42.217	89.79%	0.866	172.217

Tables 8 and 9 present the user's and producer's accuracies [66] for all of the classifications performed. Most user's and producer's accuracies were close to if not greater than 80%. The most notable exception to this was the palustrine and scrub/shrub class which in all classifications performed had producer's accuracies below 70% for most cases and in the case of palustrine, the producer's accuracy never got above 60% with the exception of the area based assessment. User's accuracies for the two classes were better (above 60% in all cases); however, with the exception of a few cases, they were still lower than the other categories. Palustrine areas were most confused with forested classes and to a lesser extent estuarine wetlands. Scrub/shrub on the other hand was often confused with agriculture and forest classes especially evergreen forests, as well as unconsolidated shore. Forested areas, while performing better than palustrine and scrub shrub, also proved troublesome. Most user's and producer's accuracies were not above 80%; however, they do not drop below 60%. One aspect that was common for all three forest classes in all cases was that many samples were incorrectly classified as palustrine or another forested class.

Table 8. User's (UA) and producer's (PA) accuracies for pre-hurricane classifications using object-based (OBC) and pixel-based (PBC) classification methodology.

	PBC		OBC			
	Tally-based		Tally-based		Area-based	
	UA	PA	UA	PA	UA	PA
Agriculture	100.00%	84.00%	97.87%	92.00%	98.12%	92.08%
Deciduous Forest	70.31%	90.00%	74.55%	82.00%	78.48%	84.03%
Evergreen Forest	66.67%	84.00%	82.22%	74.00%	85.65%	79.55%
Mixed Forest	72.41%	84.00%	82.98%	78.00%	82.02%	80.94%
Developed	65.75%	96.00%	84.21%	96.00%	84.29%	98.81%
Open Water	100.00%	98.21%	96.49%	98.21%	99.69%	99.80%
Estuarine	90.74%	98.00%	73.77%	90.00%	84.64%	96.30%
Palustrine	60.00%	22.64%	73.17%	56.60%	68.85%	60.87%
Scrub/Shrub	81.25%	54.17%	71.43%	72.92%	75.80%	68.64%
Unconsolidated Shore	84.78%	78.00%	89.58%	86.00%	92.08%	85.33%

3.2.2. Kappa Analysis

A pairwise kappa analysis was performed after the creation of the error matrices to assess whether there was a significant difference between the pre and post hurricane images (Landsat 8 *vs.* Landsat 5) as well as the object-based and pixel-based classifications. For the pre and post hurricane map comparison, the results were compared for both the object-based and pixel-based classifications, including the error matrices for the area-based assessment for the object based classification. For the object-based *vs.* pixel-based comparison, only the tally-based error matrices were compared for each method since an area-based assessment is not performed on a pixel-based map. Tables 10 and 11 present the results of the pre *vs.* post hurricane analysis and the object *vs.* pixel-based analysis respectively.

Table 9. User's (UA) and producer's (PA) accuracies for post-hurricane classifications using object-based (OBC) and pixel-based (PBC) classification methodology.

Land Cover Class	PBC		OBC			
	Tally-based		Tally-based		Area-based	
	UA	PA	UA	PA	UA	PA
Agriculture	90.38%	88.68%	89.29%	94.34%	91.41%	95.31%
Deciduous Forest	74.07%	80.00%	72.22%	78.00%	79.94%	78.65%
Evergreen Forest	69.09%	76.00%	78.26%	72.00%	75.64%	76.17%
Mixed Forest	73.47%	72.00%	64.58%	62.00%	66.84%	68.51%
Developed	71.93%	82.00%	85.45%	94.00%	87.88%	97.95%
Open Water	100.00%	98.33%	100.00%	98.33%	100.00%	99.61%
Estuarine	88.68%	94.00%	83.93%	94.00%	90.36%	96.08%
Palustrine	60.87%	50.91%	66.00%	60.00%	71.53%	68.85%
Scrub/Shrub	80.00%	64.00%	80.00%	64.00%	79.68%	55.12%
Unconsolidated Shore	92.45%	98.00%	90.74%	98.00%	94.44%	99.04%

The kappa analysis comparing the pre and post hurricane images indicates that there is no significant difference between the two maps for both the pixel and object-based classifications since all Z test statistics were less than 1.96. The kappa analysis comparing the object-based and pixel-based classifications also indicates no significant difference between the two methods for either map.

Table 10. Kappa analysis comparing pre and post hurricane error matrices.

Method	Accuracy Assessment Type	Z Test Statistic
Object	Tally-based	0.413
	Area-based	0.950
Pixel	Tally-based	0.640

* Significant at 95% confidence level.

Table 11. Kappa analysis comparing object-based *vs.* pixel-based classifications error matrices.

Classification	Z Test Statistic
Pre-Hurricane	1.52
Post-Hurricane	1.11

* Significant at 95% confidence level.

3.3. Land Cover Change Detection and Classification

With the vector results of the post-hurricane classification included to act as boundaries, a within-class segmentation was performed with the full weight placed on the results of the image differencing between the post-hurricane and pre-hurricane NDVI. It was then possible to develop class specific change threshold values using the class specific change means and standard deviations of the resulting segments. Change thresholds were calculated by multiplying the standard deviation by an adjustable parameter and then adding and subtracting that value from the mean. The appropriate adjustable parameter and thus change thresholds for each class was determined by testing several parameter values and visually assessing how well the resulting thresholds detected change within each class. A parameter value of 1.5 was found to work best for all classes except for unconsolidated shore and scrub/shrub which required a value of 1 to detect all changes. Using the calculated thresholds, the objects within the image were classified as either change or no-change. Areas labeled as change were then classified with a pre-hurricane class and merged with the post-hurricane results.

Following the completion of the UID methodology, a PCC was also carried out between the two single date maps. The pre and post hurricane map were unioned and areas with two different land cover labels were classified as change. Figure 5 shows the difference between the areas classified as change and no change using the two change detection methods. The PCC classified 8.98% of the study area as change while the UID classified 0.339% of the area. It is important to note that while developing change thresholds for the UID, change was not detected for estuarine wetlands in order to reduce errors associated with the tide. For the PCC, the two maps were simply overlaid as they were. The results from both change detection methodologies were overlaid to identify and measure spatial agreement between areas labeled as change (Figure 6). Combined, 17275.59 ha were classified as change and of this area, 523.53 ha were identified as change by both the PCC and UID (shown in green in Figure 6). The area of agreement includes 80.63% of the area identified as change by the UID methodology and 3.05% of the area identified as change by the PCC.

Given the high percentage of area identified as change by the PCC, some supplementary evaluations were conducted to assess the results of the PCC. The PCC conducted previously was performed using the object based maps which, when overlaid, can create slivers due to differences in object boundaries between dates. A PCC was performed on the results of the pixel-based classifications, eliminating slivers, and was compared to the object-based PCC. The percentage of the study area classified as change by the pixel-based PCC was 8.59%, only a 0.3% drop compared to the object-based PCC. Additionally, the effect of aggregating land cover classes on the amount of area classified as change was investigated. Two aggregations and PCCs were performed. The first involved aggregating mixed, evergreen, and deciduous forest up one level and reclassifying them as forest and then performing a PCC with the reclassified maps. The second involved aggregating estuarine and palustrine up to wetland in addition to the aggregation of the forest classes and performing a PCC for a second time. Class aggregation was done for the object-based and pixel-based classifications. The results are shown in Table 12. It should be noted that the percent difference in change area represents the percent difference in total area classified as change after performing an aggregation and PCC compared to total area classified as change using the PCC with the original land cover maps (all 10 classes with no aggregation). Thus there is no percent difference in change area for the original 10 classes (shown as NA in table)

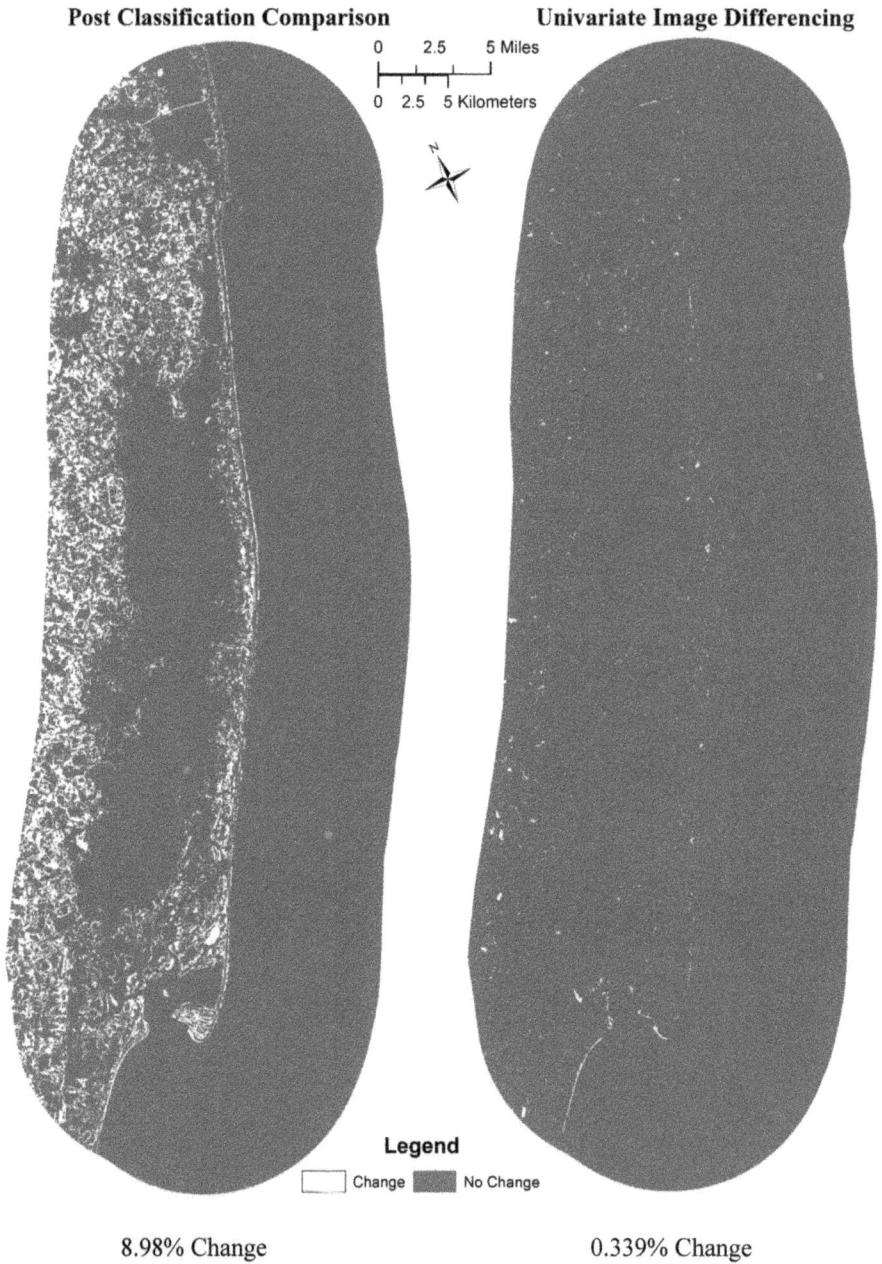

Figure 5. Areas of change and no-change as detected by the univariate image difference and post classification comparison (PCC). Percent change under each map represents percentage of study area classified as change by each method.

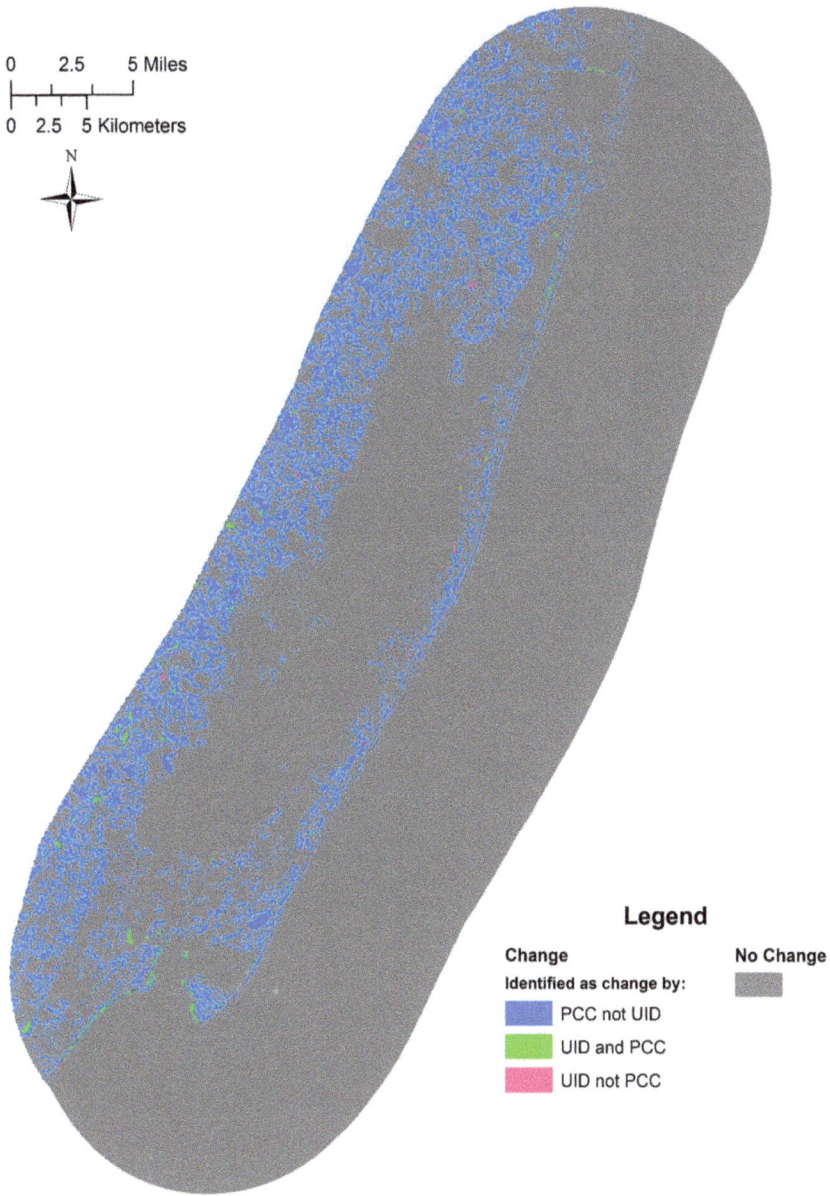

0   2.5   5 Miles

0   2.5   5 Kilometers

N

Legend

Change   No Change

Identified as change by:

PCC not UID

UID and PCC

UID not PCC

Figure 6. Overlay of the areas classified as change by the Post Classification Comparison (PCC) and Univariate Image differencing (UID). Areas in green represent agreement between methods while blue and pink are areas identified as change by one method but not the other.

Table 12. Results of aggregating land cover classes on PCC change detection.

Object-Based PCC			
Class Aggregation	Total Area Classified As Change (ha)	Percentage of Study Area Classified As Change	Percent Difference in Change Area *
Original Classes	1,7149.86	8.98%	NA
Aggregate Forest Classes	1,5030.63	7.87%	12.36%
Aggregate Forest and Wetland Classes	1,4151.87	7.41%	17.48%
Pixel-Based PCC			
Class Aggregation	Total Area Classified As Change (ha)	Percentage of Study Area Classified As Change	Percent Difference in Change Area *
Original Classes	1,6409.16	8.59%	NA
Aggregate Forest Classes	1,4236.65	7.45%	13.24%
Aggregate Forest and Wetland Classes	1,3515.3	7.07%	17.64%

* Compared to the total area classified as change using the original 10 land cover classes.

For both the object and pixel-based PCC, by aggregating the classes, the total area classified as change dropped. There was a 12.36% drop in the total area classified as change just by aggregating the forest classes together for the object-based PCC. A similar drop was seen with the pixel-based PCC. The drop was slightly higher by adding in the aggregated wetland classes. While there was a drop in the total area classified as change, the percent change still remains higher than that see with the UID.

Table 13. Land cover totals and area difference in hectares for the entire study area after using univariate image differencing to generate the pre-hurricane map. A negative difference indicates a decrease in the total area for that class between dates.

	Pre-Hurricane	Post-Hurricane	Difference (ha)	Percent Change
Deciduous Forest	5.67	5.67	0.00	0.000%
Developed	199.53	199.80	0.27	0.001%
Estuarine	4043.71	3966.58	−77.13	0.417%
Evergreen Forest	273.24	274.50	1.26	0.007%
Open Water	1,0457.08	1,0500.19	43.11	0.233%
Palustrine	1040.67	1048.59	7.92	0.043%
Scrub/Shrub	1100.70	1088.28	−12.42	0.067%
Unconsolidated Shore	1368.77	1405.76	36.99	0.200%
			Total =	0.969%

Tables 13 and 14 give the land cover totals and area differences on Assateague Island using the UID and PCC respectively. Most of these changes within the island are expected to have been a result of Hurricane Sandy since development is not common here. As can be seen in Table 13, the greatest impact was to estuarine, scrub/shrub, unconsolidated shore, and open water. Both scrub/shrub and estuarine experienced a decrease in area while unconsolidated shore experiences a considerable increase. The results after performing the PCC (Table 14) exhibit similar trends with the addition of palustrine wetlands that saw the largest increase between dates and unconsolidated shore which saw a decrease. There is very little agreement on the direction (increase or decrease) and magnitude of change.

Table 14. Land cover totals and area difference in hectares for Assateague Island using PCC. A negative difference indicates a decrease in the total area for that class between dates.

	Pre-Hurricane	Post-Hurricane	Difference (ha)	Percent Change
Deciduous Forest	6.39	5.67	−0.72	0.004%
Developed	193.68	199.80	6.12	0.033%
Estuarine	4000.57	3966.58	−33.99	0.184%
Mixed Forest	9.90	0.00	−9.90	0.054%
Evergreen Forest	284.85	274.50	−10.35	0.056%
Open Water	1,0644.9	1,0500.19	−144.71	0.783%
Palustrine	674.28	1048.59	374.31	2.024%
Scrub/Shrub	1112.58	1088.28	−24.30	0.131%
Unconsolidated Shore	1562.22	1405.76	−156.46	0.846%
			Total =	4.11%

4. Discussion and Conclusions

The main goal of this study was to quantify and qualify the type and amount of land cover change that occurred on Assateague Island as a result of Hurricane Sandy. With hurricane change detection established as the overarching goal, it was possible to perform several additional assessments; comparing the performance of Landsat 8 to Landsat 5, and object *vs.* pixel-based classification. Several findings resulted from this study and are discussed below.

4.1. Landsat 5 vs. Landsat 8

In this study the addition of the coastal blue band as well as the improvements in the Landsat OLI sensor (radiometric resolution and spectral bandwidths) were tested. First, the difference in accuracy between classifications was compared with and without the coastal blue band. The coastal blue band was the only new band assessed as the cirrus band is comprised of wavelengths that are almost completely absorbed by atmospheric moisture, providing little additional information. Additionally, the quality control band does not contain any spectral information and is instead used to assess the quality of the imagery. The coastal blue band was not found to improve the accuracy of the classification. The result is understandable given the placement of the band in the electromagnetic spectrum. In upland land cover classification, the visible bands often do not provide important information necessary to distinguish one cover type from another; the infrared bands are better suited to that end. Liu *et al.* [67] also found that both the coastal blue band and the new cirrus band had little effect on the derivations of the tassel cap components and their ability to classify land cover types. However, because this band is comprised of very narrow wavelengths, it would better be able to penetrate water and improve bathymetric studies or studies investigating subaquatic vegetation. In addition, its placement will make it important for detecting atmospheric conditions and improving atmospheric correction because it is highly scattered.

With the increase in radiometric resolution, it was expected that the classification accuracy for the post-hurricane map (Landsat 8) would be significantly better than the pre-hurricane (Landsat 5) maps due to the improved separability of spectrally similar classes and improvement in bandwidth ranges. Using several pairwise kappa analyses, however, it was found that there was no significant difference in classification accuracy for either the object-based or pixel-based classifications.

It is possible that the differences in the training and accuracy samples between maps may have had an impact. Each image underwent a separate segmentation and then the segments were intersected with the reference unit centroids. It is not likely that the same size and shaped segments were generated, thus each classification was trained with and applied to slightly different segments which could alter the results. Furthermore, it is possible that the classification scheme used was not detailed enough for the higher radiometric resolution to make a difference in the classification accuracy. The NOAA C-CAP classification scheme used here was designed to be used with 8 bit Landsat data collected by the TM and Enhanced Thematic Mapper (ETM+) sensors, and has been successfully used to classify these cover types for a number of years with a high level of accuracy. There may be enough information stored in the 8 bit imagery to distinguish the spectrally similar classes. Additional studies would need to be conducted using more detailed schemes to determine when the improved radiometric resolution would be a benefit for land cover classification. It is important to keep in mind though that the 30 m spatial resolution may limit how detailed one can get. Furthermore, the enhanced radiometric resolution may benefit studies where biological parameters are being measured from the imagery as demonstrated in Dube and Mutanga [20] who found Landsat 8 OLI data provided better estimates of above ground biomass compared to Landsat 7 ETM+.

4.2. Object vs. Pixel-Based Classification

While object-based classification has been shown to improve the classification accuracy of high-resolution imagery, with moderate resolution imagery the benefits have not been clear. One of the objectives of this study was to further investigate this relationship by comparing a pixel-based and object-based classification using the same training data and classification algorithm, which has not commonly been done in previous investigations. No significant difference between the object-based and pixel-based classifications for both the pre-hurricane and post-hurricane imagery was found which is similar to conclusions reached in other studies utilizing medium resolution imagery [39,40,68].

One of the main advantages of OBIA that allows it to perform well is its ability to reduce the spectral variability within a class and thus confusion between classes. If there was not much within class spectral variability to begin with, however, then the benefits of OBIA would be diminished. It was pointed out in Campbell *et al.* [38] that a 30 m Landsat pixel is sizable and could be considered an image object in-and-of itself and if the classification scheme was more detailed, then a single pixel may reduce the spectral noise enough to allow the pixel to be classified accurately. On the other hand, if the classes were very broad, then one would expect significantly more within-class spectral variability. For example, the class forest would include the spectral characteristics of both deciduous and evergreen forest or developed would include the characteristics of a multitude of objects such as grass, trees, roof tops, pavement, *etc.* With broad classes, the reduction in noise may actually increase the classification accuracy, especially for more heterogeneous classes.

4.3. Land Cover Change

One of the major objectives of this study was to quantify the land cover change that occurred on the island as a result of Hurricane Sandy. The UID approach was implemented to detect changes and then update the post-hurricane map in order to generate a more consistent pre-hurricane map. From there, the land cover totals and differences were calculated. The PCC was also implemented to detect where change had occurred and the type of change using the information provided in both single-date maps.

Changes across the entire study area were minimal. By percent change, the UID labeled 0.336% of the study area as change while the PCC labeled 8.98%. A minimal amount of change was expected given the time frame of the study; however, the PCC method detected significantly more. There was some agreement between the two methods. The UID methodology relies on the change in spectral reflectance to detect change, not a difference in thematic labeling which is dependent on the quality of the classification. Of the area classified as change using the UID, 80% of it was detected as change by the PCC method indicating that the PCC was able to detect real land cover changes in the study area, however, a great deal of false change was detected as well.

Two factors could explain why such a substantial difference was seen between the change detection methodologies. First, an object-based PCC suffers from the effects of slivers. When different sets of imagery undergo separate segmentations, it is rare that the resulting segments will exhibit the same geometry due to differences in the environmental conditions when the images were captured or misregistration [69]. When overlaid, differences in the edges of the segments create slivers which can lead to additional change errors. The second factor is thematic errors. The PCC method relies upon the accuracy of the single-date maps. Errors in the maps will appear as erroneous changes and can occur in both object-based and pixel-based PCC. In this case, two segments representing the same feature may have different class labels because of the classification process, which results in a change error.

The object-based PCC was compared to the pixel-based PCC to assess the possible contribution of slivers to the total change area detected by the PCC. Both methods detected a similar amount of change with the pixel-based detecting slightly less, so slivers were perhaps a minimal component to the object-based PCC. The differences in thematic labeling were investigated further by aggregating classes prior to performing the PCC (see Table 12). By simply aggregating mixed, deciduous, and evergreen forest together, the total area classified as change dropped by 12.36% and 13.24% for the object-based and pixel-based maps respectively. One would not expect the composition of a forest to change within two years, so this difference in area was due to differences in the labeling of these forested polygons and pixels between maps. By aggregating the wetland classes, an even greater drop was observed. The result of this investigation suggests that change detection errors could be reduced by simply aggregating classes up to a higher level in the classification scheme. Nevertheless, even after aggregating the wetland and forest classes, the percentage of the study area classified as change was still over 7% for both PCCs indicating that a number of labeling errors still exists.

The object-based univariate image differencing method applied here has the advantage of detecting changes due to spectral differences between dates and not because of differences in thematic labeling. Furthermore, by enforcing the boundaries of the post-hurricane map during the segmentation of the difference image, internal adjustments could be made to those segments that experienced real change. Had the difference image been segmented without the enforced boundaries, when updating the post-hurricane map, differences between the changed objects and the base map would result in silvers as demonstrated in McDermid *et al.* [70]; however, there would be significantly fewer compared to the PCC since only the overlap between the change objects and the base map occurs, Enforcing the boundaries did not eliminate spurious slivers altogether. Boundary errors in the post-hurricane map still resulted in slivers, some only a few pixels in size. Fuzzy boundaries between cover types also make it difficult to accurately delineate features

While change was measured for the entire study area, this paper focuses mainly on the results seen within Assateague Island. While the PCC estimated a greater amount of change within the island, there were some consistent results. Estuarine, scrub/shrub, and unconsolidated shore experiences the greatest changes on the island itself. Estuarine and scrub/shrub both saw decreases. Surprisingly, unconsolidated shore saw a significant decrease with the PCC compared to the UID that showed an increase. It is probable that this is due to the decrease in shoreline detected by the PCC and not with the UID. It is difficult to say whether this was real change or not as it could have been explained by differences in the tide.

While changes on the island were minimal, where they occurred is consistent with the structure of the island itself. Most of the change detected within the northern 10 km of the island, known as the North End, as well as the southernmost hook known as Tom's Cove Hook (Figure 7). The center of the island experienced almost no change with the exception of a few small overwash fans. The North End much lower relative to the rest of the island and does not have a continuous, high dune to protect the vegetation on the side closest to the bay. These characteristics leave the area very susceptible to overwash which carries sand from the ocean onto the mainland. The North End saw an increase in unconsolidated shore due to a large overwash fan burying estuarine wetland and scrub/shrub areas that are close to the ocean due to this area being so narrow. Tom's Cove Hook on the other hand is an accretionary spit. Long shore transport carries sand from the northern portion of the island south where it accumulates at the spit. The hook saw a significant increase in the southern tip of the spit following the hurricane as well as a loss of a very small amount of estuarine and scrub/shrub habitat that was buried by a small overwash fan. In general the location and type of change detected on the island is consistent with its ecology and topography.

It is important to note that these results should be taken as estimates. First, it was not possible to collect reference data on the actual changes that occurred within this time span thus it was not possible to assess the accuracy of these results. Second, a visual assessment was undertaken to choose change threshold parameters. While more objective methods exist to develop change thresholds, since change was expected to be minimal and the study area was small, a visual assessment was conducted instead. The goal was to reduce omission, thus overestimation of change was preferred rather than under estimation. It is possible that more change was detected than had actually occurred. Differences in tide between the two dates may have also played a role in the final results. While estuarine habitat was not included in the change detection process in order to reduce possible errors, a great deal of unconsolidated shore and open water changes were classified as estuarine in the pre-hurricane map leading to a higher decrease in estuarine wetlands than was actually experienced.

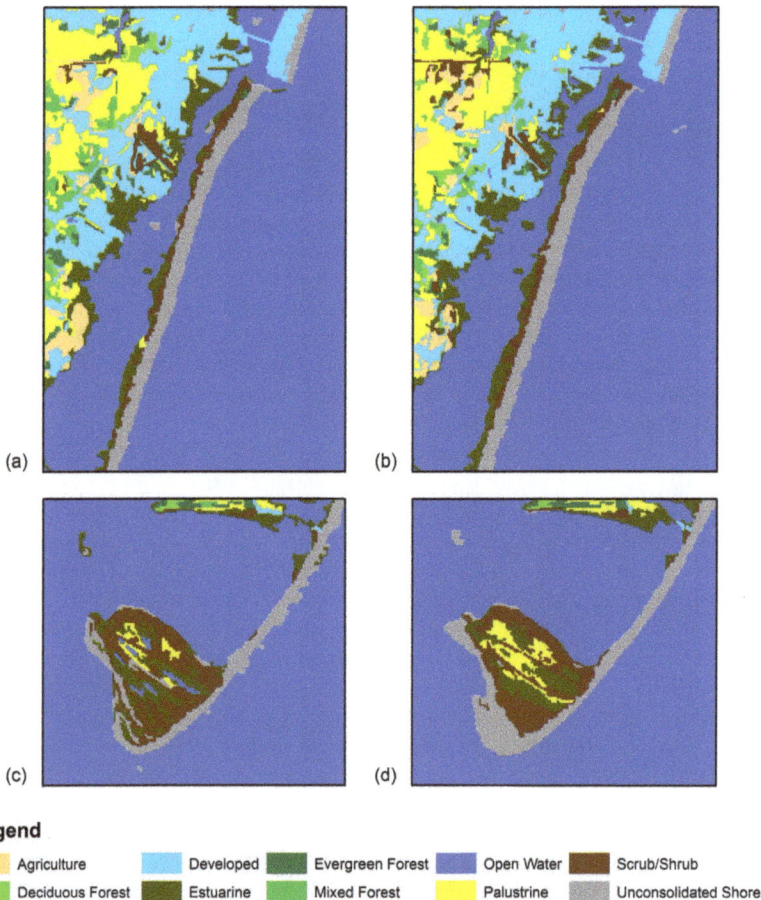

Figure 7. The North End before (**a**) and after (**b**) Hurricane Sandy. Tom's Hook Cove before (**c**), and after (**d**) the hurricane.

This study was limited to detecting large area changes. Given the spatial resolution of the Landsat data, it was not possible to detect changes smaller than 30 m × 30 m in size. Higher spatial resolution data would be needed to detect smaller changes but it was not available for the entire study area. Additionally, the use of Landsat data allows for this analysis to be integrated with pre-existing land cover maps and with the data being freely available, these maps can be updated as needed.

Acknowledgments: Partial funding was provided by the New Hampshire Agricultural Experiment Station. This is Scientific Contribution Number 2628. This work was supported by the USDA National Institute of Food and Agriculture McIntire-Stennis Project 1002519. This project was also supported in part by Grant Number G14AP00002 from the Department of the Interior, United States Geological Survey to AmericaView. Its contents are solely the responsibility of the authors; the views and conclusions contained in this document are those of the authors and should not be interpreted as representing the opinions or policies of the U.S. Government. Mention of trade names or commercial products does not constitute their endorsement by the U.S. Government. Furthermore, we wish to extend our thanks and appreciation to the staff at Assateague Island National Seashore for their support and assistance during this project.

Author Contributions: Heather Grybas conducted the research presented in this study and wrote the paper. Russell G. Congalton contributed to the development of the overall research design, provided guidance along the way, and aided in the writing of the paper.

Conflicts of Interest: The authors declare no conflict of interest

References

1. Klemas, V.V. The role of Remote Sensing in predicting and determining coastal storm impacts. *J. Coast. Res.* **2009**, *25*, 1264–1275. [CrossRef]
2. Bianchette, T. A.; Liu, K.B.; Lam, N.S.; Kiage, L.M.; Ecological impacts of hurricane ivan on the gulf coast of Alabama: A remote sensing study. *J. Coast. Res.* **2009**, *2*, 1622–1626.
3. Zhang, X.; Wang, Y.; Jiang, H.; Wang, X. Remote-sensing assessment of forest damage by Typhoon Saomai and its related factors at landscape scale. *Int. J. Remote Sens.* **2013**, *34*, 7874–7886. [CrossRef]
4. Ramsey III, E.W.; Chappell, D.K.; Baldwin, D.G. AVHRR imagery used to identify hurricane damage in a forested wetland of Louisiana. *Photogramm. Eng. Remote Sens.* **1997**, *63*, 293–297.
5. Steyer, G.D.; Couvillion, B.R.; Barras, J.A. Monitoring vegetation response to episodic disturbance events by using multitemporal vegetation indices. *J. Coast. Res.* **2013**. [CrossRef]
6. Rodgers, J.C.; Murrah, A.W.; Cooke, W.H. The impact of hurricane katrina on the coastal vegetation of the weeks bay reserve, alabama from NDVI data. *Estuar. Coast.* **2009**, *32*, 496–507. [CrossRef]
7. Knutson, T.R.; McBride, J.L.; Chan, J.; Emanuel, K.; Holland, G.; Landsea, C.; Held, I.; Kossin, J.P.; Srivastava, A.K.; Sugi, M. Tropical cyclones and climate change. *Nat. Geosci.* **2010**, *3*, 157–163. [CrossRef]
8. Webster, P.J.; Holland, G.J.; Curry, J.A.; Chang, H.-R. Changes in tropical cyclone number, duration, and intensity in a warming environment. *Science* **2005**, *309*, 1844–1846. [CrossRef] [PubMed]
9. Lam, N.S.-N.; Liu, K.-B.; Liang, W.; Bianchette, T.A.; Platt, W.J. Effects of Hurricanes on the Gulf Coast ecosystems: A remote sensing study of land cover change around Weeks Bay, Alabama. *J. Coast. Res.* **2011**, 1707–1711.
10. Part A: global and sectoral aspects. contribution of working group ii to the fifth assessment report of the intergovernmental panel on climate change. In *IPCC Climate Change 2014: Impacts, Adaptation, and Vulnerability*; Field, C.B.; Barros, V.R.; Dokken, D.J.; Mach, K.J.; Mastrandrea, M.D.; Bilir, T.E.; Chatterjee, M.; Ebi, K.L.; Estrada, Y.O.; Genova, R.C.; *et al.* (Eds.) Cambridge University Press: Cambridge, United Kingdom and New York, NY, USA, 2014; pp. 361–409.
11. Wang, Y.; Christiano, M.; Traber, M. Mapping salt marshes in Jamaica Bay and terrestrial vegetation in Fire Island National Seashore using QuickBird satellite data. In *Remote Sensing of Coastal Environments*; Weng, Q., Ed.; CRC Press: Boca Raton, FL, USA, 2010; pp. 191–208.
12. Lu, D.; Weng, Q. A survey of image classification methods and techniques for improving classification performance. *Int. J. Remote Sens.* **2007**, *28*, 823–870. [CrossRef]
13. Ramsey III, E.W.; Jacobs, D.M.; Sapkota, S.K.; Baldwin, D.G. Resource management of forested wetlands: Hurricane impact and recovery mapped by combining Landsat TM and NOAA AVHRR data. *Photogramm. Eng. Remote Sens.* **1998**, *64*, 733–738.
14. Ayala-Silva, T.; Twumasi, Y.A. Hurricane Georges and vegetation change in Puerto Rico using AVHRR satellite data. *Int. J. Remote Sens.* **2004**, *25*, 1629–1640. [CrossRef]

15. Wang, F.; Xu, Y.J. Hurricane Katrina-induced forest damage in relation to ecological factors at landscape scale. *Environ. Monit. Assess.* **2009**, *156*, 491–507. [CrossRef] [PubMed]

16. Wulder, M.A.; White, J.C.; Goward, S.N.; Masek, J.G.; Irons, J.R.; Herold, M.; Cohen, W.B.; Loveland, T.R.; Woodcock, C.E. Landsat continuity: Issues and opportunities for land cover monitoring. *Remote Sens. Environ.* **2008**, *112*, 955–969. [CrossRef]

17. Wulder, M.A.; Masek, J.G.; Cohen, W.B.; Loveland, T.R.; Woodcock, C.E. Opening the archive: How free data has enabled the science and monitoring promise of Landsat. *Remote Sens. Environ.* **2012**, *122*, 2–10. [CrossRef]

18. Roy, D.P.; Wulder, M.A.; Loveland, T.R.; C.E., W.; Allen, R.G.; Anderson, M.C.; Helder, D.; Irons, J.R.; Johnson, D.M.; Kennedy, R.; *et al.* Landsat-8: Science and product vision for terrestrial global change research. *Remote Sens. Environ.* **2014**, *145*, 154–172. [CrossRef]

19. USGS Frequently Asked Questions about the Landsat Missions. Availble Online: http://landsat.usgs.gov/ldcm_vs_previous.php (accessed on 30 September 2015).

20. Dube, T.; Mutanga, O. Evaluating the utility of the medium-spatial resolution Landsat 8 multispectral sensor in quantifying aboveground biomass in uMgeni catchment , South Africa. *ISPRS J. Photogramm. Remote Sens.* **2015**, *101*, 36–46. [CrossRef]

21. Flood, N. Continuity of Reflectance Data between Landsat-7 ETM+ and Landsat-8 OLI, for Both Top-of-Atmosphere and Surface Reflectance: A Study in the Australian Landscape. *Remote Sens.* **2014**, *6*, 7952–7970. [CrossRef]

22. NASA Landsat 8 Overview. Availble Online: http://landsat.gsfc.nasa.gov/?page_id=7195 (accessed on 30 September 2015).

23. Irons, J.R.; Dwyer, J.L.; Barsi, J.A. The next Landsat satellite: The Landsat data continuity mission. *Remote Sens. Environ.* **2012**, *122*, 11–21. [CrossRef]

24. Jia, K.; Wei, X.; Gu, X.; Yao, Y.; Xie, X.; Li, B. Land cover classification using Landsat 8 Operational Land Imager data in Beijing, China. *Geocarto Int.* **2014**, *29*, 941–951. [CrossRef]

25. Poursanidis, D.; Chrysoulakis, N.; Mitraka, Z. Landsat 8 *vs.* Landsat 5: A comparison based on urban and peri-urban land cover mapping. *Int. J. Appl. Earth Obs. Geoinf.* **2015**, *35*, 259–269. [CrossRef]

26. Ferguson, R.L.; Korfmacher, K. Remote sensing and GIS analysis of seagrass meadows in North Carolina, USA. *Aquat. Bot.* **1997**, *58*, 241–258. [CrossRef]

27. Vogelmann, J.E.; Sohl, T.; Howard, S.M. Regional characterization of land cover using multiple sources of data. *Photogramm. Eng. Remote Sens.* **1998**, *64*, 45–57.

28. Lawrence, R.L.; Wright, A. Rule-based classification systems using classification and regression tree (CART) analysis. *Photogramm. Eng. Remote Sensing* **2001**, *67*, 1137–1142.

29. Lathrop, R.G.; Montesano, P.; Haag, S. A multi-scale segmentation approach to mapping seagrass habitats using airborne digital camera imagery. *Photogramm. Eng. Remote Sens.* **2006**, *72*, 665–675. [CrossRef]

30. Yu, Q.; Gong, P.; Clinton, N.; Biging, G.; Kelly, M.; Schirokauer, D. Object-based detailed vegetation classification with airborne high spatial resolution Remote Sensing imagery. *Photogramm. Eng. Remote Sens.* **2006**, *72*, 799–811. [CrossRef]

31. Conchedda, G.; Durieux, L.; Mayaux, P. An object-based method for mapping and change analysis in mangrove ecosystems. *ISPRS J. Photogramm. Remote Sens.* **2008**, *63*, 578–589. [CrossRef]

32. Johansen, K.; Arroyo, L.A.; Phinn, S.; Witte, C. Comparison of geo-object based and pixel-based change detection of riparian environments using high spatial resolution multi-spectral imagery. *Photogramm. Eng. Remote Sens.* **2010**, *76*, 123–136. [CrossRef]

33. Blaschke, T. Object based image analysis for remote sensing. *ISPRS J. Photogramm. Remote Sens.* **2010**, *65*, 2–16. [CrossRef]

34. Blaschke, T.; Hay, G.J.; Kelly, M.; Lang, S.; Hofmann, P.; Addink, E.; Queiroz Feitosa, R.; van der Meer, F.; van der Werff, H.; van Coillie, F.; *et al.* Geographic object-based image analysis—Towards a new paradigm. *ISPRS J. Photogramm. Remote Sens.* **2014**, *87*, 180–191. [CrossRef]

35. Whiteside, T.G.; Boggs, G.S.; Maier, S.W. Comparing object-based and pixel-based classifications for mapping savannas. *Int. J. Appl. Earth Obs. Geoinf.* **2011**, *13*, 884–893. [CrossRef]

36. Flanders, D.; Hall-Beyer, M.; Pereverzoff, J. Preliminary evaluation of eCognition object-based software for cut block delineation and feature extraction. *Can. J. Remote Sens.* **2003**, *29*, 441–452. [CrossRef]

37. Yan, G.; Mas, J.-F.; Maathuis, B.H.P.; Xiangmin, Z.; Van Dijk, P.M. Comparison of pixel-based and object-oriented image classification approaches—A case study in a coal fire area, Wuda, Inner Mongolia, China. *Int. J. Remote Sens.* **2006**, *27*, 4039–4055. [CrossRef]

38. Campbell, M.; Congalton, R.G.; Hartter, J.; Ducey, M. Optimal land cover mapping and change analysis in northeastern oregon using Landsat imagery. *Photogramm. Eng. Remote Sens.* **2015**, *81*, 37–47. [CrossRef]

39. Robertson, L.D.; King, D.J. Comparison of pixel- and object-based classification in land cover change mapping. *Int. J. Remote Sens.* **2011**, *32*, 1505–1529. [CrossRef]

40. Duro, D.C.; Franklin, S.E.; Dubé, M.G. A comparison of pixel-based and object-based image analysis with selected machine learning algorithms for the classification of agricultural landscapes using SPOT-5 HRG imagery. *Remote Sens. Environ.* **2012**, *118*, 259–272. [CrossRef]

41. Hussain, M.; Chen, D.; Cheng, A.; Wei, H.; Stanley, D. Change detection from remotely sensed images: From pixel-based to object-based approaches. *ISPRS J. Photogramm. Remote Sens.* **2013**, *80*, 91–106. [CrossRef]

42. Coppin, P.; Jonckheere, I.; Nackaerts, K.; Muys, B.; Lambin, E. Review ArticleDigital change detection methods in ecosystem monitoring: A review. *Int. J. Remote Sens.* **2004**, *25*, 1565–1596. [CrossRef]

43. Dobson, J.E.; Bright, E.A.; Ferguson, R.L.; Field, D.W.; Wood, L.; Haddad, K.D.; Iredale III, H.; Jensen, J.R.; Klemas, V.V.; Orth, J.R.; *et al.* *NOAA Coastal Change Analysis Program (C-CAP): Guidance for Regional Implementation*; NOAA Technical Report NMFS 123: Seattle, WA, USA, 1995.

44. Schupp, C. *Assateague Island National Seashore Geologic Resources Inventory Report*; Natural resource report NPS/NRSS/GRD/NRR—2013/708: Fort Collins, CO, USA, 2013.

45. Carruthers, T.; Beckert, K.; Dennison, B.; Thomas, J.; Saxby, T.; Williams, M.; Fisher, T.; Kumer, J.; Schupp, C.; Sturgis, B.; *et al.* *Assateague Island National Seashore Natural Resource Condition Assessment Maryland Virginia*; Natural Resource Report NPS/ASIS/NRR—2011/405: Fort Collins, CO, USA, 2011.

46. Carruthers, T.; Beckert, K.; Schupp, C.A.; Saxby, T.; Kumer, J.P.; Thomas, J.; Sturgis, B.; Dennison, W.C.; Williams, M.; Fisher, T.; *et al.* Improving management of a mid-Atlantic coastal barrier island through assessment of habitat condition. *Estuar. Coast. Shelf Sci.* **2013**, *116*, 74–86.

47. Krantz, D.E.; Schupp, C.; Spaur, C.C.; Thomas, J.; Wells, D. Dynamic systems at the land-sea interface. In *Shifting Sands: Environmental and Cultural Change in Maryland's Coastal Bays*; Dennison, W.C., Thomas, J., Cain, C.J., Carruthers, T., Hall, M.R., Jesien, R.V., Wazniak, C.E., Wilson, D.E., Eds.; IAN Press: Cambridge, MD, USA, 2009; pp. 193–230.

48. Intergraph. *ERDAS Field Guide*; Intergraph Corporation: Huntsville, AL, USA, 2013.

49. Chavez, P.S. Image-based atmospheric corrections—Revisited and improved. *Photogramm. Eng. Remote Sens.* **1996**, *62*, 1025–1036.

50. Chander, G.; Markham, B.L.; Helder, D.L. Summary of current radiometric calibration coefficients for Landsat MSS, TM, ETM+, and EO-1 ALI sensors. *Remote Sens. Environ.* **2009**, *113*, 893–903. [CrossRef]

51. Crist, E.P.; Laurin, R.; Cicone, R. Vegetation and soils information contained in transformed Thematic Mapper data. In Proceedings of IGARSS' 86 Symposium, Zurich, Switzerland, 8–11 September 1986; pp. 1465–1470.

52. Baig, M.H.A.; Zhang, L.; Shuai, T.; Tong, Q. Derivation of a tasselled cap transformation based on Landsat 8 at-satellite reflectance. *Remote Sens. Lett.* **2014**, *5*, 423–431. [CrossRef]

53. Congalton, R.G.; Green, K. *Assessing the Accuracy of Remotely Sensed Data: Principles and Practices*, 2nd ed.; CRC Press: Boca Raton, FL, USA, 2009.

54. Congalton, R.G. Using spatial autocorrelation analysis to explore the errors in maps generated from remotely sensed data. *Photogramm. Eng. Remote Sens.* **1988**, *54*, 587–592.

55. Trimble. *eCognition Developer 9.0 User Guide*; TrimbleGermany GmbH: Munich, Germany, 2014.

56. Kim, M.; Warner, T.A.; Madden, M.; Atkinson, D.S. Multi-scale GEOBIA with very high spatial resolution digital aerial imagery: Scale, texture and image objects. *Int. J. Remote Sens.* **2011**, *32*, 2825–2850. [CrossRef]

57. Kim, M.; Madden, M.; Warner, T. Estimation of optimal image object size for the segmentation of forest stands with multispectral IKONOS imagery. In *Object-Based Image Analysis—Spatial Concepts for Knowledge-Driven Remote Sensing Applications*; Blaschke, T., Lang, S., Hay, G.J., Eds.; Springer-Verlag: Berlin, Heidelberg, 2008; pp. 291–307.

58. Drăguţ, L.; Tiede, D.; Levick, S.R. ESP: A tool to estimate scale parameter for multiresolution image segmentation of remotely sensed data. *Int. J. Geogr. Inf. Sci.* **2010**, *24*, 859–871. [CrossRef]

59. Drăguţ, L.; Csillik, O.; Eisank, C.; Tiede, D. Automated parameterisation for multi-scale image segmentation on multiple layers. *ISPRS J. Photogramm. Remote Sens.* **2014**, *88*, 119–127. [CrossRef] [PubMed]

60. Espindola, G.M.; Camara, G.; Reis, I.A.; Bins, L.S.; Monteiro, A.M. Parameter selection for region-growing image segmentation algorithms using spatial autocorrelation. *Int. J. Remote Sens.* **2006**, *27*, 3035–3040. [CrossRef]

61. Breiman, L. Random forests. *Mach. Learn.* **2001**, *45*, 5–32. [CrossRef]

62. Rodriguez-Galiano, V.F.; Ghimire, B.; Rogan, J.; Chica-Olmo, M.; Rigol-Sanchez, J.P. An assessment of the effectiveness of a random forest classifier for land-cover classification. *ISPRS J. Photogramm. Remote Sens.* **2012**, *67*, 93–104. [CrossRef]

63. Congalton, R.G.; Oderwald, R.G.; Mead, R.A. Assessing Landsat classification accuracy using discrete multivariate analysis statistical techniques. *Photogramm. Eng. Remote Sens.* **1983**, *49*, 1671–1678.

64. MacLean, M.G.; Congalton, R.G. Map accuracy assessment issues when using an object-oriented approach. In Proceedings of American Society of Photogrammetry & Remote Sensing 2012 Annual Conference, Sacramento, CA, USA, 19–23 March 2012; p. 5.

65. Xian, G.; Homer, C.; Fry, J. Updating the 2001 national land cover database land cover classification to 2006 by using Landsat imagery change detection methods. *Remote Sens. Environ.* **2009**, *113*, 1133–1147. [CrossRef]

66. Story, M.; Congalton, R.G. Accuracy assessment: A user's perspective. *Photogramm. Eng. Remote Sens.* **1986**, *52*, 397–399.

67. Liu, Q.; Liu, G.; Huang, C.; Xie, C. Comparison of tasselled cap transformations based on the selective bands of Landsat 8 OLI TOA reflectance images. *Int. J. Remote Sens.* **2015**, *36*, 417–441. [CrossRef]

68. Baker, B.A.; Warner, T.A.; Conley, J.F.; McNeil, B.E. Does spatial resolution matter? A multi-scale comparison of object-based and pixel-based methods for detecting change associated with gas well drilling operations. *Int. J. Remote Sens.* **2013**, *34*, 1633–1651. [CrossRef]

69. Chen, G.; Hay, G.J.; Carvalho, L.M.T.; Wulder, M.A. Object-based change detection. *Int. J. Remote Sens.* **2012**, *33*, 4434–4457. [CrossRef]

70. McDermid, G.J.; Linke, J.; Pape, A.D.; Laskin, D.N.; McLane, A.J.; Franklin, S.E. Object-based approaches to change analysis and thematic map update: Challenges and limitations. *Can. J. Remote Sens.* **2008**, *34*, 462–466. [CrossRef]

Journal of
Imaging

MDPI

Article

Automated Soil Physical Parameter Assessment Using Smartphone and Digital Camera Imagery

Matt Aitkenhead *, Malcolm Coull, Richard Gwatkin and David Donnelly

The James Hutton Institute, Craigiebuckler, Aberdeen AB15 8QH, UK; malcolm.coull@hutton.ac.uk (M.C.);
richard.gwatkin@hutton.ac.uk (R.G.); david.donnelly@hutton.ac.uk (D.D.)
* Correspondence: matt.aitkenhead@hutton.ac.uk; Tel.: +44-01224-395-257

Academic Editors: Gonzalo Pajares Martinsanz and Francisco Rovira-Más
Received: 29 September 2016; Accepted: 7 December 2016; Published: 13 December 2016

Abstract: Here we present work on using different types of soil profile imagery (topsoil profiles captured with a smartphone camera and full-profile images captured with a conventional digital camera) to estimate the structure, texture and drainage of the soil. The method is adapted from earlier work on developing smartphone apps for estimating topsoil organic matter content in Scotland and uses an existing visual soil structure assessment approach. Colour and image texture information was extracted from the imagery. This information was linked, using geolocation information derived from the smartphone GPS system or from field notes, with existing collections of topography, land cover, soil and climate data for Scotland. A neural network model was developed that was capable of estimating soil structure (on a five-point scale), soil texture (sand, silt, clay), bulk density, pH and drainage category using this information. The model is sufficiently accurate to provide estimates of these parameters from soils in the field. We discuss potential improvements to the approach and plans to integrate the model into a set of smartphone apps for estimating health and fertility indicators for Scottish soils.

Keywords: soil; structure; mobile phone; visual soil assessment

1. Introduction

Soil structure is the term given to the arrangement of differently-sized mineral particles and organic material into larger, more complex arrangements in the soil. Good structure improves soil hydraulic properties, allows air and roots to move freely through the soil and provides a suitable habitat for microbes, fungi and soil fauna. Poor structure implies smaller overall pore space, reduced numbers of large pores and a lack of internal spaces for the physical and chemical protection of organic matter and living organisms. Soil texture is the proportional distribution of particles of different sizes, commonly divided into sand (>50 μm), silt (2–50 μm) and clay (<2 μm) particles (these size range definitions vary in the literature). Both soil texture and structure have important impacts on several factors including plant growth, nutrient availability and soil hydrology.

Assessment of soil structure and texture is an important component of determining physical condition and degradation of soils. The soil structure and its impact on soil health and resilience to disturbance can be measured using a number of different approaches, including physical and visual, while estimates of soil texture are normally carried out by hand or using a number of laboratory approaches. It was demonstrated [1] that the geometric characteristics of soil pores varied under different levels of compaction and structural damage. Visual soil assessment approaches give results that in many cases link closely to direct physical measurements of structure-related soil parameters [2].

It has been pointed out [3] that laboratory-based assessment of soil sample structure does not produce a fair indication of soil structure, and that soil structure does not lend itself to quantification. The same is not true of soil texture, which can be measured directly in the laboratory but which still relies on fairly subjective "hand-texturing" approaches in the field. One alternative could be to determine if image analysis of the visible structure of soils can lead to automated soil structure assessment in the field. A clear statistical link has been drawn [4] between the spatial organisation and forms of soil components at the microscopic scale, and the character and stability of soil structure. If this is also true at larger scales such as those visible to digital cameras, then this automated image analysis might be possible.

Changes in soil micromorphology, specifically relating to pore size distribution and hence to soil structure, can be detected following changes to tillage practices [5]. The same management of soils with different textural proportions (sand, silt, clay) will not result in the same soil structure. Therefore, it is also worthwhile investigating whether soil texture can be linked to image analysis of soil profiles, and estimated using the same approaches.

Mineralogical analysis may traditionally have provided better information on soil structure at microscopic scales than micromorphological analysis [6]. However, improvements in characterising 2D image of complex systems such as soil may now mean that image analysis is on an equal footing with mineralogy in this area. Early work on digital information of pore structure (e.g., [7,8]) demonstrated links between soil structure and image structure, but required careful preparation of soil samples (i.e., impregnation with epoxy resin, sawing, polishing).

A study by [9] demonstrated a link between morphological measurements of soil structure in the field and soil compaction. Visualisation of soil aggregate structure and pore size/connectivity using X-ray tomography reveals links between structural parameters related to bulk density and water movement [10]. It was also demonstrated [11] that the use of 2D tomography for assessing soil structure properties could provide sufficient information to assess structural quality.

This implies that soil structural assessment, including information about bulk density, can be automated through appropriate characterisation of photographs of soil profiles. The term "soil profile" here is taken to mean a vertical section of the soil, not the description of the sequence and character of the individual soil horizons, as it is sometimes taken to mean.

An approach has been described [12] for visual assessment of soil structure that gives a numerical score to the structure. The soil features described in [12] are amenable to visual discrimination, implying that a direct estimate of structural parameters may be possible. The benefit to farmers and other land managers of a rapid, low-cost method of assessing soil structure and soil degradation is emphasised in [13].

The Visual Evaluation of Soil Structure (VESS) approach developed by [12] (which was developed for agricultural soils commonly found in northern Europe) can also be applied successfully to Oxisols and potentially other tropical soils under different land use systems [14] and to tropical soils under no-tillage agricultural systems [15]. Arguably, a single method of automated soil structure assessment could potentially be applied in many parts of the world and to soils under many different management systems.

Investigation between visually measurable pore-structure attributes and levels of compaction show clear links that imply the ability to measure soil structure from digital imagery with mathematically defined parameters [16]. A correlation was also shown [17] between visual soil structure assessment scoring and the slope of the line between soil wetness and penetration resistance. This demonstrated a clear link between visual assessment scoring and physically-relevant soil structure measurements.

An assessment of different methods for visually assessing soil structure by [18] showed that many approaches gave similar results and linked closely to shape and size of aggregates. These aggregates are visible and may be amenable to detection through image analysis approaches. This is an approach that has similarities to the method of [12] and that indicates that aggregate size in particular is something that links both visual assessment and structure in soils.

Several methods exist for field-based assessment of soil structure, including a number that rely on visual interpretation [19]. The VESS approach of [12] by Bruce Ball, however, provides an approach for determining soil structure visually, and provides information that relates well to measurements from more instrument-reliant approaches for determining soil physical structure [20].

Our aim is to show that in the last two and a half decades technological, statistical and modelling developments now mean that because it is possible to measure soil structure in the field rather than the laboratory, the quantification of soil structure and soil texture can now be achieved through image analysis directly linked to field-based soil structure assessment parameters and environmental conditions. We make extensive use of Ball's approach, including the numerical scoring system and an image analysis method that utilises some of the descriptive features in this method.

Aggregate soil stability, a key parameter in soil structure, is influenced by a number of factors including climate, vegetation growth and soil management amongst other parameters (including soil texture) [21]. Our assessment also includes several of these environmental factors, although it does not include land management history. In addition to structure, we estimate soil texture and drainage class using the same image analysis approach. Building on previous work linking soil colour and environmental factors to soil properties [22], we also include LOI (Loss On Ignition) as an estimate of soil organic matter content, and soil pH. The overall method is related to current work in Digital Soil Mapping, which links spatial covariates (and occasionally remote sensing) to soil properties.

2. Materials and Methods

2.1. Soil Photography

Using a smartphone camera, a total of 103 images were acquired of topsoil (upper 30 cm) from sites across the north-east of Scotland. The sites were selected to provide variability in land cover, soil type, topography and parent material. Images were captured using an iPhone 2, with the site location also captured using the smartphone's GPS. A colour correction card was placed within each shot to allow post-capture image colour correction, as soil colour is one of the factors that is considered important in mineralogy (and hence soil texture and structure). Natural lighting was used in each case, with no artificial sources. All photographs were taken under at least reasonable lighting conditions, without a flash being required. Colour correction was carried out using the approach described in [23], and involved calculating the ratio between actual and expected RGB values at various intensity levels (0–255 in intensity bands 16 units across). This ratio was then applied to all pixels across the image that lay within the specific intensity level. Figure 1 shows two examples of the images that were captured.

Figure 1. Two examples of topsoil profile imagery captured using the smartphone camera. The colour correction card in each image is approximately 10 cm long and 6 cm wide.

A total of 54 images were selected from the profile photographs taken during the NSIS2 (National Soil Inventory of Scotland 2007–2009) field campaign. The protocol for how these profiles was prepared and sampled is given in [24] and two examples of the imagery are given in Figure 2. The soil profile face was cleaned and photographed prior to sampling, and where possible the photograph was taken to clearly show the horizons and any important features of the soil. Clearly labelled signs with the Site ID (National Grid Intersect) were used in the photographs.

Figure 2. Two examples of soil profile imagery captured during the NSIS2 field campaign. A tape measure was placed against each profile to provide scale. In each image, the tape length is 75 cm. Photographs credited to John Bell and Willie Towers.

For the smartphone imagery, structural assessment was carried out in the field using the approach outlined by [12] and detailed in the instructions available at the website of Scotland's Rural College (SRUC). Prior to this work being carried out, the researchers involved spent time in the field testing the VESS approach and ensuring that they were confident in producing replicable estimates of structure. VESS scoring estimates were produced for the upper (top 15 cm) and lower (bottom 15 cm) of the exposed soil, to capture variation in structure with depth. For the NSIS2 imagery, the same procedure was carried out but only for the topsoil. In addition, the structural assessment for NSIS2 samples was only carried out on agricultural soils, meaning that the imagery selected is restricted to this land use/land cover class. Throughout this work, reference to soil structure is taken to mean assessment of the VESS score.

Each of the soil samples was analysed for sample size according to the specified ranges (<2 μm—clay; 2–50 μm—silt; >50 μm—sand). This analysis was carried out using a Mastersizer 2000 (Malvern Instruments). The same procedure was applied to samples relating to both types of imagery. Loss On Ignition (LOI) was carried out for each sample at 450 °C for 24 h in a muffle furnace. Bulk density was calculated from the volume and mass of an oven-dried core, dried at 105 °C for 24 h. pH was measured using a 0.01 M solution of $CaCl_2$. Structure was assessed in the field using the method described by [12] and involves a manual and visual assessment of how easily the soil breaks up and the shapes/sizes of the soil fragments. Drainage is an assessment of the drainage characteristics of the whole soil profile, and so depends also on features below the topsoil; values given include very poor (1); poor (2); imperfect (3); moderate (4) and freely-draining (5).

2.2. Image Analysis

Each soil image (lossless JPEG format) was converted to digital numbers and stored as three arrays (one each for red, green and blue pixel values) with values ranging from 0 to 255. Following colour correction (on the smartphone imagery—colour correction cards were not used in the NSIS2 protocol), non-soil pixels were removed by selection of pixels with high ratios of green or blue to red (ratio greater than 1), or that had been identified as belonging to the colour correction card. Figure 3 shows two examples of this pixel elimination process using the same images as in Figure 1, with the resulting images containing only pixels identified as belonging to soil or plant roots.

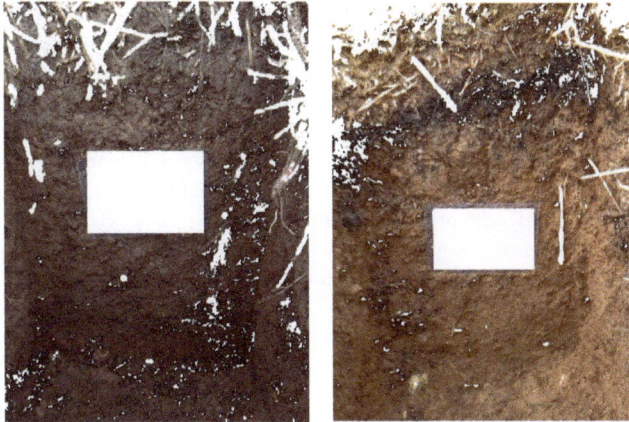

Figure 3. Topsoil profile imagery following removal of non-soil pixels (smartphone imagery). Note that some soil pixels have also been removed as they were too dark to provide useful information in the image analysis.

The average colour ratios for the upper and lower half of each image were then determined, with a value given to the mean proportion of each total pixel intensity (i.e., not the absolute, but the proportion of the summed RGB values) that was red, blue or green. Each image was therefore represented by six values (mean R, mean G, mean B for top and bottom of image). These values have been found useful in estimating topsoil organic matter content [22] and in estimating soil texture in ongoing unpublished work by this research group. Each image does not represent the same soil profile depth and so the upper and lower halves of the image are not representative of the same depths each time.

For structural assessment, a greyscale version of the image was derived by taking the average of the red, green and blue pixel values for each pixel. In order to assess the range of structures within each image, a moving window centred on the pixel of interest was passed over each pixel to normalise the intensity of that pixel in relation to its surroundings, followed by thresholding and boundary density counting. Pixels that had been removed due to colour or intensity thresholding were not included in this analysis. The following steps were carried out on each pixel of the greyscale image for a number of moving window sizes:

1. Identify the minimum (Min) and maximum (Max) pixel values within the moving window;
2. Normalise the target pixel value P_0 within the moving window range, defining the new pixel value P_1 as $P_1 = \text{INT}(256 \times (P_0 - \text{Min})/(\text{Max} - \text{Min}))$.
3. Following normalisation of each pixel, threshold the pixel value so that all values below 128 are set to 0, and all values equal to or above 128 are set to 255.
4. For each horizontal row of pixels in the image, calculate the number of boundaries between 0 and 255 for neighbouring pixels in any direction (soil pixels only). Divide this number by the number of soil pixels in the row.

In practice, identification of the minimum and maximum pixel values within the moving window was estimated by selecting 100 pixels at random within the moving window; this prevented the processing time from becoming prohibitive as the moving window size increased. It was assumed that selecting this number of pixels would provide a good estimation of the range of values within the moving window. The following moving window sizes were used: 1, 2, 4, 8, 16, 32, 64, 128 and 256 pixels. In each case, the moving window included all pixels within this distance in the x or y directions (i.e., a square window of side length $2N + 1$, where N is the moving window size). For moving window sizes of 8 or less, all pixels were used as the total number of pixels in the window was less than 100.

The result of the above steps for each moving window size is to produce a set of images that represents the local granularity of the image at a number of scales. This local granularity was assumed to reflect the physical structure within the soil. Counting the proportion of boundaries between thresholded pixels gives an indication of the distribution of aggregates within the image. The boundary proportion value was recorded for all rows in the image and average values were recorded for the upper and lower half of the image (discounting those pixels that contained non-soil pixels, the colour correction card or that were within one moving window size of it).

2.3. Environmental Characteristics

The GPS-recorded location of each sample site was used to determine a number of environmental characteristics. The parameters here were derived from a number of spatial datasets, including the following:

- The Ordnance Survey landform PANORAMA digital terrain model at 50 m resolution [25].
- Annual mean rainfall and temperature for 1460 Meteorological Office Stations from 1941 to 1970 interpolated to a 100 m resolution across Scotland [26,27]. The digital data was obtained directly from the Met Office by the Macaulay Institute under license agreement prior to 1994.
- Major soil subgroup (and soil parent material) from the 1:250,000 Soil Survey of Scotland at 100 m resolution [28]. This data was digitised at the Macaulay Institute for Soil Research (now the James Hutton Institute/JHI). Within this dataset, each map unit can contain a mixture of major soil subgroups, and we used the percentage estimate of each given within each map unit.
- The Land Cover of Scotland 1988 (LCS88) survey was carried out based on interpretation of aerial photographs and "ground-truth" assessment [29]. The work involved medium scale (1:24,000) aerial photography acquisition for the whole of Scotland over a period from 1987 to 1989 and interpretation of this imagery to obtain a 1:250,000 digital map of land cover that could be used with GIS software. The resulting datasets, which were validated and found to have a mapping accuracy of approximately 98%, were generated at the Macaulay Institute in 1989–1991 and continue to form part of the Institute (now the James Hutton Institute) data holdings.
- Geology—21 classes derived from descriptions of parent material in the Scottish Soil Map.

2.4. Modelling Soil Characteristics

An artificial neural network (ANN) model was used to estimate soil structural and texture parameters. The ANN was trained using 10-fold cross-validation. This method involves producing 10 subsamples of the training data, and 10 ANN models being developed, each trained with 9 out of the 10 randomly-generated subsamples and validated with the "missing" subsample. Each ANN was validated using a different subsample. The structure of the ANN used 104 inputs, including the 80 derived from environmental parameters and the 24 derived from image analysis (textural measurement at nine different moving window sizes, red green and blue colour average for both upper and lower image).

A single hidden layer in each ANN contained 10 nodes. There were 2 output nodes for the smartphone imagery (LOI and VESS structure each normalised in the range 0 to 1) and 7 for the NSIS2 imagery (drainage class, LOI, bulk density, sand, silt, clay and pH each normalised to the range 0 to 1). Each ANN was trained using the backpropagation gradient descent algorithm with a learning rate of 0.05. Training was carried out over 20,000 training steps each with random input selection, and validated every 1000 training steps to allow optimal training to be selected.

In addition to the above ANN design, we also tested the model using information derived only from (A) location information (80 inputs, same hidden node and output architecture) and (B) image analysis (24 inputs, same hidden node and output architecture). The same training regime was also used as for above. These secondary models were developed to allow us to determine the relative importance of spatial covariates or image characteristics in the input model. It was anticipated that image analysis would be more important for soil structural assessment and that spatial covariates would be more important in estimating soil texture because: (1) the image resolution is not sufficient to capture particles within the necessary size range to describe soil texture; and (2) the spatial covariates used would have less impact on soil structure than local management conditions.

In terms of estimating drainage, LOI and pH, it was also suspected that the spatial covariates would provide more useful information than image analysis, as visual examination of soil profiles does not allow an obvious way of estimating these while environmental conditions are known to have a strong influence on these soil characteristics.

Statistical evaluation of model performance was carried out using a number of metrics. These included r^2, RMSE (Root Mean Square Error), MAE (Mean Absolute Error), RPD (Ratio of Performance to Deviation) and RPIQ (Ratio of Performance to Interquartile distance). RPD is the ratio of standard error of prediction to the standard deviation of the variable's actual values, and is commonly used to provide an indication of model quality, with values greater than 2 often being taken to indicate good model performance. However, [30] showed that RPD is closely related to r^2, particularly for normally distributed data and large sample size, with the relationship RPD $= (1 - r^2)^{-0.5}$ at this limit. They proposed instead that RPIQ should be used, as described by [31], as it better represents the spread of values in a dataset. The values given of RPIQ are usually within a factor of 2 of RPD values, and so it is possible (but not to be encouraged) to take RPIQ values of greater than 2 to indicate good model performance.

For structure and drainage, which are categorical variables rather than being continuous, the use of r^2 and some of the other statistical metrics is acknowledged to be suspect; the use of these is only really appropriate for continuous variables. A more appropriate measure in this case is one of several possible pseudo-r^2 regression methods. However, as we are dealing here with variables that have multiple possible values rather than being binary, it was felt that the standard r^2 calculation could in this case be used as a metric of performance.

3. Results

For topsoil structure, the range of values is from 1 (best) to 5 (worst) on a numerical scale described in [32]. LOI values in Scottish topsoils can range between 0% and 100%, but the examples used here were all mineral soils and so the values were lower. Drainage values ranged from 1 (worst) to 5 (best) and were skewed towards good drainage (over 75% of the data points were given scores of 3 or above). Bulk density values were skewed towards higher values, with only 36% of the values lying below 0.75 g·cm^{-3}. Sand percentage values were approximately evenly distributed between 0% and 80%, while silt values were skewed towards higher values (a majority of Scottish soils are sandy loams or similar). Clay values were evenly distributed between 0% and 6%, with one value at 12%. The values for pH were mostly clustered between 4.5 and 7, with a small proportion of values lying above or below this range.

3.1. Model Results Using All Parameters

The accuracy with which each of the eight soil variables was estimated varied greatly in the models using image analysis and spatial covariate data, as shown in Table 1.

Table 1. Statistical evaluation of an artificial neural network (ANN) model using image analysis and spatial covariate information. For each variable, the range of values seen in the data is also given, to allow the Root Mean Square Error (RMSE) and Mean Absolute Error (MAE) values to be better understood. RPD—Ratio of Performance to Deviation; RPIQ—Ratio of Performance to Interquartile distance.

Variable	Range	r^2	RMSE	MAE	RPD	RPIQ
Structure	1 to 5	0.59	0.83	0.67	1.58	1.20
LOI (%)	0 to 25	0.64	3.62	2.87	1.66	1.34
Drainage	1 to 5	0.50	1.02	0.89	1.42	0.98
Bulk density (g·cm^{-3})	0 to 1.5	0.57	0.25	0.21	1.52	1.66
Sand (%)	0 to 80	0.58	13.60	11.46	1.50	1.57
Silt (%)	0 to 80	0.63	11.16	8.64	1.62	1.69
Clay (%)	0 to 12	0.62	1.59	1.37	0.95	1.60
pH	3 to 8	0.61	0.45	0.36	1.52	1.55

3.2. Model Results Using Image Analysis Only

As expected, all variables were worse estimated using only image analysis parameters, but the variation was not consistent (Table 2). Structure and bulk density estimates were not so strongly influenced by the removal of spatial covariate data from the model, and soil texture estimation accuracy was not reduced as greatly as expected. LOI and pH however saw relatively large drops in r^2 and increases in RMSE and MAE.

Table 2. Statistical evaluation of ANN model using image analysis data only.

Variable	Range	r^2	RMSE	MAE	RPD	RPIQ
Structure	1 to 5	0.50	0.97	0.82	1.51	1.10
LOI (%)	0 to 25	0.48	4.81	3.34	1.32	0.94
Drainage	1 to 5	0.43	1.50	1.26	1.21	0.89
Bulk density (g·cm^{-3})	0 to 1.5	0.52	0.29	0.25	1.40	1.48
Sand (%)	0 to 80	0.52	17.36	14.04	1.33	1.51
Silt (%)	0 to 80	0.55	13.45	10.13	1.40	1.57
Clay (%)	0 to 12	0.54	2.00	1.52	0.84	1.52
pH	3 to 8	0.49	0.68	0.50	1.32	1.35

3.3. Model Results Using Spatial Covariates Only

For the model using only spatial covariates, all parameters were again modelled with less accuracy than with all inputs (Table 3). However, this reduction in model performance was different from that seen with image analysis only; VESS structure and bulk density estimates saw poorer performance while LOI and pH were estimated more accurately than with image analysis alone. Drainage class and topsoil texture variables were estimated with approximately the same accuracy using the two types of input.

Table 3. Statistical evaluation of ANN model using spatial covariate data only.

Variable	Range	r^2	RMSE	MAE	RPD	RPIQ
Structure	1 to 5	0.47	1.08	0.87	1.43	1.00
LOI (%)	0 to 25	0.60	4.01	3.06	1.50	1.17
Drainage	1 to 5	0.41	1.52	1.26	1.19	0.85
Bulk density (g·cm^{-3})	0 to 1.5	0.48	0.33	0.28	1.35	1.43
Sand (%)	0 to 80	0.53	16.27	13.25	1.40	1.51
Silt (%)	0 to 80	0.58	13.79	9.92	1.45	1.59
Clay (%)	0 to 12	0.54	1.93	1.44	0.91	1.47
pH	3 to 8	0.58	0.76	0.56	1.39	1.46

4. Discussion

In absolute terms, none of the soil variables was estimated with an accuracy that would enable their estimation in the field to replace laboratory or field analysis by a trained observer. None achieved an RPD of greater than 2 (a value commonly taken to indicate "good" estimation) even with the best model. Drainage and clay percentage had RPD values of less than 1.5. Clay in particular had extremely low RPD values and did not follow the relationship given in [30] linking this with r^2. The distribution of values for clay may have affected this result, as it did not have a normal distribution, and the values given for RPIQ were also below 2. However, for estimating values for land management decision making or rapid condition assessment, the levels of accuracy achieved (as expressed by RMSE and MAE) are sufficient to give a "low/medium/high" estimate of every variable estimated.

The best model developed for estimating VESS structure gives an RMSE value of less than 1, and so if given a value of 2, for example, the user could be confident that the soil examined was not a 4 or 5. For pH similarly, the user can safely assume that the value given is within 1 pH unit of the actual value. This level of confidence is suitable for calculating liming rates. Drainage, bulk density and soil texture parameters are given with sufficient accuracy to detect compaction or evaluate candidate management options. Drainage in particular is an indication of the whole soil profile's drainage characteristics and so being able to make even a coarse assessment using information only from the upper soil is potentially very useful.

The neural network models used spatial covariates for Scotland's landscape. As such, the exact models would provide poorer performance if applied elsewhere, particularly in geographical regions with significantly different climate, soil or vegetation types. In order to achieve corresponding levels of accuracy, models would need to be developed that were specific to the regions of interest, which for many parts of the world may be difficult due to a lack of mapped information. The methodology used here is strongly dependent on available data, both from soil surveys and from spatial datasets with appropriate accuracy and resolution. As such, while globally applicable it would not provide the same level of accuracy in parts of the world that were poorly surveyed and mapped, particularly if these areas are spatially heterogeneous. This is an ongoing challenge for all areas of soil mapping, particularly in the digital age. There is also a distinction to be made between natural/seminatural and cultivated soils, where management is likely to play a much stronger role in soil characteristics. Management was not a factor included in this work, and would be necessary to include for some soils, particularly where agriculture has been present for hundreds of years.

The models are relatively small and computationally inexpensive, and could be implemented within a smartphone app quite easily. An existing app that provides a template for this is the SOCIT app, which estimates topsoil organic matter content for mineral soils in Scotland. This app links to similar spatial covariates as are used in this work, by sending the topsoil image with user coordinates to a server at the James Hutton Institute where the necessary large datasets are held. One limitation that we cannot improve on within our own work is the positional accuracy achieved by smartphones and tablets equipped with GPS; depending on the device and the quality of the signal acquired, this ranges between approximately 5 and 50 m. In linking positional information to spatial covariates, we therefore must be aware that in landscapes that are spatially heterogeneous, positional inaccuracy may increase error rates for variables of interest.

Ongoing research in this area indicates that LOI, pH and bulk density in particular can be estimated with better accuracy than demonstrated here, if sufficient data is provided. Spectroscopy in particular can provide good estimates of many soil variables non-destructively. However, this work was intended to produce a method of estimating all of the variables of interest from a dataset of available imagery, reducing the need to gather additional information. We intend to expand the resource of available data and imagery to produce a more robust and accurate model capable of improving on the levels of accuracy demonstrated here. The final objective of this work is to produce an app that can provide rapid assessments of several topsoil characteristics for Scottish soils.

The working hypothesis, that information derived from imagery would be more important for estimating structure while spatial covariates would provide more useful for estimating the other parameters, was largely borne out. Structure is no doubt affected by environmental conditions but is also strongly influenced by management, which is spatially variable but was not incorporated as an input to the modelling. Characteristics relevant to structure are often visible (e.g., cracking, solid blocky peds, macropores), implying that with appropriate image metrics, it should be possible to measure indicators of these and other structural features.

We also saw that in addition to structure, bulk density modelling was strongly affected by removing the imagery data from the modelling. It is possible that image features indicative of densely-packed soils are correlated to the measured image characteristics, and this deserves further study.

Soil texture variable estimation was approximately equal for the two data types, while LOI and pH appear to more strongly linked to spatial covariates. It is not surprising that pH is more strongly linked to environmental character than image metrics, as the visual appearance of topsoil does not often provide clues to pH levels even to trained field surveyors. Local conditions such as vegetation or parent material however are known to have a strong influence on soil pH levels.

An unexpected result was that the drop in model performance when removing one data type was not as great as expected for any of the variables modelled. It appears that both image characteristics and environmental conditions can contribute useful information to all variables that we studied here. It is possible that the characteristics measured in the imagery are in some way connected to the environmental conditions at each site, which would make it harder than expected to discriminate between the relative importance of different inputs. For example, particularly wet and cold climate conditions could result in soils with a specific visual appearance, which would be reflected in the image analysis data produced.

Further work on the linkages between environmental conditions and the visual appearance of the soil would help in resolving this. In order to achieve this, it is necessary to further investigate the possible image metrics that could be measured, as those selected for this work may not be the most appropriate. This would also benefit any further modelling work in the area of estimating soil characteristics from soil profile image analysis. It would also help to improve the spatial resolution of some of the parameters, particularly topography, as the current 100 m resolution that was used in this work does not reflect the spatial heterogeneity of Scotland's landscape in some areas, particularly the fragmented mountainous regions in the north and west. Current work on digital soil mapping at the James Hutton Institute includes the development of a range of 10 m resolution topography products, to be incorporated into further research. As mentioned above, this will correspond to the approximate resolution achieved by current smartphone/tablets equipped with GPS.

Currently, we are developing a collection of imagery for Scottish soils that covers a wider range of soils and environmental conditions than was used in this work. We are also working on the identification of image analysis metrics that provide better representation of the soil under field conditions.

Acknowledgments: This paper was funded by the Rural & Environment Science & Analytical Services Division of the Scottish Government.

Author Contributions: Matt Aitkenhead conceived and designed the experiments and wrote the paper; Malcolm Coull and David Donnelly provided spatial datasets and spatial analysis; Richard Gwatkin carried out the field work.

Conflicts of Interest: The authors declare no conflict of interest.

Abbreviations

The following abbreviations are used in this manuscript:

GPS	Global Positioning System
VESS	Visual Evaluation of Soil Structure
NSIS2	National Soil Inventory of Scotland (2)
SRUC	Scotland's Rural College
LOI	Loss On Ignition
JHI	James Hutton Institute
LCS88	Land Cover of Scotland 1988
GIS	Geographic Information System
ANN	Artificial Neural Network
RMSE	Root Mean Squared Error
MAE	Mean Absolute Error
RPD	Ratio of Performance to Deviation
RPIQ	Ratio of Performance to Inter-Quartile distance

References

1. Skvortsova, E.B. Changes in the geometric structure of pores and aggregates as indicators of the structural degradation of cultivated soils. *Eurasian Soil Sci.* **2009**, *42*, 1254–1262. [CrossRef]
2. Mueller, L.; Kay, B.D.; Deen, B.; Hu, C.; Zhang, Y.; Wolff, M.; Eulenstein, F.; Schindler, U. Visual assessment of soil structure: Part II. Implications of tillage, rotation and traffic on sites in Canada, China and Germany. *Soil Tillage Res.* **2009**, *103*, 188–196. [CrossRef]
3. Letey, J. The study of soil structure—Science or art. *Aust. J. Soil Res.* **1991**, *29*, 699–707. [CrossRef]
4. Rampazzo, N.; Blum, W.E.H.; Wimmer, B. Assessment of soil structure parameters and functions in agricultural soils. *Bodenkultur* **1998**, *49*, 69–84.
5. Vanden-Bygaart, A.J.; Protz, R.; Tomlin, A.D. Changes in pore structure in a no-till chronosequence of silt loam soils, southern Ontario. *Can. J. Soil Sci.* **1999**, *79*, 149–160. [CrossRef]
6. Rampazzo, B.; Blum, W.E.H.; Strauss, P.; Curlik, J.; Slowinska-Jurkiewicz, A. The importance of mineralogical and micromorphological investigations for the assessment of soil structure. *Int. Agrophys.* **1993**, *7*, 117–132.
7. McBratney, A.B.; Moran, C.J.; Stewart, J.B.; Cattle, S.R.; Koppi, A.J. Modifications to a method of rapid assessment of soil macropore structure by image analysis. *Geoderma* **1992**, *53*, 255–274. [CrossRef]
8. Moran, C.J.; McBratney, A.B. Acquisition and analysis of 3-component digital images of soil pore structure. I. Method. *J. Soil Sci.* **1992**, *43*, 541–549. [CrossRef]
9. Slowinska-Jurkiewicz, A.; Mikosz, A.I. The application of morphological studies for the assessment of various soil structure conditions. *Pol. J. Soil Sci.* **1995**, *28*, 1–9.
10. Sander, T.; Gerke, H.H.; Rogasik, H. Assessment of Chinese paddy-soil structure using X-ray computed tomography. *Geoderma* **2008**, *145*, 303–314. [CrossRef]
11. Pires, L.F.; Bacchi, O.O.S.; Reichardt, K. Assessment of soil structure repair due to wetting and drying cycles through 2D tomographic image analysis. *Soil Tillage Res.* **2007**, *94*, 537–545. [CrossRef]
12. Ball, B.C.; Batey, T.; Munkholm, L.J. Field assessment of soil structural quality—A development of the Peerlkamp test. *Soil Use Manag.* **2007**, *23*, 329–337. [CrossRef]
13. McGarry, D.; Sharp, G. A rapid, immediate, farmer-usable method of assessing soil structure condition to support conservation agriculture. In *Conservation Agriculture*; Garcia-Torres, L., Benites, J., Martinez-Vilela, A., Holgado-Cabrera, A., Eds.; Springer: Dordrecht, The Netherlands, 2003; pp. 375–380.
14. Balarezo, G.; Neyde, F.T.; Cassio, A.S.; Alvaro, P.; Ball, B. Visual assessment soil quality structure methodology applied to Oxisol under different soil use and management. *Cienc. Rural* **2009**, *39*, 2531–2534.
15. Balarezo, G.N.F.; da Silva, A.P.; Tormena, C.A.; Ball, B.; Rosa, J.A. Visual soil structure quality assessment on Oxisols under no-tillage system. *Sci. Agric.* **2010**, *67*, 479–482.
16. Schaeffer, B.; Stauber, M.; Mueller, R.; Schulin, R. Changes in the macro-pore structure of restored soil caused by compaction beneath heavy agricultural machinery: A morphometric study. *Eur. J. Soil Sci.* **2007**, *58*, 1062–1073. [CrossRef]

17. Cotching, W.E.; Belbin, K.C. Assessment of the influence of soil structure on soil strength/soil wetness relationships on Red Ferrosols in north-west Tasmania. *Aust. J. Soil Res.* **2007**, *45*, 147–152. [CrossRef]
18. Mueller, L.; Kay, B.D.; Hu, C.; Li, Y.; Schindler, U.; Behrendt, A.; Shepherd, T.G.; Ball, B.C. Visual assessment of soil structure: Evaluation of methodologies on sites in Canada, China and Germany Part I: Comparing visual methods and linking them with soil physical data and grain yield of cereals. *Soil Tillage Res.* **2009**, *103*, 178–187. [CrossRef]
19. McKenzie, D.; Batey, T. Recent trends in rapid assessment of soil structure in the field. *Aust. Cottongrower* **2006**, *27*, 58–61.
20. Ball, B.C.; Crawford, C.E. Mechanical weeding effects on soil structure under field carrots (*Daucus carota* L.) and beans (*Vicia faba* L.). *Soil Use Manag.* **2009**, *25*, 303–310. [CrossRef]
21. Jirku, V.; Kodesova, R.; Muhlhanselova, M.; Zigova, A. Seasonal variability of soil structure and soil hydraulic properties. In Proceedings of the 19th World Congress of Soil Science: Soil Solutions for a Changing World, Brisbane, Australia, 1–6 August 2010.
22. Aitkenhead, M.J.; Coull, M.; Towers, W.; Hudson, G.; Black, H.I.J. Prediction of soil characteristics and colour using data from the National Soils Inventory of Scotland. *Geoderma* **2013**, *200*, 99–107. [CrossRef]
23. Aitkenhead, M.J.; Donnelly, D.; Coull, M.; Hastings, E. Innovations in environmental monitoring using mobile phone technology—A review. *Int. J. Interact. Mob. Technol.* **2014**, *8*, 50–58. [CrossRef]
24. Lilly, A.; Bell, J.S.; Hudson, G.; Nolan, A.J.; Towers, W. *National Soil Inventory of Scotland 2007–2009: Profile Description and Soil Sampling Protocols. (NSIS_2)*; Technical Bulletin; James Hutton Institute: Aberdeen, UK, 2011.
25. Ordnance Survey OpenData. Available online: http://www.ordnancesurvey.co.uk/oswebsite/products/os-opendata.html (accessed on 10 June 2014).
26. Matthews, K.B.; MacDonald, A.; Aspinall, R.J.; Hudson, G.; Law, A.N.R.; Paterson, E. Climatic soil moisture deficit—Climate and soil data integration in a GIS. *Clim. Chang.* **1994**, *28*, 273–287. [CrossRef]
27. MacDonald, A.M.; Matthews, K.B.; Paterson, E.; Aspinall, R.J. The impact of climate change on the soil/moisture regime of Scottish mineral soils. *Environ. Pollut.* **1994**, *83*, 245–250. [CrossRef]
28. Macaulay Institute for Soil Research. *Organization and Methods of the 1:250,000 Soil Survey of Scotland*; University Press: Aberdeen, UK, 1984.
29. Macaulay Land Use Research Institute. *The Land Cover of Scotland: Final Report*; University Press: Aberdeen, UK, 1993.
30. Minasny, B.; McBratney, A.B. Why you don't need to use RPD. *Pedometron* **2013**, *33*, 14–15.
31. Bellon-Maurel, V.; Fernandez-Ahumada, E.; Palagos, B.; Roger, J.M.; McBratney, A. Critical reviews of chemometric indicators commonly used for assessing the quality of the prediction of soil attributes by NIR spectroscopy. *Trends Anal. Chem.* **2010**, *29*, 1073–1081. [CrossRef]
32. Ball, B. Visual Evaluation of Soil Structure. SRUC (Scotland's Rural College), 2016. Available online: http://www.sruc.ac.uk/info/120625/visual_evaluation_of_soil_structure (accessed on 22 September 2016).

MDPI
St. Alban-Anlage 66
4052 Basel
Switzerland
Tel. +41 61 683 77 34
Fax +41 61 302 89 18
www.mdpi.com

Journal of Imaging Editorial Office
E-mail: jimaging@mdpi.com
www.mdpi.com/journal/jimaging